T0309112

Heat and Mass Transfer

Heat and Mass Transfer

Editor: Celine Kennedy

www.murphy-moorepublishing.com

MURPHY & MOORE

Cataloging-in-Publication Data

Heat and mass transfer / edited by Celine Kennedy.
p. cm.
Includes bibliographical references and index.
ISBN 978-1-63987-749-2
1. Heat. 2. Mass transfer. 3. Heat--Transmission. I. Kennedy, Celine.
QC254.2 .P75 2023
536--dc23

Murphy & Moore Publishing
1 Rockefeller Plaza,
New York City,
NY 10020, USA

ISBN 978-1-63987-749-2

Contents

Preface

The main aim of this book is to educate learners and enhance their research focus by presenting diverse topics covering this vast field. This is an advanced book which compiles significant studies by distinguished experts in the area of analysis. This book addresses successive solutions to the challenges arising in the area of application, along with it; the book provides scope for future developments.

Heat transfer refers to the movement of heat across the border of a system caused due to difference between the temperature of the system and its surroundings. Mass transfer refers to the physical phenomenon that involves the observation of a net movement of generic particles from one place to another. These phenomena are the basis of various mechanisms and processes such as distillation of alcohol, the evaporation of water, and purification of blood in the liver and kidneys. Heat and mass transfer is a significant and well-established branch of engineering and physics. It also finds application in multiple manufacturing procedures. This book aims to shed light on the role of heat and mass transfer in the energy systems. It will also provide interesting topics for research, which interested readers can take up. With state-of-the-art inputs by acclaimed experts of this field, this book targets students and professionals.

It was a great honour to edit this book, though there were challenges, as it involved a lot of communication and networking between me and the editorial team. However, the end result was this all-inclusive book covering diverse themes in the field.

Finally, it is important to acknowledge the efforts of the contributors for their excellent chapters, through which a wide variety of issues have been addressed. I would also like to thank my colleagues for their valuable feedback during the making of this book.

Editor

1

Optimization Method for the Evaluation of Convective Heat and Mass Transfer Effective Coefficients and Energy Sources in Drying Processes

Marcin Stasiak [1], Grzegorz Musielak [2,*] and Dominik Mierzwa [2]

1 Institute of Mathematics, Poznań University of Technology, 60-965 Poznań, Poland; Marcin.Stasiak@put.poznan.pl
2 Institute of Technology and Chemical Engineering, Poznań University of Technology, 60-965 Poznań, Poland; Dominik.Mierzwa@put.poznan.pl
* Correspondence: Grzegorz.Musielak@put.poznan.pl

Abstract: A new optimization method for the assessment of the coefficients existing in a model of drying kinetics is developed and presented in this article. This method consists of matching the drying kinetics resulting from the mathematical model with the drying kinetics resulting from the experiments. Both the heat and mass transfer coefficients, the critical relative humidity, and the additional ultrasound energy (heat) source are included in the optimization procedure. The Adams–Bashforth multistep method of solving nonlinear ordinary differential equations is used. The inverse problem of model parameter estimation is solved by the Rosenbrock optimization method. The methodology is illustrated by the example of the ultrasound-assisted convective drying of apple and carrot. A high level of agreement between the results obtained experimentally and numerically was found. The obtained results confirmed the great influence of ultrasound on the drying kinetics. It was found that ultrasound application improved the mass transfer by 20–80% and heat transfer by 30–90%. It was also found that the heating effect caused by the ultrasound's absorption was very small, with a value below 1%.

Keywords: mass and heat transfer coefficients; heat source; evaluation; ultrasound

---

1. Introduction

Biological materials require very gentle drying methods because they are extremely sensitive to a high temperature and long process time [1,2]. Simple convection drying often leads to the degradation of these materials. For this reason, hybrid methods which combine convective drying with ultrasound assistance have been developed in recent decades [3]. The results of this development may have a positive effect on the production and storage of fruit and vegetables. In the drying technology of biological products, new drying methods applying ultrasound and microwave enhancement are required [4]. This can significantly reduce the energy consumption, shorten the drying time, and improve the quality of food by preserving its valuable nutrients (proteins, carbohydrates, vitamins, and minerals) and visual aspects (color and shape).

The published literature [5–10] indicates that a high-intensity ultrasound is able to improve the mass and heat transfer processes of dried products. This is especially true for drying materials such as vegetables and fruits, which are biological materials sensitive to high temperatures. The positive attributes of such methods have recently led to an innovative ultrasound apparatus being developed [11]. New investigations and possibilities of ultrasound-assisted drying as a result of new dryer application using ultrasound have been proposed [9,12–16]. A broad review of ultrasound use in drying processes is presented in [3].

To describe ultrasound-assisted drying, a number of drying models enabling the assessment of the drying kinetics and process effectiveness have been developed. The simplest models are the empirical ones, which interpolate experimental data using exponential functions, e.g., the Weibull model [13]. The models of this type allow the effective diffusion coefficient to be determined. Another group of models includes the diffusion ones. Some of these models neglect external resistance [17,18], whilst others consider this resistance [19,20]. These models only describe the mass transfer, and do not describe the heat transfer. A model that describes simultaneous mass and heat transfer and takes into account external resistance is described in [21]. Unfortunately, this model does not account for material shrinkage. This is a drawback, especially when describing drying food. A model taking into account heat and mass transfer, together with external resistance and material shrinkage, has been developed by Kowalski and Pawłowski [10,22]. In our article, this model is used to describe the drying kinetics. Mathematical modeling of the drying process needs to determine the process parameters [23]. To evaluate the drying kinetics according to the model proposed above, it is necessary to determine the model parameters, namely, the effective coefficients of heat h_T and mass h_m transfer, as well as the energy source due to ultrasound absorption ΔQ and the critical moisture content X_{cr}. The main aim of this work is to present a numerical algorithm for the evaluation of these parameters. In modeling and simulations of the heat and mass transfer processes in the available literature, the Runge–Kutta method and ready-made solvers (e.g., implemented in MatLab) are often used [24–26]. This method of the ordinary differential equations solution is not stable for long-term simulation. Therefore, the Adams–Bashforth method was used in this work [27]. The model parameters were estimated on the basis of the ultrasound-assisted drying kinetics realized experimentally. The estimation was based on the inverse problem solution, obtained by the use of optimization techniques. The non-gradient Rosenbrock optimization method was chosen because of its efficiency [28].

2. Mathematical Model

Here, drying is considered as a coupled heat and mass transfer process described by a global model of drying kinetics, that is, a model expressing the variation of the drying material moisture content and temperature over time. The drying kinetics is described by a system of coupled ordinary differential equations [10,22]:

$$m_s \frac{dX}{dt} = -A_m h_m \ln \frac{\varphi|_{\partial B}\, p_{vs}(T)}{\varphi_a p_{vs}(T_a)}, \tag{1}$$

$$m_s \frac{d}{dt}[(c_s + c_l X)T] = A_T h_T (T_a - T) - A_m l h_m \ln \frac{\varphi|_{\partial B}\, p_{vs}(T)}{\varphi_a p_{vs}(T_a)} + \Delta Q, \tag{2}$$

where X is the moisture content (dry basis); T denotes the temperature; m_s is the dry mass; A_m and A_T denote the surfaces of mass and heat exchange, respectively; φ_a is the relative air humidity in ambient air; $\varphi\big|_{\partial B}$ is the air humidity close to the dried sample surface; p_{vs} is the temperature-dependent saturated vapor pressure; c_s and c_l denote the specific heat of dry solid and moisture, respectively; l is the latent heat of evaporation; h_m and h_T denote the mass and heat transfer coefficients, respectively; and ΔQ is the heat in the material sample due to the absorption of ultrasonic waves. Both equations are determined as a result of the mass and energy balances. Equation (1) describes the mass exchange: The mass accumulation of moisture (left hand side) is equal to the convective moisture mass flux (right hand side). Equation (2) describes the heat transfer between the dried material and the surrounding air: The heat accumulation (left hand side) is equal to the heat flux delivered by convection (the first term to the right hand side) minus the heat consumed by evaporation (the second term to the right hand side) plus the heat of ultrasonic wave absorption (the third term to the right hand side).

This model of drying kinetics is used to describe the drying of strongly deformable biological materials, such as fruit and vegetables. Therefore, it is necessary to take into account that such materials undergo a large amount of shrinkage during drying. It is assumed that the dried material undergoes linear volumetric shrinkage according to the following formula:

$$V = V(X) = [1 - \alpha_V(X_0 - X)]V_0, \tag{3}$$

where α_V denotes the volumetric shrinkage coefficient, X_0 is the initial moisture content, and X represents the actual moisture content. The surfaces A_m and A_T (Equations (1) and (2)) denote the surfaces of mass and heat exchange, respectively. If the dried material is placed on a moisture-impermeable support, the surfaces A_m and A_T are different. If, on the other hand, the material experiences constant motion during drying, the evaporation area is not limited. In this case, the heat and mass exchange occurs through the whole material surface. Then, the surfaces A_m and A_T are equal. In the rotary dryer used in our tests (Figure 1), the material is not supported and experiences constant motion, and both heating and drying take place over the whole surface. Hence, a simple transformation of Equation (3) gives the area of both mass and heat exchange as a function of humidity, described by the following equation:

$$A_m = A_T = A(X) = [1 - \alpha_V(X_0 - X)]^{2/3}A_0. \tag{4}$$

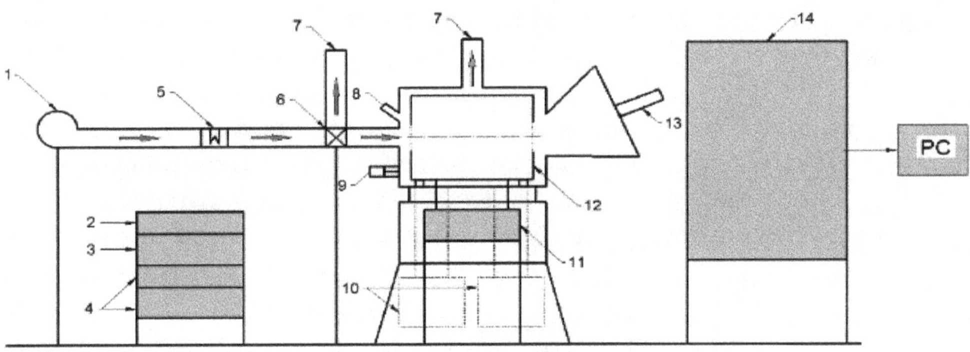

Figure 1. The scheme of the hybrid dryer: 1—blower (fan); 2—AUS controller; 3—AUS preamplifier; 4—microwave feeders; 5—heater; 6—pneumatic valve; 7—air outlet; 8—pyrometer; 9—drum drive; 10—microwave generators; 11—balance; 12—rotatable drum; 13—AUS ultrasound transducer; and 14—control unit.

At the beginning of drying, the material is heated. Then, usually, two periods can be distinguished in the drying chart [29] (see Figure 2a).

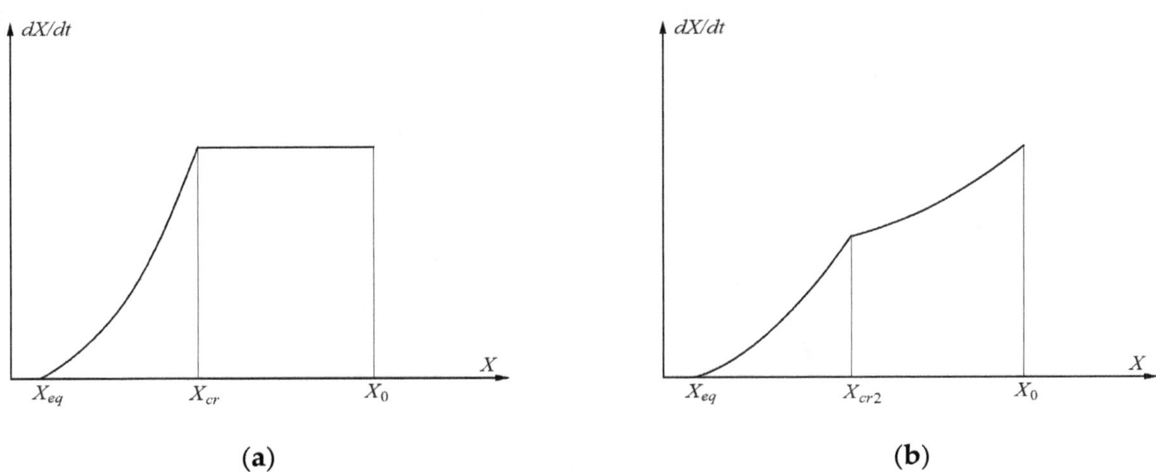

Figure 2. Schematic presentation of the drying rate function: (**a**) Existence of the constant drying period, and (**b**) existence of the two stages of the falling drying rate period.

The first one is called the constant drying rate period (CDRP). During this period, a film of free moisture exists on the surface of the dried material. The drying rate only depends on external conditions. During this period, the temperature of the dried material does not change. When the average moisture content reaches its critical value X_{cr}, the free moisture film is reduced and the drying rate decreases. The falling drying rate period (FDRP) starts. During this period, the temperature of the dried material rises to the equilibrium temperature.

The proposed model enables the description of the material heating, the constant (CDRP), and the falling (FDRP) drying rate periods. One of the possibilities for describing the above periods using the mathematical model presented above (Equations (1) and (2)) is the adoption of a discontinuous function describing the dependence of the relative air humidity at the surface of the dried material $\varphi|_{\partial B}$ on the moisture content in the dried material X [30]:

$$\varphi|_{\partial B} = \begin{cases} 1 & \text{for } X_0 \geq X \geq X_{cr} \\ \varphi_a + (1-\varphi_a)\frac{X-X_{eq}}{X_{cr}-X_{eq}} & \text{for } X_{cr} \geq X \geq X_{eq} \\ \varphi_a \frac{X}{X_{eq}} & \text{for } X_{eq} \geq X \end{cases}, \tag{5}$$

where X_0, X_{cr}, and X_{eq} are the initial, critical, and equilibrium values of the moisture contents, respectively and φ_a is the relative humidity of the drying medium. The critical moisture point X_{cr} describes the transition between the constant drying rate period (CDRP) and the falling drying rate period (FDRP).

Based on the experimental tests, one can state that during the drying of fruits and vegetables, no CDRP exists [1]. In some cases, the FDRP period can be divided into two stages [29] (see Figure 2b). The first stage could be interpreted as a result of a decrease in the dried material surface, which causes a decrease in the drying rate. During the second drying stage, the drying rate only depends on the internal mass transfer resistance. To describe these two stages of the FDRP, the following relation is proposed [30]:

$$\varphi|_{\partial B} = \begin{cases} \varphi_{cr2} + (1-\varphi_{cr2})\frac{X-X_{cr2}}{X_0-X_{cr2}} & \text{for } X_0 \geq X \geq X_{cr2} \\ \varphi_a + (\varphi_{cr2}-\varphi_a)\frac{X-X_{eq}}{X_{cr2}-X_{eq}} & \text{for } X_{cr2} \geq X \geq X_{eq} \\ \varphi_a \frac{X}{X_{eq}} & \text{for } X_{eq} \geq X \end{cases}, \tag{6}$$

where the second critical point (X_{cr2}, φ_{cr2}) describes the transition between the two aforementioned stages of the FDRP.

At a constant total pressure (equal to 0.1 MPa) of humid air, the pressure of saturated vapor p_{vs} is a function of temperature:

$$\begin{aligned} p_{vs}(T) = {} & 9.61966 \times 10^{-4}T^4 - 1.08405264 \times T^3 + 4.61325529 \times 10^{-2}T^2 \\ & -8.77803513 \times 10^{-4}T + 6.29588464 \times 10^6, \end{aligned} \tag{7}$$

where T is the absolute temperature ranging between 273 and 373 K. This function is obtained by the interpolation of experimental data presented in [31].

3. Solution Method

The drying kinetics of Equations (1) and (2) should be rearranged to produce the following forms:

$$\frac{dX}{dt} = -\frac{A_m h_m}{m_s} \ln \frac{\varphi|_{\partial B}\, p_{vs}(T)}{\varphi_a p_{vs}(T_a)}, \tag{8}$$

$$\frac{dT}{dt} = \frac{1}{m_s(c_s + c_l X)}\left[A_T h_T (T_a - T) + (c_l T - l) A_m h_m \ln \frac{\varphi|_{\partial B}\, p_{vs}(T)}{\varphi_a p_{vs}(T_a)} + \Delta Q\right]. \tag{9}$$

This transformation allows us to present the initial problem in the form of a system of two nonlinear ordinary differential equations with appropriate initial conditions. This system can be written as

$$\begin{cases} \frac{dX}{dt} = \Psi(t, X, T) \text{ for } t > 0 \\ \frac{dT}{dt} = \Phi(t, X, T) \text{ for } t > 0 \\ X(t=0) = X_0 \; T(t=0) = T_0 \end{cases} . \tag{10}$$

Drying is a long process. Therefore, the solution of the above system of equations requires the use of an appropriate numerical method for such problems. Therefore, the Adams–Bashforth method was used to solve the system of Equation (10) [27]. This is an explicit method of solving nonlinear ordinary differential equations. It is based on Lagrangian polynomial interpolation of the solution. The method is a non-self-starting multistep one. This means that, in order to use this method, four starting points must be specified with another numerical method. These points were obtained using the fourth row Runge–Kutta method [32].

4. Model Parameter Determination

In order to determine the values of the process parameters in the mathematical model, the solution of the inverse problem is used. It consists of searching for the direct problem solution that is closest to the experimental results describing the kinetics of the process. The best fit between the experimental and numerical results is obtained using optimization techniques. Four parameters, which describe the drying process, are introduced in the mathematical model (Equations (1)–(7)): The heat h_T and mass h_m transfer coefficients; the heat source ΔQ describing the absorption of ultrasonic waves; and the critical relative humidity φ_{cr2}.

The optimal numerical solution is sought in the quadratic norm. This means that the best fit occurs when the sum of the squares of the residuals of the experimental and numerical values is the smallest. However, the numerical values of these differences in the moisture content and the temperature are of different orders. Therefore, the differences are normalized by dividing them by their maximum values. As a result, the share of both parameters—the moisture content and the temperature—is of the same order in the objective function which is defined as follows:

$$f(h_m, h_T, \Delta Q, \varphi_{cr2}) = \sum_{i=1}^{N} \left[\left(\frac{X_{\text{num},i} - X_{\text{exp},i}}{X_{\max} - X_{\min}} \right)^2 + \left(\frac{T_{\text{num},i} - T_{\text{exp},i}}{T_{\max} - T_{\min}} \right)^2 \right]. \tag{11}$$

The search for the minimum of the objective function (11) is a four-parameter optimization problem. Calculation of the numerical value of this function does not require a long computation time, but its derivatives in relation to independent variables cannot be determined. Therefore, an appropriate non-gradient optimization method should be used. The Rosenbrock optimization method was chosen as the appropriate one to solve the formulated optimization problem [28].

5. Materials and Methods

The proposed method for determination of the model parameters is illustrated for the ultrasound-assisted convective drying of apple and carrot samples, as examples.

There are two material parameters used in the model, namely, the specific heat of dry solid c_s and the volumetric shrinkage coefficient α_V. Their values should be determined before starting the optimization procedure. The value of the material specific heat c_s as a function of temperature was calculated based on formulas described in the literature [33]. The procedure employed for calculating the specific heat of the moist material was as follows. First, the mass fractions of the components contained in the apples and carrots were determined on the basis on the data from the National Food Institute of the Technical University of Denmark [34]. The individual specific heats of all constituents as temperature functions (polynomials) were determined on the basis of [33]. Then, the specific heat of dry matter was obtained as the sum of the products of the individual specific heats and the mass fractions of the components. The specific heat of the wet material is the sum of the specific heat of dry matter and the specific heat of water (temperature-dependent) multiplied by the moisture content X (see Equation (2)). The volumetric shrinkage coefficient, α_V, was determined on the basis of an additional set of experiments. During the experiments, the materials were subject to slow convection drying. The volumes of fresh material samples were measured before drying and in given time intervals during the process. The measurement was carried out with the use of the gravimetrical method based on Archimedes' law. In this method, the volume is calculated on the basis of weight measurement in air and water. Due to the amount of water removed by the sample, its weight in water is less than in air. The weight difference enables the sample volume measurement.

The drying kinetics was determined from the experimental tests carried out in the rotary hybrid dryer, which allows drying with three different techniques, i.e., hot air, microwaves, and ultrasound used separately and/or simultaneously. Microwaves were not used during these experiments. The scheme of the dryer is shown in Figure 1.

The material under first investigation was an apple (*Malus domestica* cv. Ligol), bought at a local market and stored in a refrigerator at a temperature of 5 °C prior to the experiments. For each test, fresh fruits were cut into 10 mm size cubes, and 200 g of such material was taken for drying experiments. The second material under investigation was a carrot (*Daucus carota* L. cv. Nantes), bought at the same place and stored in the same way as apples. The vegetable portion used for drying tests had a similar shape to the previous one. For each test, fresh carrots were cleaned and cut into slices with a 5 mm thickness. The amount of material taken for drying experiments was equal to 200 g. All of the tests were carried out according to the schemes and process parameters listed in Table 1. The air velocity was controlled by the flow sensor placed in the pipeline between the fan and the drying chamber. The diameters of the pipe and the drying chamber were different, so the effective air velocity in the chamber was five times smaller than the velocity in the pipe. The values of both velocities are described in Table 1.

Table 1. Drying schemes and parameters.

Scheme No. and Abbreviation	Convective Drying Parameters	Ultrasound Power If Used
	Apples	
1—CV_404	40 °C, air flow 4 m/s (effective 0.8 m/s)	–
2—CVUD_404200	40 °C, air flow 4 m/s (effective 0.8 m/s)	200 W
	Carrots	
1—CV_502	50 °C, air flow 2 m/s (effective 0.4 m/s)	–
2—CV_702	70 °C, air flow 2 m/s (effective 0.4 m/s)	–
3—CVUD_502200	50 °C, air flow 2 m/s (effective 0.4 m/s)	200 W
4—CVUD_702200	70 °C, air flow 2 m/s (effective 0.4 m/s)	200 W

The tests for all drying schemes were carried out in triplicate. The initial moisture contents of the investigated samples were determined using a Precisa XM120 moisture analyzer, and through drying at 70 °C for 24 h in the chamber drier. The initial moisture content for carrots was equal to 0.88 ± 0.005 kg/kg_{wb}, whereas for apples, it was 0.83 ± 0.01kg/kg_{wb}. The drying processes were performed until the final moisture content of the carrots reached 0.05 kg/kg_{wb} and that of the apples reached 0.1 kg/kg_{wb}. During drying tests, the controller maintained the set process parameters, and collected all the data in its internal memory at constant time intervals. Only the material's temperature was measured independently using a Dwyer HTDL-30 wireless temperature data logger, which allowed the collection of this parameter for material placed in a rotating drum. The wireless Dwyer sensor is a small electronic tube which ends with an elastic thermocouple. The setup was placed in a drier drum where the thermocouple was placed inside the samples in its center. The sample was secured against slipping from thermocouple by a very thin string.

Figure 3 shows the drying kinetics, i.e., the drying curves and the curves of temperature evolution, obtained experimentally. It is convenient to analyze the drying kinetics with the use of temperature curves. The drying kinetics of both materials show the presence of a heating period for the material, the absence of the first drying period (no constant temperature), and the presence of a second drying period. The drying kinetics of apples shows the existence of the single FDRP curve. In contrast, the carrot kinetics, especially the change in the temperature rise rate, indicates the existence of two stages of the FDRP curves.

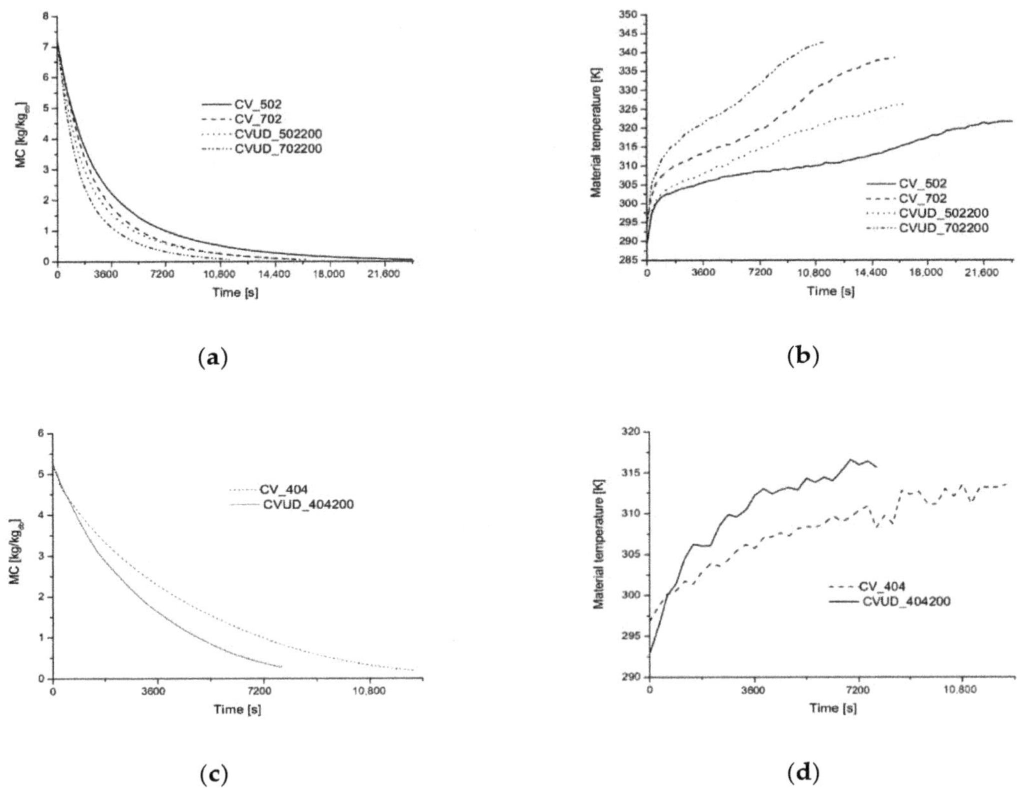

Figure 3. Drying kinetics of the ultrasound-assisted convective drying of carrot samples: (**a**) Drying curves and (**b**) temperature evolution curves. Drying kinetics of the ultrasound-assisted convective drying of apple samples: (**c**) Drying curves and (**d**) temperature evolution curves.

The temperature of the drying medium, T_a, was about 40, 50, and 70 °C. When ultrasound was applied in drying schemes CVUD_404200, CVUD_502200, and CVUD_702200, the material temperature T increased above the drying medium temperature T_a due to the absorption of ultrasonic waves. This phenomenon follows from the additional heat source ΔQ.

6. Numerical Results

The proposed method of parameter estimation in the drying model was tested using the experimental data described in the previous paragraph. According to the experimental results, all of the processes showed no first drying period. Therefore, the values of the critical moisture content X_{cr} for all drying schemes were assumed to be equal to the initial one X_0.

The drying kinetics of apples did not demonstrate the existence of two stages of the FDRP. Therefore, Equation (5) describing the FDRP was used in the mathematical model.

Unlike the drying kinetics of apples, the drying kinetics of carrots reveals the existence of two stages of the FDRP. Therefore, Equation (6) describing these two stages was used in the mathematical model. The values of the second critical moisture content X_{cr2} were estimated on the basis of the drying kinetics (experimental data), and are given in Table 2.

Table 2. Results of the model parameters determination for carrot samples.

Scheme No. and Abbreviation	Calculated Parameters				Assumed Value	Value of the Objective Function
	h_m (kg/m^{2}s)	h_T (J/m^{2}sK)	φ_{cr2} (1)	ΔQ (W)	X_{cr2} (kg/kg)	$f(h_m,h_T,\Delta Q,\varphi_{cr2})$
1—CV_502	0.000416	24.3	0.191	–	0.3	0.6589
2—CV_702	0.00116	20.8	0.194	–	1.0	0.2682
3—CVUD_502200	0.000773	44.5	0.323	0.584	0.3	0.1048
4—CVUD_702200	0.00140	40.5	0.275	0.887	0.8	0.1219

The Rosenbrock optimization scheme was applied for the determination of the model parameters. The method requires a starting point in the optimization variable space. The optimization method is considered to be convergent if it gives the same result, regardless of the assumed starting point. Therefore, all calculations were carried out using different starting points for the optimization procedure (h_m, h_T, φ_{cr}, and ΔQ). Examples of optimization procedure paths are shown in Figure 4. The chart shows three optimization paths (red, blue, and green). Each of these paths starts at a different point in space (points 1, 2, and 3 in Figure 4). For clarity, the chart was made with the standardized variable ranging from zero to the maximum value occurring during the calculation. Each of these paths was obtained as a result of the optimization procedure starting from a different point in the optimization variable space. It was found, regardless of the starting point of the performed calculations, that the same optimal point was obtained (see Figure 4). This means that the objective function, f, had only one minimum, and that the proposed algorithm was convergent to this solution.

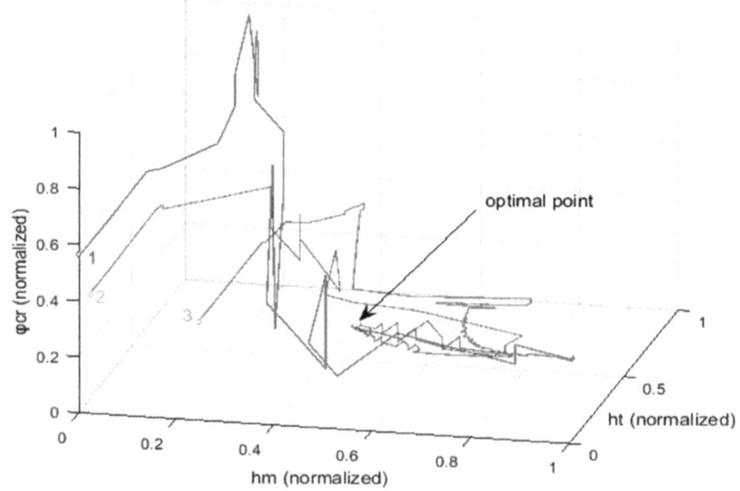

Figure 4. Optimization paths of the mass and heat transfer coefficients and the critical relative humidity (h_m, h_T, and φ_{cr}) (drying scheme CVUD_502200).

The numerical procedure was used to calculate process parameters describing the heat and mass transfer during all of the performed experimental tests. Examples of the drying kinetics obtained, both during experiments and by numerical simulations, are shown in Figure 5. The results demonstrate a good qualitative and quantitative agreement between the experimental and numerical results. It should be noted that the proposed procedure involves two tasks. The first concerns solving nonlinear equations for long-term integration of the model equations. The second is the search for the optimal solution. The performed tests exhibited a good convergence of both tasks, namely, of the initial problem solution and the optimization method.

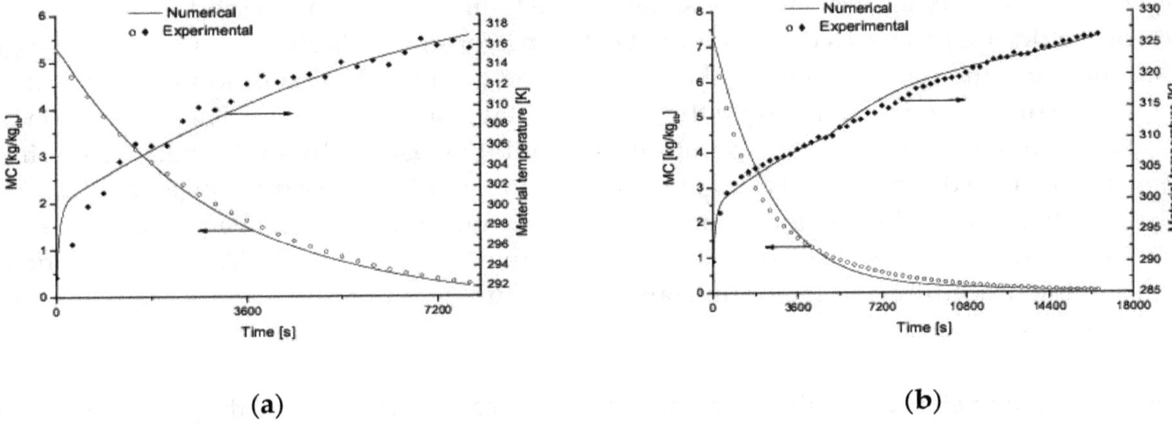

(a) (b)

Figure 5. Drying kinetics of ultrasound-assisted convective drying: (**a**) Drying of apple—drying scheme CVUD_404200, and (**b**) drying of carrot—drying scheme CVUD_502200.

Tables 2 and 3 summarize the results of the presented studies. The parameters, which describe the drying process, namely, the effective coefficients of heat h_T and mass h_m convective transfer, the additional heat source ΔQ, and the critical relative humidity φ_{cr2}, were determined using the optimization procedure described previously (in the section on the estimation of model parameters). It was found that ultrasound application improves all parameters (h_m, h_T, and φ_{cr2}) describing the convective heat and mass transfer between the samples and the drying medium. It was also found that the heating effect of ultrasound application caused by the ultrasound's absorption ΔQ was very small. The power of the energy absorbed was less than 1% of the ultrasound generator's power.

Table 3. Results of the model parameter determination for apple samples.

Scheme No. and Abbreviation	Calculated Parameters			Value of the Objective Function $f(h_m,h_T,\Delta Q,X_{cr})$
	h_m (kg/m^{2}s)	h_T (J/m^{2}sK)	ΔQ (W)	
1—CV_404	0.000501	22.5	–	0.1235
2—CVUD_404200	0.000843	29.4	0.158	0.1321

7. Discussion

The described method has been used by the authors to determine the coefficient of the mass transfer of strawberries [35], raspberries [36], and green pepper [37]. The mass transfer coefficient in the case of convective drying was equal to 8.61×10^{-5}, 1.73×10^{-4}, and 2.80×10^{-4} kg/m^2 s, for the strawberries, raspberries, and green pepper, respectively. The application of ultrasound resulted in an increase of the coefficient by 96% for the strawberries, 45% for the raspberries, and 26% for the green pepper. These results indicate that the mass transfer coefficient depends on the material to be dried. The values of the coefficient of the mass transfer obtained in the present work presented in Tables 2 and 3 are an order of greater value. This results from the differences in the movement of the material in the dryer. The pouring of the material in rotating cylinder construction improves the heat and mass transfer. The obtained results show that both estimated transfer coefficients depend on the material to be dried, the dryer construction, and the drying conditions.

8. Conclusions

In this article, an optimization procedure is proposed for determination of the coefficient in the drying model used by the authors. The coefficients of the effective heat and mass transfer by convective drying and convective drying enhanced with ultrasound are introduced. The presented drying model and the method of coefficient estimation enable the description of the single stage of the falling drying period (FDRP) (drying of apples) and two stages of the falling drying rate period (FDRP) (drying of carrots) for the drying processes of bio-products. The performed tests showed a good convergence of the solution and the optimization methods. The good quality of the solution was confirmed by the compliance of the model results with the experimental results (Figure 5). The application of the Adams–Bashforth method resulted in solution compatibility, especially in the part describing the transition between the drying periods and in the part of the solution describing the end of the process. The good quality of the applied optimization scheme was confirmed by the convergence of the solution at one point, regardless of the starting point in the variable space (Figure 4). The quality of both the equation solving procedure and the optimization procedure was quantified by the low value of the objective function (Tables 2 and 3).

The kinetic parameters determined from the numerical calculation showed the positive influence of convective drying with ultrasound enhancement for both tested materials. The results of the performed calculations confirm that absorption of the air-borne ultrasound energy into the dried material is very small. The method presented in this article was shown to be very efficient for determination of the heat and mass transfer processes based on the presented model of drying kinetics. This model could be easily extended to other complex drying processes, for example, hybrid drying based on several drying techniques.

Author Contributions: Conceptualization, G.M. and M.S.; methodology, G.M. and M.S.; software preparation, M.S.; experiments D.M.; numerical calculation, M.S.; results visualization D.M. and M.S.; writing—original draft preparation, G.M.; writing—review and editing, M.S., G.M., and D.M. All authors have read and agreed to the published version of the manuscript.

Nomenclature

Symbol	Designation	Unit
c_s	specific heat of dry solid	J/kgK
c_l	specific heat of moisture (water)	J/kgK
f	objective function	1
h_m	mass transfer coefficient	kg/m^2s
h_T	heat transfer coefficient	J/m^2sK
l	latent heat of evaporation	J/kg
m_s	dry mass	kg
p_{vs}	saturated vapor pressure	Pa
t	time	s
A	surface of sample	m^2
A_0	initial surface of sample	m^2
A_m	surface of mass exchange	m^2
A_T	surface of heat exchange	m^2
MC	moisture content dry basis	kg/kg_{db}
T	absolute temperature	K
T_0	initial temperature	K
T_a	ambient air temperature	K
T_{exp}	experimental value of temperature	K
T_{max}	maximal value of temperature	K
T_{min}	minimal value of temperature	K
T_{num}	numerical value of temperature	K
V	volume of sample	m^3
V_0	initial volume of sample	m^3

X	moisture content dry basis	kg/kg_{db}
X_0	initial moisture content dry basis	kg/kg_{db}
X_{cr}	critical moisture content dry basis	kg/kg_{db}
X_{cr2}	second critical moisture content dry basis	kg/kg_{db}
X_{eq}	equilibrium moisture content dry basis	kg/kg_{db}
X_{exp}	experimental value of moisture content	kg/kg_{db}
X_{max}	maximal value of moisture content	kg/kg_{db}
X_{min}	minimal value of moisture content	kg/kg_{db}
X_{num}	numerical value of moisture content	kg/kg_{db}
α_V	volumetric shrinkage coefficient	1
φ	relative air humidity	1
φ_a	relative air humidity in ambient air	1
$\varphi\vert_{\partial B}$	air humidity close to the dried sample surface	1
φ_{cr2}	second critical air relative humidity	1
ΔQ	heat due to absorption of ultrasonic waves	W
$\Phi(t, X, t)$	function—simplifying designation	kg/s
$\Psi(t, X, t)$	function—simplifying designation	K/s

References

1. Jayaraman, K.S.; Das Gupta, D.K. Drying of Fuits and Vegetables. In *Handbook of Industrial Drying*, 4th ed.; Mujumdar, A.S., Ed.; CRC Press: Boca Raton, FL, USA, 2014; pp. 611–636.
2. Chen, G.; Mujumdar, A.S. Drying Herbal Medicines and Tea. In *Handbook of Industrial Drying*, 4th ed.; Mujumdar, A.S., Ed.; CRC Press: Boca Raton, FL, USA, 2014; pp. 637–646.
3. Musielak, G.; Mierzwa, D.; Kroehnke, J. Food Drying Enhancement by Ultrasound—A Review. *Trends Food Sci. Technol.* **2016**, *56*, 126–141. [CrossRef]
4. Szadzińska, J.; Pashminehazar, R.; Kharaghani, A.; Tsotsas, E.; Łechtańska, J. Microwave and Ultrasound Assisted Convective Drying of Raspberries: Drying Kinetics and Microstructural Changes. *Dry. Technol.* **2018**, *37*, 1–12. [CrossRef]
5. Gallego-Juárez, J.A.; Riera, E.; de la Fuente Blanco, S.; Rodriguez-Corral, G.; Acosta-Aparicio, V.M.; Blanco, A. Application of High-Power Ultrasound for Dehydration of Vegetables: Processes and Devices. *Dry Technol.* **2007**, *25*, 1893–1901. [CrossRef]
6. Cárcel, J.A.; García-Pérez, J.V.; Benedito, J.; Mulet, A. Food Process Innovation through New Technologies. *J. Food Eng.* **2012**, *110*, 200–207. [CrossRef]
7. Bantle, M.; Hanssler, J. Ultrasonic Convective Drying Kinetics of Clipfish during the Initial Drying Period. *Dry Technol.* **2013**, *31*, 1307–1316. [CrossRef]
8. Siucińska, K.; Konopacka, D. Application of Ultrasound to Modify and Improve Dried Fruit and Vegetable Tissue—A Review. *Dry Technol.* **2014**, *32*, 1360–1368. [CrossRef]
9. Kowalski, S.J.; Mierzwa, D. Ultrasound-Assisted Convective Drying of Biological Materials. *Dry Technol.* **2015**, *33*, 1601–1613. [CrossRef]
10. Kowalski, S.J.; Pawłowski, A. Intensification of Apple Drying due to Ultrasound Enhancement. *J. Food Eng.* **2015**, *156*, 1–9. [CrossRef]
11. Gallego-Juárez, J.A.; Rodríguez-Corral, G.; San Emeterio-Prieto, J.L.; Montoya-Vitini, F. Electroacoustic Unit for Generating High Sonic and Ultrasonic Intensities in Gases and Interphases. U.S. Patent No. 5,299,175, 29 March 1994.
12. Gallego-Juárez, J.A.; Riera, E.; Rodríguez-Corral, G.; Gálvez Moraleda, J.C.; Yang, T.S. A New High-Intensity Ultrasonic Technology for Food Dehydration. *Dry Technol.* **1999**, *17*, 597–608. [CrossRef]
13. García-Pérez, J.V.; Cárcel, J.A.; de la Fuente-Blanco, S.; Riera-Franco de Sarabia, E. Ultrasonic Drying of Foodstuff in a Fluidized Bed: Parametric Study. *Ultrasonics* **2006**, *44*, e539–e543. [CrossRef]
14. Khmelev, V.N.; Shalunov, A.V.; Barsukov, R.V.; Abramenko, D.S.; Lebedev, A.N. Studies of Ultrasonic Dehydration Efficiency. *J. Zhejiang Univ. Sci. A* **2011**, *12*, 247–254. [CrossRef]
15. Kouchakzadeh, A.; Ghobadi, P. Modeling of Ultrasonic-Convective Drying of Pistachios. *Agric. Eng. Int. CIGR J.* **2012**, *14*, 144–149.

16. Konopacka, D.; Parosa, R.; Piecko, J.; Połubok, A.; Siucińska, K. Ultrasound & Microwave Hybrid Drying Device for Colored Fruit Preservation—Product Quality and Energy Efficiency. In Proceedings of the 8th Asia-Pacific Drying Conference (ADC 2015), Kuala Lumpur, Malaysia, 10–12 August 2015; pp. 252–258.
17. Cárcel, J.A.; García-Pérez, J.V.; Riera, E.; Mulet, A. Influence of Highintensity Ultrasound on Drying Kinetics of Persimmon. *Dry Technol.* **2007**, *25*, 185–193. [CrossRef]
18. García- Pérez, J.V.; Cárcel, J.A.; Benedito, J.; Mulet, A. Power Ultrasound Mass Transfer Enhancement in Food Drying. *Food Bioprod. Process.* **2007**, *85*, 247–254. [CrossRef]
19. Cárcel, J.A.; García-Pérez, J.V.; Riera, E.; Mulet, A. Improvement of Convective Drying of Carrot by Applying Power Ultrasound. Influence of Mass Load Density. *Dry Technol.* **2011**, *29*, 174–182. [CrossRef]
20. Gamboa-Santos, J.; Montilla, A.; Cárcel, J.A.; Villamiel, M.; García-Pérez, J.V. Air-Borne Ultrasound Application in the Convective Drying of Strawberry. *J. Food Eng.* **2014**, *128*, 132–139. [CrossRef]
21. Rodríguez, O.; Santacatalina, J.V.; Simal, S.; García- Pérez, J.V.; Femenia, A.; Rosselló, C. Influence of Power Ultrasound Application on Drying Kinetics of Apple and its Antioxidant and Microstructural Properties. *J. Food Eng.* **2014**, *129*, 21–29. [CrossRef]
22. Kowalski, S.J.; Pawłowski, A. Modeling of Kinetics in Stationary and Intermittent Drying. *Dry Technol.* **2010**, *28*, 1023–1031. [CrossRef]
23. Aversa, M.; Van der Voort, A.-J.; de Heij, W.; Tournois, B.; Curcio, S. An Experimental Analysis of Acoustic Drying of Carrots: Evaluation of Heat Transfer Coefficients in Different Drying Conditions. *Dry Technol.* **2011**, *29*, 239–244. [CrossRef]
24. Raszkowski, T.; Samson, A. Numerical Approaches to the Heat Transfer Problem in Modern Electronic Structures. *Comput. Sci.* **2017**, *18*, 71–93. [CrossRef]
25. Zwarycz-Makkles, K.; Majorkowska-Mech, D. Gear and Runge-Kutta Numerical Discretization Method in Differential Equations of Adsorption in Adsorption Heat Pump. *Appl. Sci.* **2018**, *8*, 2437. [CrossRef]
26. Duc, L.A.; Hyuk, K.D. Mathematical Modeling and Simulation of Rapeseed Drying on Concurrent-Flow Dryer. In *Current Drying Processes*; Palqa-Rosas, I., Ed.; IntechOpen: Rijeka, Croatia; London, UK, 2020. [CrossRef]
27. Krupowicz, A. *Numerical Methods of Initial Value Problems of Ordinary Differential Equations*; PWN: Warsaw, Poland, 1986.
28. Rosenbrock, H.H. An Automatic Method for Finding the Greatest or Least Value of a Function. *Comput. J.* **1960**, *3*, 175–184. [CrossRef]
29. Mujumdar, A.S. Principles, Classification, and Selection of Dryers. In *Handbook of Industrial Drying*, 4th ed.; Mujumdar, A.S., Ed.; CRC Press: Boca Raton, FL, USA, 2014; pp. 3–30.
30. Musielak, G.; Banaszak, J. Non-Linear Heat and Mass Transfer during Convective Drying of Kaolin Cylinder under Non-Steady Conditions. *Transp. Porous Media* **2007**, *66*, 121–134.
31. Strumiłło, C. *Foundations of the Drying Theory and Technology*, 2nd ed.; WNT: Warsaw, Poland, 1983.
32. Davis, M. *Numerical Methods and Modelling for Chemical Engineers*; Wiley: Montreal, QC, Canada, 1984.
33. Toledo, R.T. *Fundamentals of Food Process Engineering*; Heldman, D.R., Ed.; Springer: New York, NY, USA, 2007.
34. DTU Food. Available online: http://www.foodcomp.dk/v7/fcdb_details.asp?FoodId=1128 (accessed on 16 April 2016).
35. Szadzińska, J.; Kowalski, S.J.; Stasiak, M. Microwave and ultrasound enhancement of convective drying of strawberries: Experimental and modeling efficiency. *Int. J. Heat Mass Transf.* **2016**, *103*, 1065–1074. [CrossRef]
36. Kowalski, S.J.; Pawłowski, A.; Szadzińska, J.; Łechtańska, J.; Stasiak, M. High power airborne ultrasound assist in combined drying of raspberries. *Innov. Food Sci. Emerg. Technol.* **2016**, *34*, 225–233. [CrossRef]
37. Szadzińska, J.; Łechtańska, J.; Kowalski, S.J.; Stasiak, M. The effect of high power airborne ultrasound and microwaves on convective drying effectiveness and quality of green pepper. *Ultrason. Sonochem.* **2017**, *34*, 531–539. [CrossRef]

Revisiting the Role of Mass and Heat Transfer in Gas–Solid Catalytic Reactions

Riccardo Tesser [1] and Elio Santacesaria [2,*]

1 NICL—Naples Industrial Chemistry Laboratory, Department of Chemical Science, University of Naples Federico II, 80126 Naples, Italy; riccardo.tesser@unina.it
2 CEO of Eurochem Engineering Ltd., 20139 Milan, Italy
* Correspondence: elio.santacesaria@eurochemengineering.com

Abstract: The tremendous progress in the computing power of modern computers has in the last 20 years favored the use of numerical methods for solving complex problems in the field of chemical kinetics and of reactor simulations considering also the effect of mass and heat transfer. Many classical textbooks dealing with the topic have, therefore, become quite obsolete. The present work is a review of the role that heat and mass transfer have in the kinetic studies of gas–solid catalytic reactions. The scope was to collect in a relatively short document the necessary knowledge for a correct simulation of gas–solid catalytic reactors. The first part of the review deals with the most reliable approach to the description of the heat and mass transfer outside and inside a single catalytic particle. Some different examples of calculations allow for an easier understanding of the described methods. The second part of the review is related to the heat and mass transfer in packed bed reactors, considering the macroscopic gradients that derive from the solution of mass and energy balances on the whole reactor. Moreover, in this second part, some examples of calculations, applied to chemical reactions of industrial interest, are reported for a better understanding of the systems studied.

Keywords: gas–solid catalytic reactions; chemical kinetics; heat and mass transfer

---

1. Introduction

When a reaction occurs inside a catalytic particle, the reagents are consumed, giving rise to products and, in the meantime, heat is released or absorbed according to whether the enthalpy of the reaction is positive or negative. Inside and around the particles, gradients of respective concentration and temperature are generated as a consequence. Then, if the particles are put inside a tubular reactor (see Figure 1), macroscopic gradients (both in axial and radial directions) also arise as a consequence of the average rate of reaction in any single catalytic particle and the regime of mass and heat flow developed in the specific reactor. In Figure 1, all the possible gradients related to both temperature and concentration occurring in a tubular gas–solid catalytic reactor are illustrated.

These macroscopic (or "long-range") gradients can be vanished by employing "gradientless" reactors that are isothermal CSTRs (continuous stirred tank reactors) normally used in laboratory kinetic studies (see Figure 2A,B).

Moreover, each particle inside a reactor has its own history, and microscopic gradients are developed in conditions at the particle surface that are generally different from the internal particle conditions.

At the industrial scale, gas–solid catalytic processes are usually carried out in very large capacity equipment represented by packed bed reactors with productivity of thousands of tons per year.

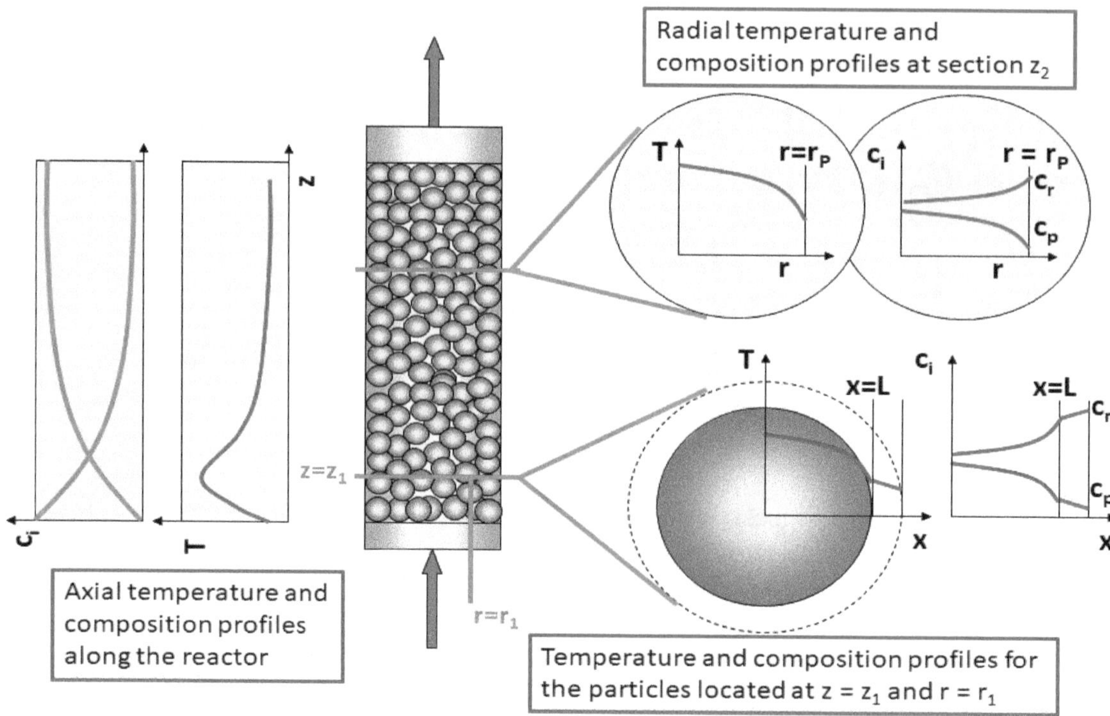

Figure 1. Overview of temperature and concentration macroscopic and microscopic gradients in packed bed reactor (taken and adapted from [1]).

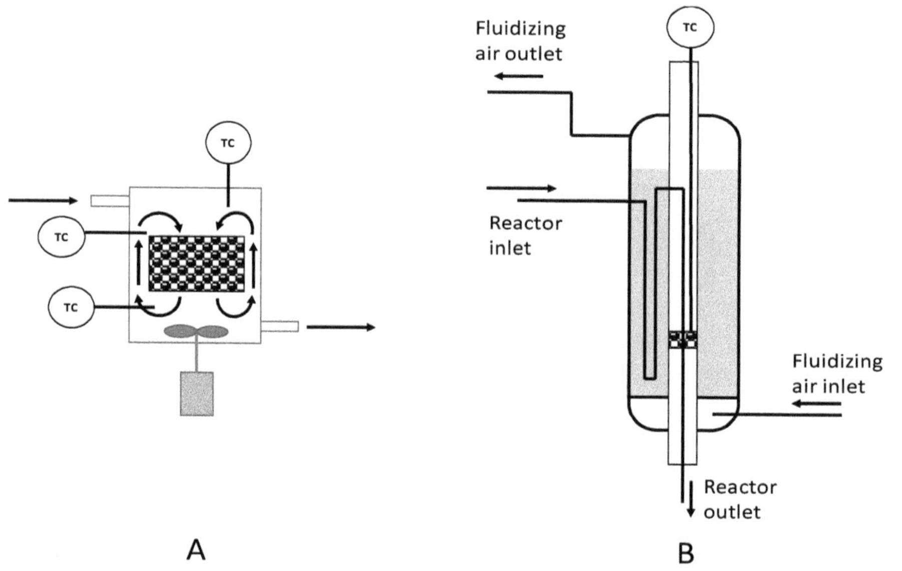

Figure 2. Examples of "gradientless" laboratory reactors. (**A**) Continuous stirred tank reactor (CSTR); (**B**) fixed-bed reactor.

Such reactors are arranged in a complex scheme also containing all the auxiliary equipment necessary for feeding, cooling, heating, or pressurizing operations. The necessity of supplying or removing heat according to the enthalpy of the reaction is the main reason for which reactors with multiple tubes (in many cases thousands of tubes) are preferred. The heat removal is obtained by circulating an opportune fluid externally to the tubes in order to limit the temperature rise (or drop) of the reactive mixture.

Normally, the goal is to obtain isothermal conditions, however, very frequently, these ideal conditions cannot be reached. On the contrary, when an equilibrium reaction is involved in the reaction scheme, such as for example:

$$SO_2 + \frac{1}{2} O_2 \rightleftarrows SO_3$$

a single reactor with large diameter, in which structurally different catalytic packed beds are contained and operating in adiabatic conditions, is preferred, because this type of reactor allows for the control of the overall conversion through the temperature of the outlet flow stream. The heat removal, in this case, is obtained by cooling the flow stream between two different catalytic stages of the reactor.

Two ideal limit conditions can be recognized, the isothermal and adiabatic, realized thanks to a more or less efficient system of heat exchange. However, a condition not isothermal and not adiabatic is more frequently encountered in practice. This implies the development of more complex models to describe the system in which the mentioned limit conditions are considered as particular cases.

Some other aspects are important in the design of fixed-bed reactors, such as pressure drop, safe operating protocol (to avoid runaway problems), temperature range, and catalyst packing modality.

From a general point of view, the design approach of catalytic fixed-bed reactors consists in correctly defining and then solving the mass and energy balance equations. Normally, the solution of such equations must be achieved only numerically, especially when the kinetic systems are characterized by a complex reaction scheme. The problem must to be solved simultaneously both at a microscopic local level, with the obtainment of the reagents and product concentration particle profiles, as well as of the effectiveness factor for all the occurring reactions, and at a macroscopic level, reproducing all the long-range concentration and temperature profiles. This specific situation requires an evaluation of the catalyst effectiveness factor in each position in the catalytic bed, considering the conditions we have at any instance in that point. This subject has been previously described in many books, papers, and reviews [1–17]. A modern and comprehensive approach to the problem, with many solved exercises, can be found in [1]. On the basis of all the examined literature, the scope of this review is to give, in a concise way, all the information necessary to the researchers to correctly face the study of the gas–solid reactions. In the following paragraphs, we consider, first of all, the mass and heat transfer occurring in a single catalytic particle, and then we will treat the macroscopic gradients related to the whole fixed-bed reactor.

2. Mass and Heat Transfer in a Single Catalytic Particle

When a reaction occurs inside a catalytic particle, the reagents are consumed for giving products and a certain amount of heat is consumed or released according to the thermal characteristic of the reaction (exothermic or endothermic). The concentration of the reagents decreases from the external geometric surface of the particles toward the center. The concentration of the products, on the contrary, increases. The temperature changes as a consequence of the heat consumed or released by the reaction, increasing or decreasing from the external surface to the center of the catalytic particle. In other words, the reaction is responsible of the concentration and temperature gradients originating inside the particle that act as driving forces for both the mass and heat transfer inside the catalyst particle. The faster the reaction, the steeper the gradients. In the case of high reaction rate, this effect is propagated toward the external part of the catalyst particle, generating other gradients of concentration and temperature between the catalyst surface and the bulk fluid. When the fluid flow regime is turbulent, as normally occurs in industrial reactors, the external gradients are confined to very thin layer, named the boundary layer, that surrounds the solid surface. The boundary layer is quiescent, and consequently mass and heat transfer occur through it, with a relatively slow process characterized by the molecular diffusion mechanism. The effects of reaction and diffusion rates are concentration and temperature profiles, respectively, such as the ones reported in Figure 3. External diffusion and chemical reaction are consecutive steps, and their contributions to the overall reaction rates can be considered separately. A similar approach cannot be adopted for the internal diffusion as it occurs simultaneously with the chemical reaction. To describe the influence of internal diffusion on reaction rate requires solving the mass and heat balance equations related to any single particle for evaluating the concentration and temperature profiles inside the pellet.

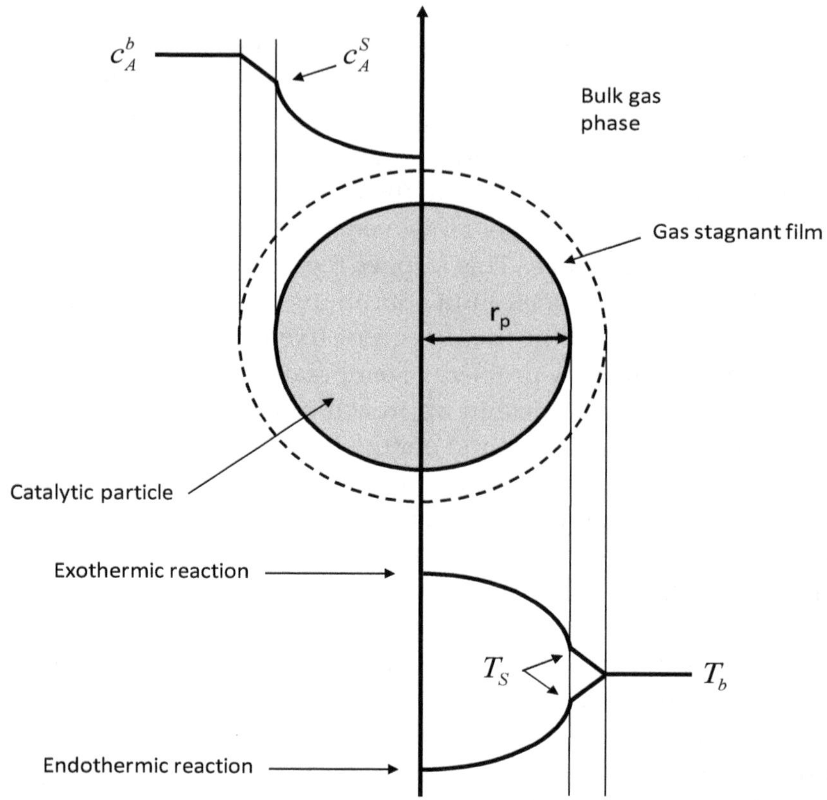

Figure 3. Profiles of concentration and temperature in a spherical catalytic particle.

2.1. *Diffusion with Reaction in a Single Catalytic Particle: Mass and Heat Balance Equations*

For a spherical particle, the mass balance can be written by considering the inlet, outlet, reaction, and accumulation terms related to a spherical shell of thickness dr and radius r (see Figure 4):

$$\begin{bmatrix} \text{diffusionrate} \\ \text{inwardatx} = \text{x} + \text{dx} \end{bmatrix} - \begin{bmatrix} \text{diffusionrate} \\ \text{outwardatx} = \text{x} \end{bmatrix} - \begin{bmatrix} \text{reactionrate} \\ \text{intotheshell} \end{bmatrix} = [\text{accumulation}] \quad (1)$$

Assuming steady state conditions, it results null the accumulation term and then

$$4\pi(x+dx)^2 N_{x+dx} - 4\pi x^2 N_x - 4\pi x^2 dx \cdot v_r = 0 \quad (2)$$

By introducing the Fick's law $N = D_{eff}\frac{dc}{dx}$ for the internal diffusive flux and a generic power law for the reaction rate $v = S_v k_S c^n$, related to a single reaction, and through rearranging we obtain

$$\frac{D_{eff}}{x^2}\frac{d}{dx}\left(x^2\frac{dc}{dx}\right) - S_v k_S c^n = 0 \quad (3)$$

with the boundary conditions

$$\begin{aligned} &\text{for } x = r_P \rightarrow c = c_S \\ &\text{for } x = 0 \rightarrow \tfrac{dc}{dx} = 0 \end{aligned}$$

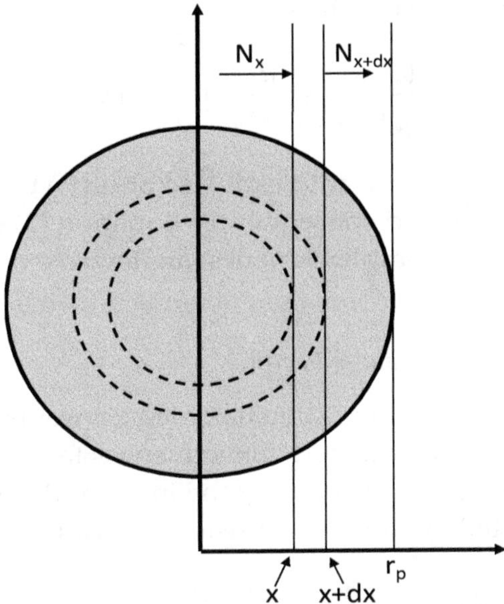

Figure 4. Reference scheme for mass and heat balance related to a catalyst particle.

For the heat balance, it is possible to follow a similar approach by introducing Fourier's law $q = -k_{eff}\frac{dT}{dx}$ instead of Fick's law, obtaining the following equation:

$$\frac{k_{eff}}{x^2}\frac{d}{dx}\left(x^2\frac{dT}{dx}\right) - (-\Delta H)S_V k_S c^n = 0 \tag{4}$$

with boundary conditions

$$\text{for } x = r_P \rightarrow T = T_S$$
$$\text{for } x = 0 \rightarrow \frac{dT}{dx} = 0$$

Considering the common terms of Equations (3) and (4), it is possible to write

$$(T - T_S) = \frac{D_{eff}}{k_{eff}}(c - c_S)(-\Delta H) \tag{5}$$

From this equation, we can conclude that for any concentration profile, inside the particle, a corresponding profile of temperature can easily be determined by using Equation (5). Alternatively, a full energy balance on the particle must be solved. A maximum temperature gradient $\Delta T_{\max}$ can be obtained when the concentration at the center of the particle can be assumed near to zero; in this case, $\Delta c \approx c_S$, and hence

$$\Delta T_{\max} = \frac{D_{eff}}{k_{eff}} c_S(-\Delta H) \tag{6}$$

Referring $\Delta T_{\max}$ to T_S, the temperature at the catalyst surface, the Prater's number is obtained, defined as $\beta = \Delta T_{\max}/T_S$.

As the thermal conductivity of solid catalyst particles is normally much higher than those of the gaseous reaction mixture, in steady state conditions, internal temperature gradients are rarely important in practice.

The evaluation of the internal profiles of both concentration and temperature requires the solution of Equations (3) and (4). For this purpose, it is opportune to introduce some dimensionless terms such as

$$\varepsilon_{dr} = x/r_P \; \gamma_{dr} = c/c_S \; \psi_{dr} = v_r(c)/v_r(c_S) \; \phi = r_P\sqrt{\frac{S_V k_S c_S^{n-1}}{D_{eff}}} \tag{7}$$

and Equation (3) for mass conservation becomes

$$\frac{1}{\varepsilon_{dr}^{2}}\frac{d}{d\varepsilon_{dr}}\left(\varepsilon_{dr}^{2}\frac{d\gamma_{dr}}{d\varepsilon_{dr}}\right)=\phi^{2}\psi_{dr} \tag{8}$$

where ϕ is called Thiele modulus [18]. It is interesting to observe that for $n = 1$, the Thiele modulus is independent of the concentration, and consequently Equation (3) or (8) can be solved analitically, while for different reaction orders or complex kinetics, an iterative numerical solution strategy must be adopted.

2.2. Definition and Evolution of the Effectiveness Factor

If the internal concentration profile of γ (dimensionless concentration) is known, it is possible to evaluate another dimensionless term η, named "effectiveness factor", defined as the ratio between the observed reaction rate, more or less affected by the internal diffusion, and the rate occurring in chemical regime, that is, not limited by internal diffusion. We can write

$$\eta=\frac{\text{effective reaction rate}}{\text{reaction rate from kinetic law}} \tag{9}$$

and can write accordingly

$$\eta=\frac{\int_{0}^{r_p}4\pi x^{2}v_{r}(c)dx}{\frac{4}{3}\pi r_{p}^{3}v_{r}(c_{S})}=3\int_{0}^{1}\varepsilon_{dr}^{2}\psi_{dr}(\gamma_{dr})d\varepsilon_{dr} \tag{10}$$

Therefore, η is a dimensionless factor directly giving the effect of the internal diffusion on the reaction rate. For a reaction rate of a single reaction of n-th order, affected by internal diffusion, we can simply write

$$v_{r}=\eta k_{S}S_{V}c_{S}^{n}=\eta k_{V}c_{S}^{n} \tag{11}$$

The effectiveness factor η can also be determined by considering that, in steady state conditions, the overall reaction rate in a particle is equal to the rate of external mass transfer from bulk to the surface. Equation (10) can be rewritten as

$$\eta=\frac{-4\pi r_{P}^{2}D_{eff}\left(\frac{dc}{dx}\right)_{rp}}{\frac{4}{3}\pi r_{P}^{3}v_{r}(c_{S})}=\frac{-3D_{eff}\left(\frac{dc}{dx}\right)_{rp}}{r_{P}k_{S}c_{S}} \tag{12}$$

As mentioned, for reaction order $n = 1$, the concentration profile can be analytically determined, and this can be done with Equation (13).

$$-\frac{dc}{dx}=\frac{c_{S}}{x}\left[\frac{\phi}{\tanh\phi}-\frac{1}{\phi}\right] \tag{13}$$

from which the following expression for the effectiveness factor can be derived by assuming a particle with spherical geometry:

$$\eta=\frac{3}{\phi}\left[\frac{1}{\tanh\phi}-\frac{1}{\phi}\right] \tag{14}$$

This equation changes with the shape of the catalyst particles, and the Thiele modulus ϕ changes too, as the quantity r_P in Equation (7) becomes a characteristic length given by the ratio between volume and external surface area of the catalytic pellet.

In some cases, the kinetic law is unknown, even if the data of the reaction rate are available. The evaluation of the Thiele modulus and of the effectiveness factor in these cases is not possible.

For this purpose, it is useful to define another dimensionless modulus, named the "Weisz modulus" [19], through the following relation:

$$M_W = \frac{r_P{}^2 v_r}{c_S D_{eff}} = \phi^2 \eta \tag{15}$$

The Weisz modulus allows for the evaluation of the effectiveness factor when experimental data of reaction rate are available.

Different plots of η, the "effectiveness factor", as a function of ϕ or M_W can be drawn. Examples of these plots for spherical particles and first-order reactions are reported in Figure 5A,B. In these plots, we can recognize three different zones, the first, at low ϕ and M_W values, delimiting the chemical regime; the latter for high ϕ and M_W values, identifying the diffusional regime; and an intermediate zone corresponding to the gradual transition from chemical to diffusional regime. When the diffusional regime is operative, the effectiveness factor η can be calculated in an approximated way as $\eta = 1/\phi = 1/M_W$. This method of calculation can also be extended to the intermediate zone. This asymptotic approximation gives place to errors in η of less than 5%.

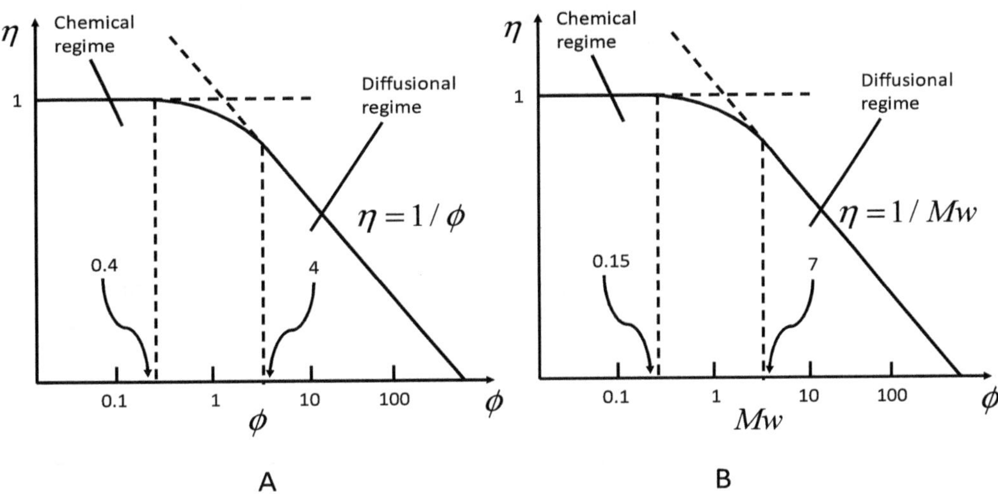

Figure 5. (**A**) Relationship between effectiveness factor and Thiele modulus; (**B**) relationship between effectiveness factor and Weisz modulus. Re-elaborated from Santacesaria [20] with the permission of Elsevier-Catalysis Today 1997.

The effect of the catalyst particle shape on $\eta = \eta(\phi)$ is quite small, while a larger influence has the reaction order, as can be appreciated in Figure 6.

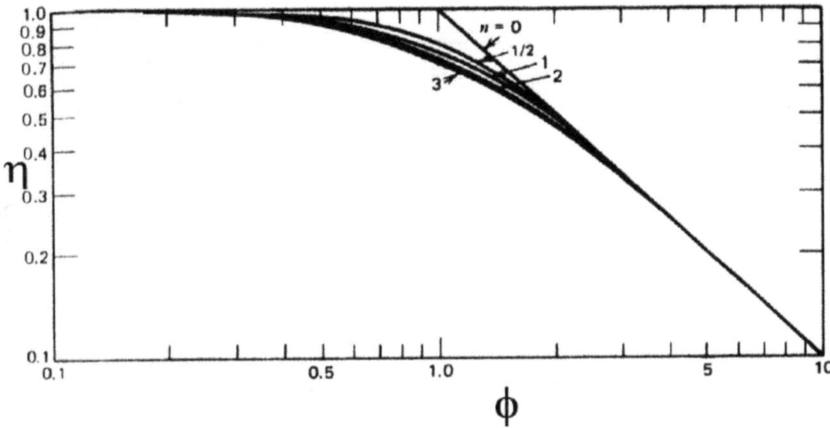

Figure 6. Effectiveness factor versus Thiele modulus for different order of reaction. Re-elaborated from Froment and Bischoff [21].

The effectiveness factor can also be evaluated experimentally by determining the reaction rate in the presence of catalyst pellets of different diameters and on finely powdered catalyst operating in chemical regime:

$$\eta = \frac{\text{rate observed for a ginen particle size}}{\text{rate observed in chemical regime on powdered catalyst}} \tag{16}$$

We have already seen that inside the catalyst particle, in correspondence to any concentration gradient, a temperature gradient is associated, determinable with Equations (5) or (6). The evolution of the effectiveness factor with the Thiele and Weisz moduli, reported in Figures 3 and 4, corresponds to isothermal conditions. When the reaction is exothermic or endothermic, the temperature inside the particle is, respectively, greater or lower than the external fluid. In these cases, the effectiveness factor can be affected by two other dimensionless factors:

(a) a heat generation parameter:

$$\beta = \frac{c_S(-\Delta H)D_{eff}}{k_{eff}T_S} = \frac{\Delta T_{\max}}{T_S} = \text{Prater's number} \tag{17}$$

(b) the reaction rate exponential parameter:

$$\alpha_E = \frac{E}{RT_S} \tag{18}$$

For exothermic reactions $\beta > 0$ while for endothermic reactions $\beta < 0$; obviously, the isothermal condition can be identified when $\beta = 0$. For exothermic reactions, the effectiveness factor can be much greater than 1, while for endothermic reactions, this value is never reached. Examples of curves η-ϕ for different β values, at any given value of α_E, are reported in Figure 7.

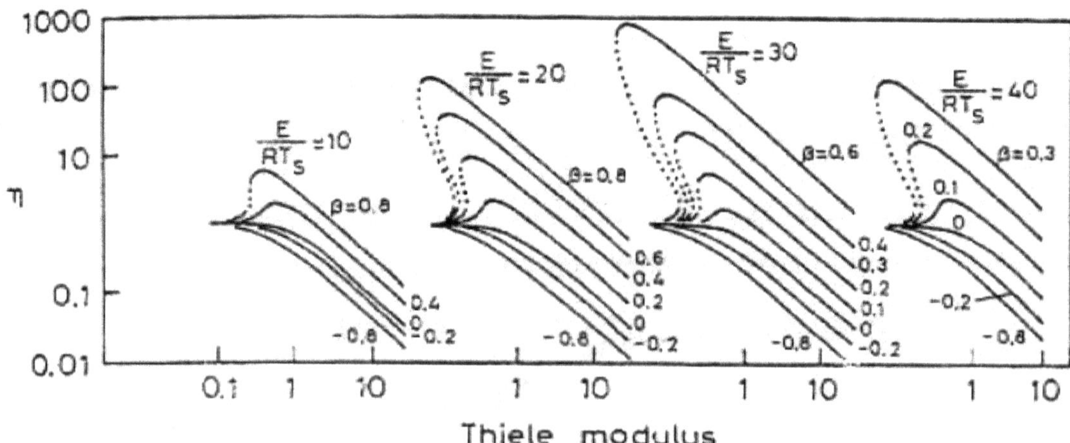

Figure 7. Effectiveness factors against the Thiele modulus in the case of both exothermic and endothermic reactions with an internal temperature gradient. Re-elaborated from Santacesaria [20] with the permission of Elsevier-Catalysis Today 1997.

2.3. *Determination of the Effective Diffusional Coefficient D_{eff} and the Effective Thermal Conductivity k_{eff}*

Effective diffusional coefficient depends on bulk diffusion coefficient D_{be}, the diffusion coefficient of the fluid in the macropores, and on the Knudsen diffusion coefficient D_{ke}, the diffusion coefficient in the micropores. We can write

$$\frac{1}{D_{eff}} = \frac{1}{D_{be}} + \frac{1}{D_{ke}} \tag{19}$$

where

$$D_{be} = \frac{D_{12}\theta}{\tau} \text{ and } D_{ke} = 1.94 \cdot 10^4 \frac{\theta^2}{\tau S_V \rho_P}\sqrt{\frac{T}{M}} \quad (20)$$

with θ being the porosity of the solid; τ being the tortuosity factor, an empirical parameter dependent on the characteristics of the pellets porosity texture with values falling in the range 0.3–10; S_V being the specific surface area; and ρ_p the catalyst particle density. D_{12}, which is normally considered equal to D_{21}, is the molecular diffusion coefficient for two components:

$$D_{12} = \frac{1.858 \cdot 10^{-3}\sqrt{\frac{T^3(M_1+M_2)}{M_1M_2}}}{P\sigma_{12}^2\Omega_D} \quad (21)$$

σ_{12} is the kinetic diameter for the molecules, while Ω_D, named "collision integral", is a function of k_BT/ε_{12}; k_B is the Boltzmann constant, while ε is a molecular interaction parameter. Both σ_{12} and ε_{12} can be determined from the Lennard–Jones intermolecular potential equation (Equation (22)):

$$\phi_{LJ}(r) = 4\varepsilon_{ij}\left[\left(\frac{\sigma_{ij}}{\rho_d}\right)^{12} - \left(\frac{\sigma_{ij}}{\rho_d}\right)^{6}\right] \quad (22)$$

$$\sigma_{ij} = (\sigma_i + \sigma_j)/2 \quad (23)$$

$$\varepsilon_{ij} = \sqrt{\varepsilon_i\varepsilon_j} \quad (24)$$

where ρ_d is the intermolecular distance, and σ_i and ε_i can be evaluated from critical temperature and volume of the molecules, that is, $\varepsilon_i/k_B = 0.75\ T_c$ and $\sigma_i = 0.833\ V_{ci}{}^{1/3}$.

When we have a mixture of more than two components, the calculation can be made by averaging the properties. Molecular diffusion coefficient D_{im} is, for example,

$$D_{im} = \frac{(1 - y_i)}{\sum\limits_j y_j/D_{ij}} \quad (25)$$

Because of the uncertainty of the tortuosity factor τ, many experimental data have been determined for D_{eff}, generally in steady-state conditions by using an apparatus such as the one schematized in Figure 8. A single pellet is put in a device in which different gases are fed above and below the catalyst particle at the same pressure. Each gas slowly flows through the pellet and is determined at the outlet. The rate of gas diffusion through the pellet is related to D_{eff} because

$$N_A = -D_{eff}\frac{dc_A}{dr} = -\frac{P}{RT}D_{eff}\frac{dy_A}{dr} = -D_{eff}\frac{P}{RT}\frac{(y_A - y_A^0)}{\Delta r} \quad (26)$$

A dynamic method can also be used by employing a pulse of a diffusing component. The response pulse is related to the value of D_{eff}.

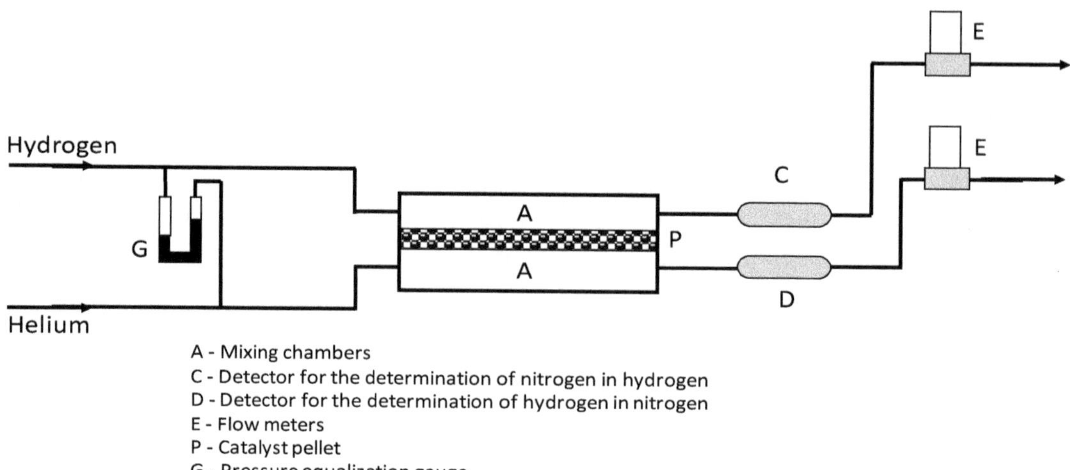

Figure 8. Scheme of the experimental device for the determination of effective diffusivity of a catalyst pellet.

The effective thermal conductivities of catalyst pellet could be surprisingly low for the numerous void spaces hindering the transport of energy. A simple but approximate approach for calculating k_{eff} has been given by Woodside and Messner [22]:

$$k_{eff} = k_{Sol}\left(\frac{k_f}{k_{Sol}}\right)^{1-\varepsilon_{Bs}} \tag{27}$$

where k_f and k_{Sol} are the thermal conductivities of the bulk fluid and of the solid phase, respectively, while ε_{Bs} is the void fraction of the solid.

Notwithstanding the difficulties in predicting k_{eff}, a reliable value can be estimated because it falls in a rather restricted range 0.1–0.4 Btu/(h ft °F) [1].

2.4. External Gradients

As before mentioned, external diffusion and reaction inside the catalytic particles can be considered as consecutive steps. Therefore, the corresponding rates can be expressed with different relationships. The external mass transfer rate expression derives from the first Fick's law and results in

$$v_{mt} = k_m a_m (c_b - c_S) \tag{28}$$

In steady state conditions, this expression must be equated to the one describing the rate of internal diffusion with reaction, that is,

$$v_r = \eta k c_S^n = k_m a_m (c_b - c_S) \tag{29}$$

For $n = 1$, after the elimination of c_S, it is possible to write

$$v_r = \frac{c_b}{\frac{1}{k_m a_m} + \frac{1}{\eta k}} \tag{30}$$

where the contribution of the resistance to the reaction rate, by external and internal mass transfer rate, clearly appears at the denominator of Equation (30). External diffusion strongly affects the kinetics, as the transport phenomena weakly depend on the temperature, and for a great contribution of the external diffusion on the reaction rate, the activation energy observed is about one-half of the true value observable in a chemical regime.

As mass transfer is originated by the reaction, it is always accompanied by heat transfer due to the heat absorbed or released by the reactions inside the particle. Therefore, for the rate of heat transfer, we can write

$$Q = k_m a_m (c_b - c_S)(-\Delta H) = h a_m (T_b - T_S) \tag{31}$$

Again, we can derive the temperature gradient from the corresponding concentration gradient

$$\Delta T = \Delta c(-\Delta H)\frac{k_m}{h} \tag{32}$$

that is, temperature and concentration gradients are strictly related, but the behavior of exothermic and endothermic reactions is quite different. It is useful to observe that both concentration and temperature gradients can fall between two limits:

$$\Delta c_{\text{min}} \cong 0 \text{ when } c_b \cong c_S \text{ and } \Delta c_{\text{max}} \cong c_b \text{ when } c_S \cong 0 \tag{33}$$

$$\Delta T_{\text{min}} \cong 0 \text{ when } T_b \cong T_S \text{ and } \Delta T_{\text{max}} \cong c_b(-\Delta H)k_m/h$$

It is possible to estimate mass and heat transfer coefficients from fluid dynamic correlations. As mentioned before, concentration and temperature gradients external to the particles are located in a thin layer (the boundary layer) surrounding the particle. The molar flow rate for each component will be

$$N_i = k_c(c_b - c_S) = k_g(p_b - p_S) \tag{34}$$

k_c and k_g are related to the molecular diffusion coefficient D_{12}, that is,

$$k_c = \frac{D_{12}}{\delta}\; k_g = \frac{D_{12}}{\delta RT} \tag{35}$$

where δ is the thickness of the boundary layer. Similarly, the heat flow through the boundary layer will be

$$q = h(T_b - T_S) \;\; (\text{heat/time} \times \text{surface area}) \tag{36}$$

Again, h is related to the thermal conductivity of the fluid, and k_f is a molecular property given by

$$k_f = 1.989 \times 10^{-4} \frac{\sqrt{T/M}}{\sigma^2 \Omega}\left(\frac{\text{cal}}{\text{cm s K}}\right) \tag{37}$$

and to the thickness of the boundary layer. This thickness depends on the fluid dynamic conditions adopted; consequently, the average transport coefficients (mass and energy) can be determined from the correlation between dimensionless groups such as Sherwood, Schmidt, and Reynolds numbers. Much experimental data have been correlated, and the following empirical relationship has been obtained for tubular reactors:

$$J_D = S_h \cdot S_c^{2/3} = \frac{\alpha_D}{\varepsilon_D} R_e^{-\beta_D} \tag{38}$$

$$S_h = \frac{k_c \rho}{G}\; S_c = \frac{\mu}{D\rho}\; R_e = \frac{G d_p}{\mu} \tag{39}$$

For $R_e > 10$, it results in $\alpha_D = 0.458$ and $\beta_D = 0.407$. For heat transfer coefficient, a quite similar approach is possible, giving place to

$$J_H = \frac{h}{C_p G} P_r^{2/3} = \frac{\alpha_H}{\varepsilon_H} R_e^{-\beta_H} \tag{40}$$

where $P_r = \mu C_p/k_t$ = number of Prandtl. A correlation exists between J_H and J_D, that is, $J_H \cong 1.08 J_D$. From these relations, it is possible to evaluate the heat and mass transfer coefficients. By putting in Equation (32) k_c and h derived from Equations (38) and (40), we obtain

$$\Delta T = \Delta c(-\Delta H)\frac{1}{\rho C_p}(L_e)^{2/3}\left(\frac{J_D}{J_H}\right) \tag{41}$$

L_e = Lewis number $= \frac{C_p\mu/k}{\mu/\rho D} \cong 1$, being also $J_D/J_H \cong 1$, resulting in

$$\Delta T \cong \Delta c(-\Delta H)\frac{1}{\rho C_p} \tag{42}$$

Therefore, $\Delta T_{\max}$ can also be determined as

$$\Delta T_{\max} \cong c_b(-\Delta H)\frac{1}{\rho C_p} \tag{43}$$

Equations (42) and (43) show that it is possible to have a significant temperature gradient even if the concentration gradient is very low as a consequence of the high value of ΔH. In conclusion, in steady state conditions, only two coupled equations are needed in order to quantitatively evaluate the effect of the external mass and heat transfer. These equations are

$$\begin{cases} k_m a_m (c_b - c_S) = \eta k c_S^n \\ h a_m (T_b - T_S) = \eta k c_S^n(-\Delta H) \end{cases} \tag{44}$$

In unsteady state conditions, four differential equations are needed, with these being different chemical and physical transport rates. The contribution of the external diffusion to reaction rate can then be estimated only on the basis of the fluid dynamic conditions in the system.

2.5. *Diffusion and Selectivity*

The selectivity of solid catalysts can be affected by diffusion in different ways according to the type of complex reactions involved. Consider as examples some very simple systems such as [3]

$$A \xrightarrow{k_1} B \text{ (desired)} \qquad R \xrightarrow{k_2} S$$

Independent reactions
1

$$A \to B, \quad A \to C$$

Competitive reactions
2

$$A \xrightarrow{k_1} B \xrightarrow{k_2} C \quad (B \text{ desired})$$

Consecutive reactions
3

(45)

All the reactions are considered first-order reactions for simplicity.

First Case

By considering for each reaction both external and internal diffusion contribution, in the first case, we express the overall reaction rate as reported in Equation (30). The selectivity can be expressed as the ratio between r_1 and r_2, that is,

$$S = \frac{r_1}{r_2} = \frac{[1/(k_m)_R a_m + 1/\eta_2 k_2]c_A}{[1/(k_m)_A a_m + 1/\eta_1 k_1]c_R} \tag{46}$$

In the case wherein the diffusion limitation is negligible, the selectivity becomes

$$S = \frac{k_1 c_A}{k_2 c_R} \tag{47}$$

By comparing Equations (46) and (47), we find a decrease of the selectivity to the desired product B for the effect of both external and internal mass transfer limitation. By considering predominantly the effect of internal diffusion and introducing the approximation (see Figure 5A) $\eta \cong 1/\phi$, we have

$$r_1 = \frac{1}{\phi_1} k_1 c_A = \frac{3}{r}\sqrt{\frac{k_1 (D_A)_{eff} c_A}{\rho_p}} \tag{48}$$

$$r_2 = \frac{1}{\phi_2} k_2 c_R = \frac{3}{r}\sqrt{\frac{k_2 (D_R)_{eff} c_R}{\rho_p}} \tag{49}$$

The selectivity becomes

$$S = \frac{r_1}{r_2} = \sqrt{\frac{k_1}{k_2}\frac{c_A}{c_R}} \tag{50}$$

considering $(D_A)_{eff} \cong (D_R)_{eff}$. By comparing Equations (47) and (50), we find that internal diffusion reduces the selectivity to the square root of Equation (47).

Second Case

For competitive reactions, diffusion limitations have an effect on the selectivity only when the occurring reactions have different reaction orders. Otherwise, for reactions having the same reaction order, no effect on the selectivity can be observed.

Third Case

Considering the occurrence of consecutive reactions in a chemical regime, that is, without diffusion limitation, selectivity can be written as

$$S = \frac{B\text{production}}{A\text{consumption}} = \frac{k_1 c_A - k_2 c_B}{k_1 c_A} = 1 - \frac{k_2 c_B}{k_1 c_A} \tag{51}$$

When internal diffusion resistance is operative ($\eta < 0.2$), we have to calculate concentration profiles for both A and B. Assuming the effective diffusivities to be equal, selectivity results [3]

$$S = \frac{(k_1/k_2)^{1/2}}{1 + (k_1/k_2)^{1/2}} - (k_2/k_1)^{1/2}\frac{c_B}{c_A} \tag{52}$$

As can be seen, selectivity is also consistently lowered in this case for the influence of the internal diffusion.

2.6. *Effectiveness Factor for a Complex Reaction Network*

According to the general definition of effectiveness factor introduced in Section 2.2 and expressed by Equation (10), we can extend our treatment to a more general situation represented by N_r reactions with rate equations that are generic functions of temperature and composition, regardless of the form of these kinetic expressions. For such a system, an expression of the effectiveness factor related to reaction j can be written as

$$\eta_j = \frac{\int_0^{r_p} 4\pi x^2 v_{r,j}(c_i, T)dx}{\frac{4}{3}\pi r_p^3 v_{r,j}(c_i^S, T^S)} \tag{53}$$

Evaluating the integral in Equation (53) requires solving the mass and heat balance inside the particle in order to evaluate the internal profiles of both temperatures and concentration. The balance equations, for steady state conditions, can be written with the same criteria adopted for Equations (3) and (4), but considering multiple reactions and multicomponent systems characterized by N_c chemical species:

$$D_{eff_i}\left[\frac{\partial^2 c_i^P}{\partial x^2}+\frac{2}{x}\frac{\partial c_i^P}{\partial x}\right]=\rho_P\sum_{j=1}^{N_{re}}\gamma_{i,j}v_{r,j}\quad i=1,2,\ldots,N_c \tag{54}$$

$$k_{eff}\left[\frac{\partial^2 T_P}{\partial x^2}+\frac{2}{x}\frac{\partial T_P}{\partial x}\right]=\rho_P\sum_{j=1}^{N_r}(-\Delta H_j)v_{r,j} \tag{55}$$

The simultaneous solution of this system of coupled partial differential equations (PDEs) must be accomplished using the following boundary conditions:

$$\begin{array}{l}\frac{\partial c_i^P}{\partial x}=0\,\frac{\partial T_P}{\partial x}=0\text{ at }x=0\\ c_i^P=c_i^S\;T_P=\;T_S\text{ at }x=r_P\end{array} \tag{56}$$

The described model is related to the simultaneous occurrence of both diffusion and chemical reactions inside a catalytic particle and consists of a system of coupled partial differential equations in pone dimension with boundary values. The solution can be obtained numerically with different algorithms reported in the literature (finite differences, orthogonal collocation, method of lines, etc.).

The method of lines (MOL) [19] in particular consists in converting the system of partial differential equations—Equations (54) and (55)—in an ordinary differential equations system. The first step of this method consists in considering the transient version of Equations (54) and (55) represented by the following equations:

$$\varepsilon_P\frac{\partial c_i^P}{\partial t}=D_{eff_i}\left[\frac{\partial^2 c_i^P}{\partial x^2}+\frac{2}{x}\frac{\partial c_i^P}{\partial x}\right]-\rho_P\sum_{j=1}^{N_r}\gamma_{i,j}v_{r,j} \tag{57}$$

$$\varepsilon_P\rho_P C_P^P\frac{\partial T_P}{\partial t}=K_{eff}\left[\frac{\partial^2 T_P}{\partial x^2}+\frac{2}{x}\frac{\partial T_P}{\partial x}\right]-\rho_P\sum_{j=1}^{N_r}(-\Delta H_j)v_{r,j} \tag{58}$$

The successive step consists in a discretization of the particle radial coordinate in a series of equally spaced radial nodes from $r = 0$ to $r = Rp$. Then, the spatial derivatives in Equations (57) and (58) are replaced by their finite difference approximation. The discretization scheme is reported in Scheme 1.

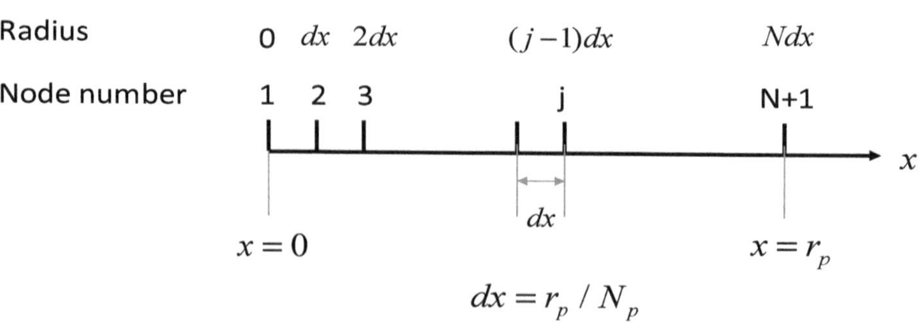

Scheme 1. Scheme of discretization.

At each node along the radius, ordinary differential equation (ODE) equations can be written in replacement of PDEs (57) and (58). The resulting set of ordinary differential equations (ODEs) can be integrated with respect to time until stationary conditions are reached. The obtained values represent the steady-state solution of Equations (54) and (55). The method of lines is largely preferred through considering, first of all, the large availability of efficient and robust ODE solvers and also for the low numerical instability related to the transformed problem. A further advantage of the MOL method can be appreciated when the system of model ODEs is "stiff", as in this case it can be treated with specifically developed ODE solvers such as, for example, GEAR and LSODE [23], or commercial solver included in MATLAB [24].

An alternative strategy to solve the particle balances for concentration and temperature internal profiles is the finite difference scheme [1] applied to Equations (54) and (55). The first step of this strategy consists, also in this case, of a nodal discretization along particle radius and then by replacing radial derivatives with a finite difference approximation formula. This method transforms the PDE system in a system of coupled nonlinear algebraic equations of the following form related to the mass balance of a generic component:

$$D_{eff}\left[\frac{2}{(i-1)\Delta x}\frac{c_A^{i+1}-c_A^{i-1}}{2\Delta x}+\frac{c_A^{i+1}+c_A^{i-1}-2c_A^i}{(\Delta x)^2}\right]-(R_{ni})=0 \tag{59}$$

In this equation, the term R_{ni} represents the reaction rate evaluated at the location of nodal point i. In this way, the original second order PDE has been transformed into a system of nonlinear algebraic system with ${c_A}^i$ as unknowns. It is worth noting that this approach is of general validity, as R_{ni} can represent any kinetic expression and can straightforwardly be extended to multiple chemical reactions by substituting the generation term with a sum of all reaction rates involving a specific component.

In the case of nonisothermal particles, heat balance must be taken into account, and the resulting finite difference nodal equation system is represented, in analogy to mass balance, by the following equation:

$$k_t\left[\frac{2}{(i-1)\Delta x}\frac{T^{i+1}-T^{i-1}}{2\Delta x}+\frac{T^{i+1}+T^{i-1}-2T^i}{(\Delta x)^2}\right]+(-\Delta H)(R_{ni})=0 \tag{60}$$

From a numerical point of view, the two numerical approaches (method of lines and method of finite differences) are quite equivocal and are both able to treat virtually any type of kinetic in a solid catalytic particle.

2.7. *An Example of Calculation of Effectiveness Factor Complex Reactions*

We considered the conversion of methanol to formaldehyde catalyzed by iron–molybdenum oxide catalyst. Two consecutive reactions occur in the process [25]:

$$\begin{aligned} CH_3OH + 1/2O_2 &\rightarrow CH_2O + H_2O \\ CH_2O + 1/2O_2 &\rightarrow CO + H_2O \end{aligned} \tag{61}$$

The conditions for the reactions, together with catalyst characteristics and other physical parameters [25] used in the calculations, are reported in Table 1.

Table 1. Physico-chemical data for the calculation.

$K_e = 2.72 \times 10^{-4}$	KJ/(s m K)	effective thermal conductivity
$D_e = 1.07 \times 10^{-5}$ exp(-672/T)	m^2/s	effective diffusivity
$\rho_P = 1180$	Kg/m^3	particle density
$C_p = 2.5$	KJ/(mole K)	particle specific heat
$P = 1.68$	atm	total pressure
$T_S = 539$	K	surface temperature
$d_P = 3.5 \times 10^{-3}$	m	particle diameter
Bulk gas composition	mol%	
CH_3OH	9.0	
O_2	10.0	
CH_2O	0.5	
H_2O	2.0	
CO	1.0	
N_2	77.5	

These reactions follow a redox mechanism, and the most reliable kinetics is the one suggested by Mars and Krevelen [26]:

$$v = \frac{k_1 k_2 P_m P^n_{O_2}}{k_1 P_m + k_2 P^n_{O_2}} \quad (62)$$

Different values of n have been suggested in the literature, generally considering $n = 1/2$ [27] or $n = 1$ [28]. The inhibition effect of water, formed in both the reactions, can also be introduced in the form of a Langmuir–Hinshelwood term [29], such as

$$v_r = \frac{k_1 k_2 P_m P^n_{O_2}}{k_1 P_m + k_2 P^n_{O_2}} \left(\frac{1}{1 + b_w P_w} \right) \quad (63)$$

Riggs [30] has proposed, on the contrary, pseudo Langmuir–Hinshelwood kinetic laws of the following type:

$$\begin{aligned} v_{r1} &= \frac{k_1 P_m}{1+a_1 P_m + a_2 P_w} \\ v_{r2} &= \frac{k_2 P_f}{1+b_1 P_m + b_2 P_w} \end{aligned} \quad (64)$$

where P_m, P_w, and P_f are, respectively, the partial pressures of methanol, water, and formaldehyde; k_1, k_2, a_1, a_2, b_1, and b_2 are parameters whose values and dependence on temperature is reported Table 2.

Table 2. Kinetic parameters for the model.

$k_1 = 5.37 \times 10^2$ exp(-7055/T)
$k_2 = 6.42 \times 10^{-5}$ exp(-1293/T)
$a_1 = 5.68 \times 10^2$ exp(-1126/T)
$a_2 = 8.37 \times 10^{-5}$ exp(7124/T)
$b_1 = 6.45 \times 10^{-9}$ exp(12,195/T)
$b_2 = 2.84 \times 10^{-3}$ exp(4803/T)
$\Delta H_1 = 37{,}480$ cal/mole
$\Delta H_2 = 56{,}520$ cal/mole

The application of the model represented by Equations (57) and (58) to this example was performed with the following assumptions:

- Catalytic particle is spherical with uniform reactivity, density, and thermal conductivity.
- The heat of reactions does not change with the temperature.
- The external diffusion resistance is negligible, and therefore the surface concentration is equal to the one of the bulk.
- The effective diffusivity has been assumed equal for all the involved chemical species.

The numerical solution of this example was achieved by discretizing the particle radius with 20 internal nodes ($N_n = 20$). As the reactive mixture is constituted by six different components, we had globally $(N_c + 1)\, N_n = 140$ ODEs to be integrated to the stationary state. A further check demonstrated that by increasing the number of internal discretization points brings a negligible variation in the effectiveness factors. As result of this calculation, we obtained the concentration profile of each component inside the catalytic particle, as shown in Figure 9A. By examining this plot, it is clear that the concentration profiles of the reagents methanol and oxygen decreased from the external surface to the center of the pellet, while the opposite occurred for products.

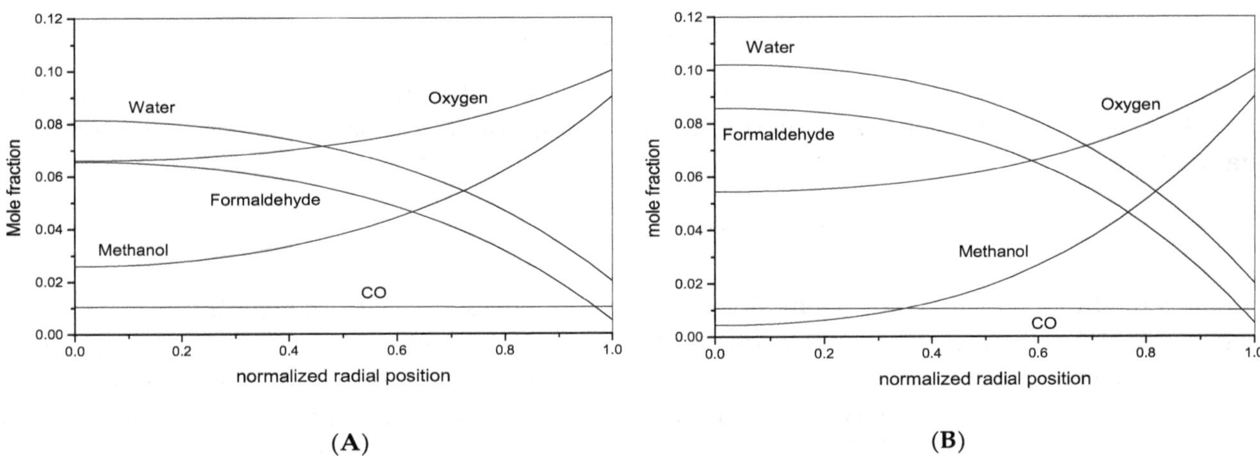

Figure 9. Mole fraction profiles inside a catalytic particle. (**A**) Kinetics of Riggs [30]; (**B**) kinetics of Dente et al. [28].

By employing Equation (53), it is possible from these profiles to evaluate the effectiveness factors for each reaction obtaining the following results: $\eta_1 = 0.778$, $\eta_2 = 8.672$.

The high effectiveness factor obtained for the second reaction was due to the low concentration of formaldehyde in the bulk gas, in comparison with the formaldehyde concentration accumulated inside the particle, which was significantly higher.

A further result of this example is related to the temperature profile reported in Figure 10. With the reactions being very exothermic, the temperature increased, as expected, from the external surface toward the center, and the overall ΔT was about 3.5 °C. In Figure 10, reported for comparison, are also the same calculations made by adopting the Mars–Krevelen model with the parameters taken from Dente et. al. [28] and Riggs [30].

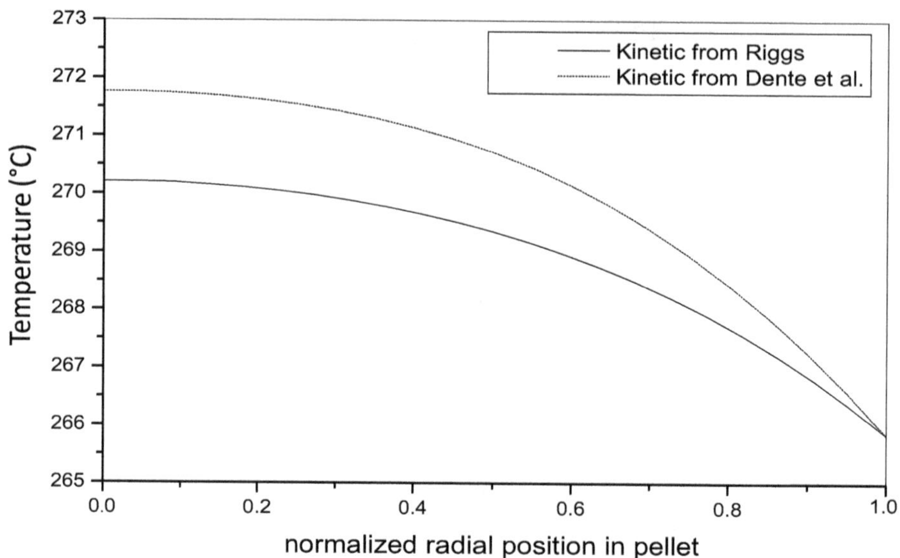

Figure 10. Internal temperature profile for a catalytic particle obtained with different kinetic models.

3. Mass and Heat Transfer in Packed Bed Reactors: Long Range Gradients

3.1. Conservation Equations for Fixed-Bed Reactors: Mass and Energy Balances

The generic mass conservation equation for a system of N_c components involved in a reaction network of N_r chemical reactions, related to the I'th component, can be written as in the following equation [16]:

$$\frac{\partial c_i}{\partial t} = -\nabla(c_i u + J_i) + \sum_{j=1}^{N_r} \gamma_{i,j} v_{r,j} \tag{65}$$

where u is the fluid velocity component along various dimensions, c_i is the concentration of a generic component, $\gamma_{i,j}$ is the stoichiometric coefficient of chemical species i in reaction j, and $v_{r,j}$ is the j-th rate of reaction based on fluid volume. The quantity J_i represents the molar flux of the i-th component originated by the concentration gradients, temperature gradients, and pressure gradients. The molar flux is in relation with the effective diffusion coefficient D_i by Fick's law, represented by the following equation:

$$J_i = -D_i \nabla c_i \tag{66}$$

Equation (65) is valid in both steady and unsteady state conditions and also contains the accumulation term resulting from the unbalanced difference between input, output, and chemical reactions terms. The overall balance is referred to a suitable control volume.

In the case of a fixed-bed reactor, the control volume assumes the shape of an annulus in a cylindrical coordinate system. By applying the conservation concepts expressed by Equation (65), assuming that only the velocity in the direction of flow ($u_z = v$) is dominant with respect to other directions and as represented in Figure 11, the general Equations (65) and (66) can be combined to give

$$\varepsilon_B \frac{\partial c_i}{\partial t} = -\frac{\partial}{\partial z}(uc_i) + \frac{\partial}{\partial z}\left[D_{a_i} \frac{\partial c_i}{\partial z}\right] + \frac{1}{r}\frac{\partial}{\partial r}\left[D_{r_i} \frac{\partial c_i}{\partial r}\right] + (1-\varepsilon_B)\sum_{j=1}^{N_{re}} \gamma_{i,j} v^G_{r,j} \tag{67}$$

where D_{ai} and D_{ri} are the effective dispersion coefficients (diffusivities), in axial and radial directions, for the i-th component. These quantities are referred to the total cross-sectional area perpendicular to the diffusion direction; u is the linear velocity in the catalyst bed and ε_B is the void fraction of the catalyst bed. The overall reaction rate $v^G_{r,j}$ is then multiplied by the factor $(1-\varepsilon_B)$ as the reaction rate is based on the catalyst particle volume.

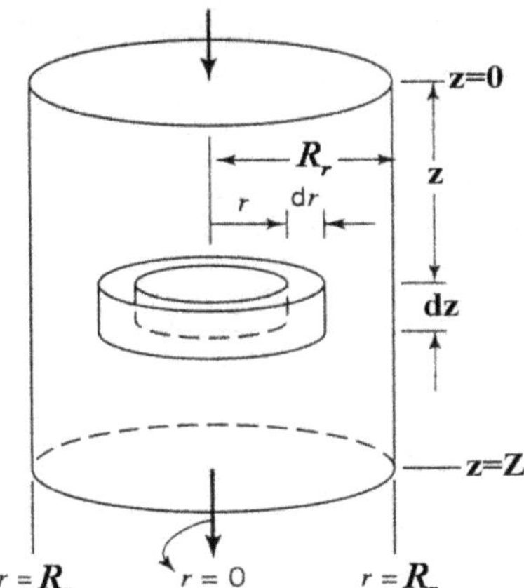

Figure 11. Scheme of the coordinate system and control volume for the fixed-bed conservation equations.

A simplification can be introduced in Equation (3) by assuming a constant linear velocity in z-direction (reactor axis) and also constant diffusivities along both z and r. Under these assumptions, Equation (67) can be reformulated as follows:

$$\varepsilon_B \frac{\partial c_i}{\partial t} + u\frac{\partial c_i}{\partial z} - D_{a_i}\frac{\partial^2 c_i}{\partial z^2} - D_{r_i}\left[\frac{\partial^2 c_i}{\partial r^2} + \frac{1}{r}\frac{\partial c_i}{\partial r}\right] = +(1-\varepsilon_B)\sum_{j=1}^{N_{re}} \gamma_{i,j} v^G_{r,j} \tag{68}$$

A similar approach can be adopted for the energy balance by replacing in Equation (68) the following quantities: the term $\rho C_p T$ instead of concentration of chemical species C_i, the effective thermal conductivities K instead of diffusivities D, and reaction enthalpy term $(-\Delta H_j)$ R_{Gj} instead of reaction rate R_{Gj}:

$$\varepsilon_B \frac{\partial T}{\partial t} + u\frac{\partial T}{\partial z} - K_a\frac{\partial^2 T}{\partial z^2} - K_r\left[\frac{\partial^2 T}{\partial r^2} + \frac{1}{r}\frac{\partial T}{\partial r}\right] = \frac{(1-\varepsilon_B)}{\rho C_p}\sum_{j=1}^{N_{re}} (-\Delta H_j) v^G_{r,j} \tag{69}$$

where ρ and C_P are the density and specific heat (average values) referred to the gas mixture, respectively.

Considering a fixed-bed reactor, bulk phase concentration and temperature can be regarded, in general, as functions of both r and z coordinates:

$$\begin{aligned} c_i^B &= f(z,r) \\ T_b &= g(z,r) \end{aligned} \tag{70}$$

In the assumptions above, the general mass and energy balance equations for the fixed-bed reactor in which N_r chemical reactions and N_c components are involved are

$$\begin{gathered} \varepsilon_B \frac{\partial c_i{}^B}{\partial t} + u\frac{\partial c_i{}^B}{\partial z} - D_{a_i}\frac{\partial^2 c_i{}^B}{\partial z^2} - D_{r_i}\left[\frac{\partial^2 c_i{}^B}{\partial r^2} + \frac{1}{r}\frac{\partial c_i{}^B}{\partial r}\right] = (1-\varepsilon_B)\sum_{j=1}^{N_R} \gamma_{i,j} v^G_{r,j} \\ i = 1,2,\ldots N_c \end{gathered} \tag{71}$$

$$\varepsilon_B \frac{\partial T_B}{\partial t} + u\frac{\partial T_B}{\partial z} - K_a\frac{\partial^2 T_B}{\partial z^2} - K_r\left[\frac{\partial^2 T_B}{\partial r^2} + \frac{1}{r}\frac{\partial T_B}{\partial r}\right] = \frac{(1-\varepsilon_B)}{\rho C_p}\sum_{j=1}^{N_R} (-\Delta H_j) v^G_{r,j} \tag{72}$$

Equations (71) and (72) represent a system of PDEs (partial differential equations) for which a solution can be obtained by imposing some suitable boundary conditions related to both variables (temperature and concentration) and their derivatives with respect to z and r. Usual boundary conditions can be written as follows:

$$\frac{\partial T_B}{\partial r} = \frac{\partial c_i^B}{\partial r} = 0 \text{atthecenterlineofthereactor}(r = 0)\text{forallz} \tag{73}$$

$$\frac{\partial c_i^B}{\partial r} = 0\,;\, h_w(T_B - T_C) = -\rho C_p K_r \frac{\partial T_B}{\partial r} \text{atthewallofreactor}(r = R)\text{forallz} \tag{74}$$

The first boundary condition (Equation (73)) can be written by considering the symmetry around the axis of the tubular reactor, while the second condition (Equation (74)) expresses the constraint that no mass transfer occurs across the reactor wall. The second part of Equation (74) expresses the zero-accumulation of energy and is related to the heat transfer boundary condition according to which the heat transferred to the cooling fluid, at a temperature T_c, is equal to the heat conducted at the wall.

The axial boundary conditions, written at the reactor inlet, consists of the following equations:

$$\begin{aligned} (uc_i^B)_{in} &= (uc_i^B - D_{a_i}\tfrac{\partial c_i^B}{\partial z})_{z=0} \\ & \qquad\qquad \text{atz} = 0 \\ (uT_B)_{in} &= (uT_B - K_a \tfrac{\partial T_B}{\partial z})_{z=0} \end{aligned} \tag{75}$$

While at the outlet

$$\frac{\partial c_i^B}{\partial z} = \frac{\partial T_B}{\partial z} = 0 \text{ at z} = Z \tag{76}$$

The boundary conditions (Equations (75) and (76)) are based on the flux continuity (both mass and heat) across a boundary, represented by the catalytic bed inlet and outlet.

3.2. External Transport Resistance and Particle Gradients

The link between macroscopical ("long-range"), concentration, and temperature gradients, described by the conservation equations for the entire reactor, and the microscopic situation locally developed around catalytic particles and inside it, is represented by a relation between the overall rate of reaction and the intrinsic kinetic. At a macroscopical level, the observed reaction rate, R_{Gi}, represents the rate of mass transfer across an interface between fluid and solid phase, which is ultimately related to the flux at the catalyst particle surface:

$$\sum_{j=1}^{N_r} \gamma_{i,j} v_{r,j}^G = \frac{k_g}{L}(c_i^B - c_i^S) = \frac{D_{ei}}{L}\frac{\partial c_i^P}{\partial x}\bigg|_{x=L} = \sum_{j=1}^{N_{re}} \gamma_{i,j}\eta_j v_{r,j} \;\; \text{j} = 1, 2, \ldots, N_r \tag{77}$$

with the following mean of the symbols:

- k_g—gas-solid mass transfer coefficient (film);
- L—characteristic length of particle (radius for spherical pellets);
- $c_i{}^S$—surface concentration of component i;
- $c_i{}^P$—particle internal concentration of component i;
- D_{ei}—effective diffusivity of component i into the particle;
- x—particle radial coordinate;
- η_j—effectiveness factor for reaction j;
- $v_{r,j}$—intrinsic rate of reaction j.

In a similar way, we can write a relation for the thermal flux:

$$\sum_{j=1}^{N_r} (-\Delta H_j) v_{r,j}^G = \frac{h}{L}(T_S - T_b) = -\frac{K_{eff}}{L}\frac{\partial T_P}{\partial x}\bigg|_{x=L} \tag{78}$$

where

- h—film heat transfer coefficient;
- T_S—temperature at the surface of the pellet;
- T_P—temperature inside the pellet;
- K_{eff}—effective thermal conductivity of the catalytic particle.

By considering Equation (77), the relationship between the rate of reaction at a macroscopic level and the intrinsic reaction rate is expressed for each chemical reaction by the effectiveness factor η or, in an equivalent way, by means of the concentration gradients measured at the particle surface. This consideration evidences the necessity to solve mass and energy balance equations related to catalytic particles to calculate local (microscopic) concentration and temperature profile. This calculation must be replicated, in principle, in each position along the reactor.

Conservation equations for the particles can be written as in the following equations (Equations (79) and (80)):

$$\varepsilon_P \frac{\partial c_i^P}{\partial t} = D_{e_i}\left[\frac{\partial^2 c_i^P}{\partial x^2} + \frac{2}{x}\frac{\partial c_i^P}{\partial x}\right] - \rho_P \sum_{j=1}^{N_r} \gamma_{i,j} r_{c_j} \quad i = 1,2,\ldots,N_c \tag{79}$$

$$\varepsilon_P \rho_P C_P^P \frac{\partial T_P}{\partial t} = K_e\left[\frac{\partial^2 T_P}{\partial x^2} + \frac{2}{x}\frac{\partial T_P}{\partial x}\right] - \rho_P \sum_{j=1}^{N_r} (-\Delta H_j) r_{c_j} \tag{80}$$

with the following meanings of the symbols:

- ε_P—catalytic particle void fraction;
- ρ_P—catalytic particle density;
- $C_P{}^P$—catalytic particle specific heat.

The simultaneous solution of PDE system represented by Equations (79) and (80) can be obtained by imposing some boundary conditions that are valid at the center and at the external surface of the catalyst particle respectively. These boundary conditions can be derived from symmetry consideration and from continuity related to both concentration and temperature:

$$\begin{array}{c} \frac{\partial c_i^P}{\partial x} = 0 \ \frac{\partial T_P}{\partial x} = 0 \text{ at } r = 0(\text{center}) \\ c_i^P = c_i^S \ T_P = \ T_S \text{ at } r = L(\text{surface}) \end{array} \tag{81}$$

As it was defined, the problem consists in a set of non-linear partial differential equations (PDEs) that must be solved at two levels: the first is a local level, related to a single catalytic particle, and the second is a long-range scale for the entire reactor. The solution of the problem in the full form, expressed by the Equations (72) to (79), is a complex task, even by adopting sophisticated numerical solution algorithms, while an analytical exact solution is impossible for the mostly practical cases. In the following part of this review, an overview of the possible simplifications is presented and some simplified equations are reported in association with problems much easier to solve.

3.3. *Conservation Equations in Dimensionless Form and Possible Simplification*

A convenient way to introduce the mentioned simplifications is in rewriting mass and energy balances for the reactor in a dimensionless form. This strategy has both the scope to emphasize some parameters of the reactor and the ability to implement a more robust procedure for the numerical solution.

We can pose

$$n_d = \frac{Z}{d_P}\ \mathrm{m}_d = \frac{R}{d_P}\ \mathrm{A}_d = \frac{Z}{R}\ \theta_d = \frac{Z}{u}\ \overline{c}_i = \frac{c_i^B}{c_i^{B(in)}}$$
$$\overline{T} = \frac{T_B}{T_{B(in)}}\ \overline{r} = \frac{r}{R_r}\ \overline{z} = \frac{z}{Z}\ \overline{t} = \frac{t}{\theta_d} \tag{82}$$

with

- d_P—particle diameter;
- R—fixed-bed reactor radius;
- Z—fixed-bed reactor length;
- ${c^{B(in)}}_i$—reactor inlet concentration;
- $T_{B(in)}$—reactor inlet temperature.

Within these assumptions, the reactor conservation equations become

$$\varepsilon_B \frac{\partial \overline{c_i}}{\partial \overline{t}} + \frac{\partial \overline{c_i}}{\partial \overline{z}} - \frac{1}{n_d P_{ma}} \frac{\partial^2 \overline{c_i}}{\partial \overline{z}^2} - \frac{1}{m_d P_{mr}} \left[\frac{\partial^2 \overline{c_i}}{\partial \overline{r}^2} + \frac{1}{\overline{r}} \frac{\partial \overline{c_i}}{\partial \overline{r}} \right] = \frac{(1-\varepsilon_B)\theta_d}{c_i^{B(in)}} \sum_{j=1}^{N_{re}} \gamma_{i,j} v_{r,j}^G \quad i = 1,2,\ldots,N_c \tag{83}$$

$$\varepsilon_B \frac{\partial \overline{T}}{\partial \overline{t}} + \frac{\partial \overline{T}}{\partial \overline{z}} - \frac{1}{n_d P_{ha}} \frac{\partial^2 \overline{T}}{\partial \overline{z}^2} - \frac{A_d}{m_d P_{hr}} \left[\frac{\partial^2 \overline{T}}{\partial \overline{r}^2} + \frac{1}{\overline{r}} \frac{\partial \overline{T}}{\partial \overline{r}} \right] = \frac{(1-\varepsilon_B)\theta_d}{\rho C_p T_{B(in)}} \sum_{j=1}^{N_{re}} (-\Delta H_j) v_{r,j}^G \tag{84}$$

In the Equations (83) and (84), we can recognize some fundamentals dimensionless groups that are related to mass dispersion, which is related to axial and radial directions, represented by Peclet's numbers expressed by the following equations:

$$P_{ma} = \frac{d_P u}{D_a}\ (\text{axial})\ \ P_{mr} = \frac{d_P u}{D_r}\ (\text{radial}) \tag{85}$$

and analogously for heat dispersion we have

$$P_{ha} = \frac{d_P u}{k_a}\ (\text{axial})\ \ P_{hr} = \frac{d_P u}{k_r}\ (\text{radial}) \tag{86}$$

The quantitative criteria that can be adopted to determine if the dispersion phenomena affect the overall reactor performances are Peclet's numbers and reactor-to-particle size ratios (n, m, and A). Moreover, these criteria can give indications to decide whether or not some simplifications are allowed. The operative conditions adopted and chemical reaction characteristics can suggest further simplifications according to which mass and energy conservation equations can be solved in a simplified form. The first and more common simplification is represented by the steady state, allowing the elimination of time variable and all its derivatives, in the left-hand sides of Equations (71), (72), (79), and (80). From an energetic point of view, the reaction enthalpy also plays a very important role. When the reaction heat is negligible or very low, the reactor can be run isothermally, and then, with the temperature being a constant, the heat balance equation can be eliminated. If the reactor is thermally insulated from the environment, it is operated in adiabatic conditions, as many reactors are in practice. In this case, radial gradients could be negligible, and therefore only a one-dimensional model is sufficient for the description of the reactor behavior. An intermediate situation, comprising these two limit cases described, is represented by a reactor working in conditions that cannot be considered isothermal nor adiabatic. This is the case of very exothermal reactions for which an external cooling system is required in order to guarantee the safety of the reactor and to preserve the catalyst durability. In this case, a numerical solution of conservation equations in full form appears to be the only feasible strategy. However, the conservation equations can still be applied in a simplified form, even if the problem remains complex to solve and is more difficult with respect to the two limit cases (isotherm and adiabatic) cited previously. Normally, for an extremely exothermic chemical reaction, the packed beds with small diameter are used for promoting the heat removal, and in this case the radial temperature profile can be neglected. The problem is again mono-dimensional in this case. In general, according to Carberry [31], the gradients along the reactor radius, for practical purposes, can be neglected when the radial aspect ratio $m = R/d_p$ is less than 3 or 4. Further guidelines can be gained by examining the values of Peclet's numbers and the reactor aspect ratios; as an example, the axial aspect ratio $n=Z/d_P$ is usually very large, and considering that P_{ma} is about 2 for gases flowing through a catalytic bed for Reynold's number (based on particle diameter) greater than 10, then the term nP_{ma} is also large, revealing that axial mass dispersion can be almost completely neglected. Table 3 [32] summarizes the general guidelines to introduce principal simplifications in the mass balance for a packed-bed reactor operating under stationary conditions; the two limit cases are also reported with concern to the isothermal and adiabatic reactor together with the intermediate situation in which the reactor cannot be considered isothermal nor adiabatic.

Table 3. Guidelines for simplifications in the left-hand side of conservation equations, with reference to stationary conditions.

Reactor Conditions	Aspect Ratio Criteria	Left-Hand Side of Equations (71) and (72)
Isothermal		$u\frac{\partial c_B^i}{\partial z}$
Adiabatic	$\left(\frac{Z}{d_P}\right)\left(\frac{d_P}{D_a}\right) > 300\ R_e > 10$	$u\frac{\partial c_B^i}{\partial z}$ $\rho C p u\frac{\partial T_B}{\partial z}$
	$\left(\frac{Z}{d_P}\right)\left(\frac{d_P}{D_a}\right) < 300\ R_e > 10$	$u\frac{\partial c_B^i}{\partial z}$ $\begin{cases} \rho C p u\frac{\partial T_B}{\partial z} & \text{or, if necessary} \\ \rho C p u\frac{\partial T_B}{\partial z} - K_a\frac{\partial^2 T_B}{\partial z^2} \end{cases}$
Non-isothermal and non-adiabatic	$\frac{R_r}{d_P} > 4$	$u\frac{\partial c_B^i}{\partial z} - D_r\left(\frac{\partial^2 c_B^i}{\partial r^2} + \frac{1}{r}\frac{\partial c_B^i}{\partial r}\right)$ $\rho C p u\frac{\partial T_B}{\partial z} - K_r\left(\frac{\partial^2 T_B}{\partial r^2} + \frac{1}{r}\frac{\partial T_B}{\partial r}\right)$
	$\frac{R_r}{d_P} \leq 4\ \mathrm{R_e} > 30$	$u\frac{\partial c_B^i}{\partial z}$ $\rho C p u\frac{\partial T_B}{\partial z}$

3.4. *Examples of Applications*

In the following sections, we examine some examples concerning fixed-bed reactors operating in the various possible thermal regimes.

3.4.1. Isothermal Conditions

Isothermal conditions are seldom obtained in industrial packed bed reactors and are only for systems with a very low heat of reaction, whereas they are most commonly encountered in slurry reactors because liquid phase has a high thermal conductivity. Therefore, in these cases, we can have only internal, and sometimes external, diffusion limitation to the reaction.

3.4.2. Adiabatic Conditions

If the reactor is operated so that heat transfer to the surrounding is negligible, the system could be considered in adiabatic conditions. For simplicity, we can consider a system of a single reaction, $A{\rightarrow}P$, in steady state adiabatic conditions, and then the material energy balances for a tubular reactor with no axial and radial dispersion could be derived from Equations (69) and (70), resulting in the following expressions:

$$\begin{aligned} \frac{dF_A}{dz} &= \rho_B A \overline{R}_1 \\ \frac{dT}{dz} &= \frac{\rho_B}{GC_P}(-\Delta H_1)\overline{R}_1 \end{aligned} \tag{87}$$

where

- G—mass velocity;
- cross section of the reactor tube;
- F_A, $F^0{}_A$ component molar flow rate;
- $\overline{R}_j$—reaction rate for reaction j based on catalyst mass.

In the above equations, it is convenient to introduce the fractional conversion, X_A, obtaining the following equations:

$$\begin{aligned} \frac{dX_A}{dz} &= -\frac{\rho_B A}{F_A^0}\overline{R}_1 \\ \frac{dT}{dz} &= \frac{\rho_B}{GC_P}(-\Delta H_1)\overline{R}_1 \end{aligned} \tag{88}$$

Dividing Equations (87) and (88) term by term, we obtain an expression relating the conversion and the temperature:

$$\frac{dX_A}{dT} = -\frac{AGC_P}{F_A^0}(-\Delta H_1) \text{ or in integrated form: } X_A = \alpha_A + \beta_A \mathrm{T} \tag{89}$$

with α_A and β_A as constants. The main result expressed by the previous equation is that a linear relationship exists between the temperature and the conversion for an adiabatic reactor.

Adiabatic reactors are frequently employed in industrial practice, especially in the case of equilibrium reactions for which the desired conversion is achieved through assembling the reactor in a series of adiabatic catalytic beds provided with intermediate heat removal or supplying system in accordance with the reaction being exothermic or endothermic. Figures 12 and 13 report a schematic reactor configuration for an exothermic and an endothermic reaction, respectively, together with temperature-conversion diagrams that show conversion equilibrium curves and straight lines resulting from balance Equation (89) and cooling or heating. With such an arrangement, it is possible to achieve good control over the final conversion of reversible reactions by controlling the temperature at the outlet of each catalytic bed. In the diagrams reported in Figures 12 and 13, dashed lines represent cooling or heating operations.

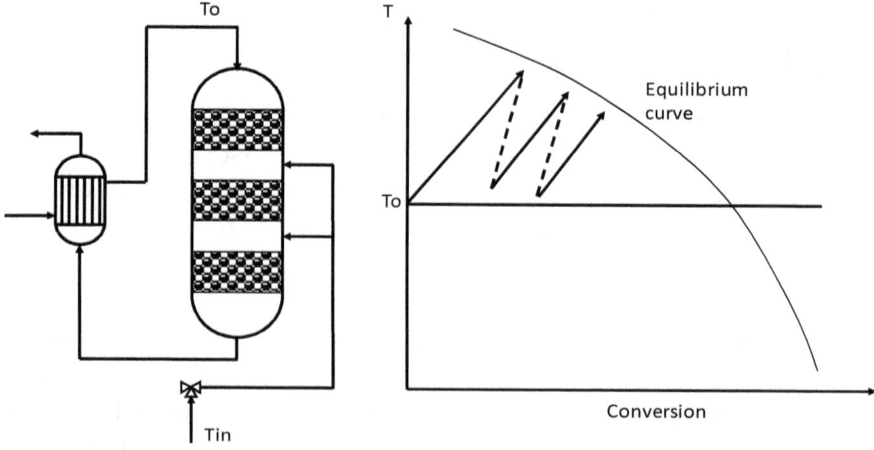

Figure 12. Scheme of a multistage adiabatic reactor (exothermic reaction). (**left**) reactor setup; (**right**) temperature profile.

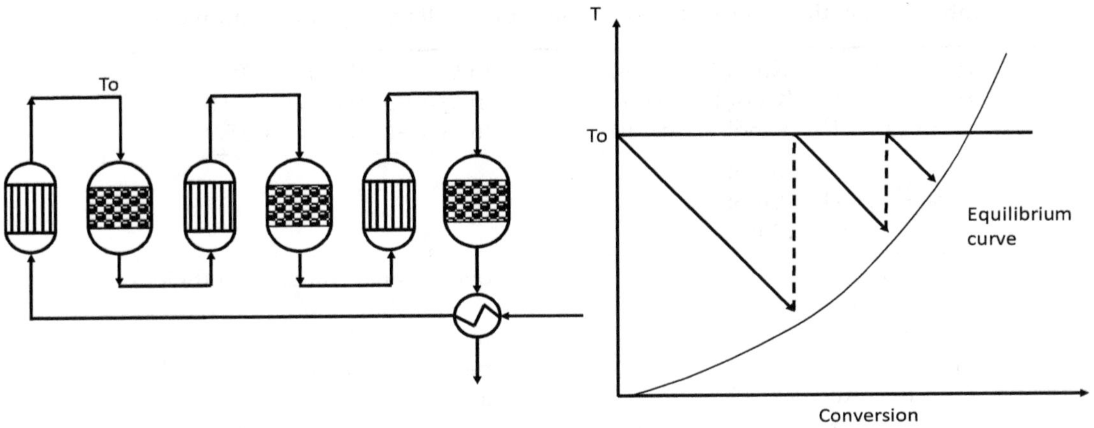

Figure 13. Scheme of a multistage adiabatic reactor (endothermic reaction). (**left**) reactor setup; (**right**) temperature profile.

4. Non-Isothermic and Non-Adiabatic Conditions

4.1. Conversion of o-Xylene to Phthalic Anhydride

Let us consider, first of all, a reaction that is performed in a packed-bed tubular reactor, operated in an modality non-isothermal and non-adiabatic that consists in the synthesis of phthalic anhydride (PA) obtained by oxidation of o-xylene (OX) with oxygen (O). A simplified scheme for this oxidation reaction can be expressed as follows:

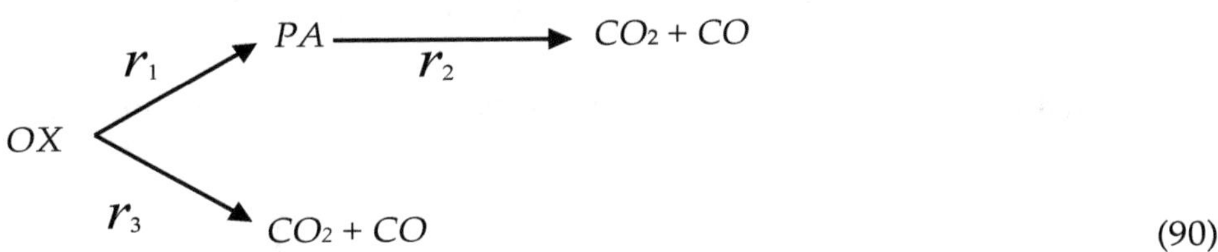

(90)

The reaction is catalyzed by vanadium pentoxide supported on α-alumina and has a high exothermal character. From Equation (90), it is evident that the reaction can lead to CO_2 and CO production, if not properly thermally controlled, giving a low yield in PA. For the reactor simulation, therefore, thermal effect must be taken into account for both the reaction and the heat exchanged with the cooling medium. The kinetic equations and related parameters for the reactions (Equation (90)) are reported in Table 4, together with the characteristics of the reactor and of the catalytic particles used in the simulations [33].

A specific characteristic of this reactor is the catalyst dilution with an inert material in the first part of the reactor (0.75 m) that is realized at the purpose of an improved temperature control.

For the model development, some basic assumptions should be stated, as in the following points:

- No axial and radial dispersion;
- No radial temperature and concentration gradients in the reactor body;
- Plug flow behavior of the reactor;
- No limitation related to internal diffusion in catalytic particles.

Table 4. Kinetic data for the conversion of o-xylene to phthalic anhydride.

$r_1 = k_1\, P_{OX}\, P_O$ (Kmol/Kg-cat h) $r_2 = k_2\, P_{PA}\, P_O$ (Kmol/Kg-cat h) $r_3 = k_3\, P_{OX}\, P_O$ (Kmol/Kg-cat h)	$\ln k_1 = -27{,}000/RT + 19.837$ $\ln k_2 = -31{,}000/RT + 20.860$ $\ln k_3 = -28{,}600/RT + 18.970$
$\Delta H_1 = -307$ Kcal/mol $\Delta H_2 = -783$ Kcal/mol $\Delta H_3 = -1090$ Kcal/mol	
$U = 82.7$ Kcal/ m^2 h °C	overall heat transfer coefficient
$D = 0.025$ m	reactor diameter
$Z = 3$ m	reactor length
$d_P = 0.003$ m	particle diameter
$C_P = 0.25$ Kcal/Kg °C	average specific heat
$\rho_B = 1300$ Kg/m^3	bulk density of the bed
Feed composition:	$y_{OX} = 0.0093$ $y_O = 0.208$
Feed molar flow rate	$F = 0.779$ moles/h
Inert dilution of the catalyst	$m_I = 0.5$ for the first quarter
Inlet temperature	$T_0 = 370$ °C

The assumptions related to radial profiles can be supported by the criteria expressed in Table 3 for radial aspect ratio $m = R/d_p$ that can be estimated as $m = 4.1$ and then slightly above the limit. By considering the assumptions and simplifications applied to this system, we can write a material balance equation directly from Equation (69) considered in the stationary state:

$$\frac{dF_i}{dz} = F\frac{dy_i}{dz} = \rho_B \frac{\pi D_r^2}{4} \sum_{j=1}^{N_r} \gamma_{i,j} \frac{\overline{R}_j}{(1+m_I)} \qquad i = 1, 2, \ldots, N_c \tag{91}$$

assuming a constant molar flow rate F, and with the following substitution:

$$u = \frac{Q}{A} \quad Qc_i = F_i \quad A = \frac{\pi D_r^2}{4} \quad F_i = y_i F \tag{92}$$

where:

- Q—volumetric overall flow rate;
- A—cross section of the reactor tube;
- D_r—reactor diameter;
- F_i—component molar flow rate;
- y_i—mole fraction of component i;
- m_I—mass of inert per unit mass of catalyst (dilution ratio);
- $\overline{R_j}$—reaction rate for reaction j based on catalyst mass.

The heat is constituted by Equation (72) and can be modified in a way similar to that adopted for mass balance and according to the absence of radial profiles and to the heat exchange of external cooling fluid in the reactor jacket. The thermal exchange with the surrounding (thermal fluid into the jacket) cannot be considered only as a boundary condition but as a separate term in the energy balance equation. A behavior similar to that of a double pipe heat exchanger (see Figure 14) can be adopted for the reactor and then, referring to a unit of reactor volume, the heat transferred across the external surface is defined as

$$q = \frac{U(T_C - T)\pi D_r dz}{A dz} = \frac{U(T_C - T)\pi D_r}{A} = \frac{4U(T_C - T)}{D_r} \tag{93}$$

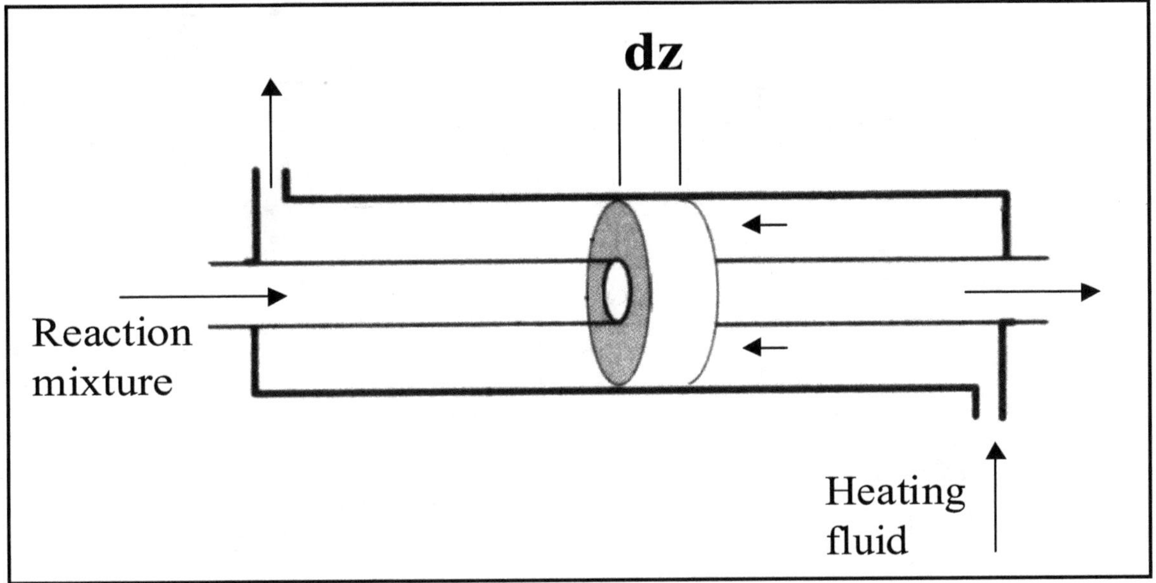

Figure 14. Double-pipe countercurrent reactor.

Equation (93) represents an additional term in the energy balance, and must be added to the heat associated with the reaction, resulting in the following overall differential equation for temperature evolution along the reactor axis:

$$\frac{dT}{dz} = \frac{\rho_B}{GC_P}\sum_{j=1}^{N_r}(-\Delta H_j)\frac{\overline{R}_j}{(1+m_I)} + \frac{4U}{DGC_P}(T_C - T) \tag{94}$$

with $G = \frac{F \cdot M_F}{A}$, with the following meanings for the symbols:

- G—mass velocity;
- M_F—average molecular weight of mixture.

The system of differential Equations (91) and (94) can be integrated in axial direction, z, for the calculation of temperature and composition profiles. The temperature profile resulting from this mono-dimensional model (axial coordinate) is reported in Figure 15 [33], with this diagram also reporting, as a comparison, the result of a more complex bi-dimensional model in which profiles in a radial direction are also taken into account.

As was shown before, the bi-dimensional model involves the solution of partial differential equations. In the considered example is the numerical strategy of finite differences method (FDM). The two models (one and two dimensions) give comparable results for what concerns axial temperature profiles. A conclusion is that the one-dimensional model can be considered sufficiently accurate for many practical purposes. The bi-dimensional model, however, foresees a slightly higher conversion to CO and CO_2, due to the higher temperature along the reactor.

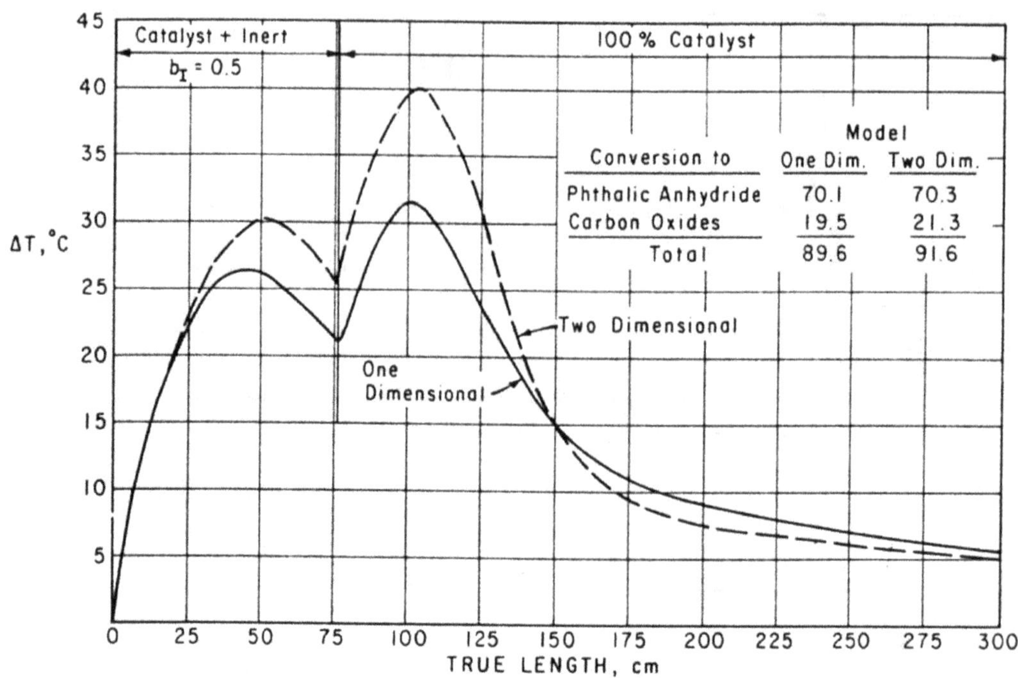

Figure 15. Comparison of the results of the one- and bi-dimensional models for reactor simulation (elaborated from data reported by Froment [33], see also [1]).

4.2. Conversion of Methanol to Formaldehyde

As a further example of a system that cannot be considered isothermal nor adiabatic, we chose the same reaction previously adopted in Section 2.7 for the evaluation of the effectiveness factor in a non-isothermal pellet, that is, the catalytic conversion of methanol to formaldehyde. Two reactions occurred as seen previously (Equation (61)).

These reactions were performed in a tubular reactor packed with catalyst and equipped with a jacket in which a heat transfer fluid is circulated with the purpose to a better temperature control. Table 5 reports the reactor operating conditions and other characteristics. A simulation was performed by using these conditions and the kinetic data from Riggs [29] (details were reported in [25]), obtaining composition and temperature profiles along the reactor axis. In this case study, a further aspect was introduced into the model, consisting in the calculation of the catalyst effectiveness factor along the reactor, considering diffusional limitations inside the particles.

Table 5. Reactor characteristic and operating conditions.

Inlet temperature	539 K
Total pressure	1.68 atm
Bulk density of the bed	0.88 Kg/m^3
Overall heat transfer coefficient U	0.171 KJ/(m^2 s K)
Heating medium temperature	544 K
Reactor diameter	2.54 x 10^{-2} m
Particles diameter	3.5 x 10^{-3} m
Reactor length	0.35 m
Gas inlet composition	mol %
CH_3OH	9
O_2	10
CH_2O	0.5
H_2O	2
CO	1
N_2	77.5

Some simplifying assumptions were introduced in the present case for the model development in a way similar to that of the example reported in the previous section:

- Negligible dispersion in axial and radial directions;
- Absence of concentration and temperature profiles along the reactor radius;
- Plug flow reactor behavior.

By applying the criteria of Table 3, radial profiles can be considered negligible as the aspect ratio in radial direction was $m = R/d_P = 3.6$, which was well below the limit value of 4. Under these assumptions, the resulting model is mono-dimensional because it only considers axial reactor profiles. At each location along the reactor axis, an effectiveness factor calculation was performed to obtain the value of the reaction rate that is related to that point, determining, in this way, an effectiveness factor axial profile. On the basis of the described assumptions and introducing molar flow rates relative to each chemical component, we can express material balance equations by the following model:

$$\frac{dF_i}{dz} = \rho_B \frac{\pi D_r^2}{4} \sum_{j=1}^{N_r} \gamma_{i,j} \overline{R}_j \quad i = 1, 2, \ldots, N_c \tag{95}$$

$$u = \frac{Q}{A} \quad Qc_i = F_i \quad A = \frac{\pi D_r^2}{4} \tag{96}$$

that can be derived upon the following substitution in Equation (71):

The energy balance, represented by Equation (72), can also be simplified, as done for the mass balance, according to the assumed absence of radial profiles and to the presence of reactor jacket with cooling fluid, as reported, for example, in Session 4.1. The heat exchanged per unit of reactor volume between the reactor and the cooling jacket can be defined as follows:

$$q = \frac{U(T_C - T)\pi D_r dz}{A dz} \tag{97}$$

This term must be added algebraically to the reaction enthalpy term in the heat balance equation, yielding the following expression:

$$\left(\sum_{i=1}^{N_c} F_i C_{P_i} \right) \frac{dT}{dz} = \frac{\pi D_r^2}{4} \rho_B \sum_{j=1}^{N_r} (-\Delta H_j) \overline{R}_j + \pi D_r U (T_C - T) \tag{98}$$

Equations (95) and (98) represent a system of N_c+1 coupled ordinary differential equations that must be integrated along the z axial direction to calculate the desired profiles of composition and temperature. At each integration step along z, a calculation of the effectiveness factor for each chemical reaction must be performed according to the procedure described in Session 2.6. A suitable integration algorithm must be adopted with a variable z step size, inversely proportional to the axial derivative dT/dz, so that a smaller step size is used when a steep temperature increase is detected in correspondence to a steeper profile. Figure 16 reports the axial temperature profile as a result of this simulation. This figure shows that the reaction mixture fed to the reactor undergoes a steep increase in gas temperature due to the strong exothermic character of this reactive system.

As methanol conversion proceeds (see composition profile reported in Figure 17), the main reaction rate also decreases, and the same trend can be appreciated for the temperature. Finally, in Figure 18, the profiles of the effectiveness factors for the two reactions is reported. It is interesting to observe that the main reaction is characterized, in the first part of the reactor, by an effectiveness factor much higher than unity, with this indicating that catalytic particles are not isothermal and a temperature profile is developed inside them.

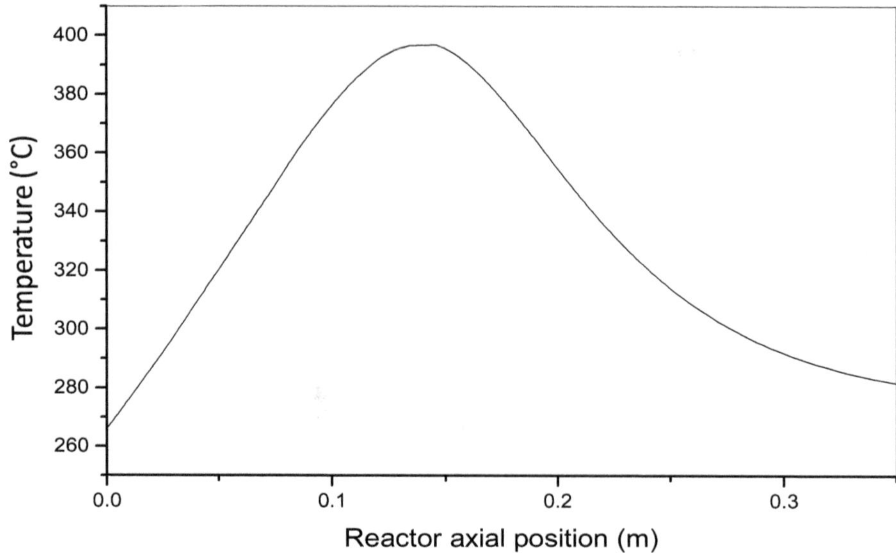

Figure 16. Axial temperature profile.

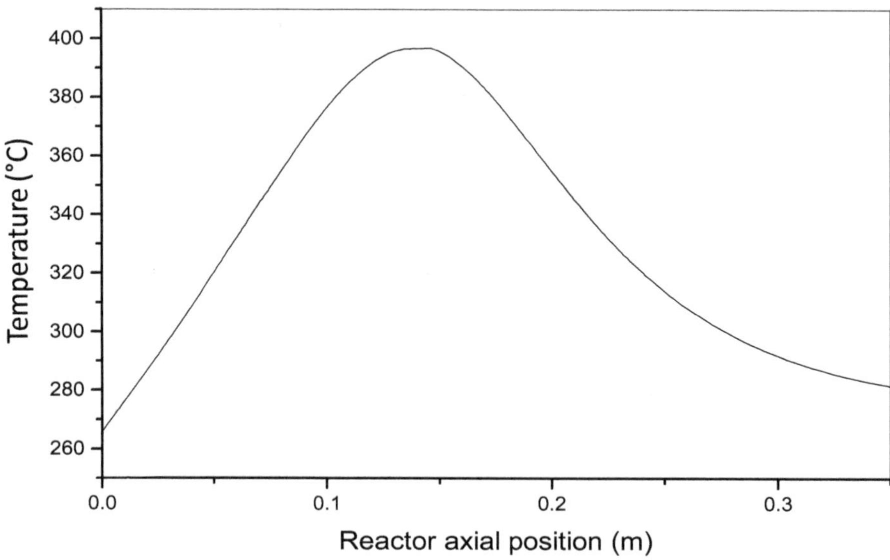

Figure 17. Axial profiles for component mole fractions.

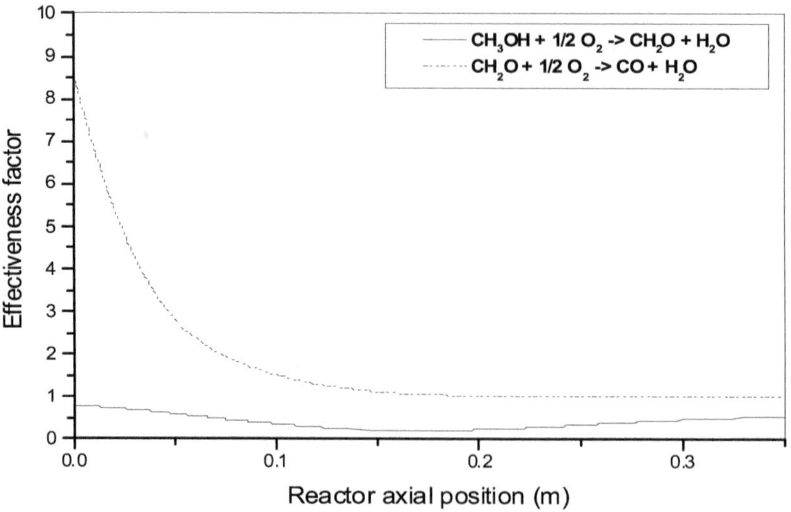

Figure 18. Axial profiles for the effectiveness factor for the two reactions.

5. Conclusions

The role of heat and mass transfer in affecting the kinetic studies in gas–solid tubular reactors was discussed in detail by surveying the abundant literature published on the subject. All the occurring phenomena were described and the equations for their interpretation were given.

Considering the enormous progress of electronic computers, many problems that were intractable in the past for their mathematical complexity can today be easily and rigorously solved with numerical approaches. For more clarity, some examples of mathematical solutions were reported.

Author Contributions: E.S. wrote the first part of the work related to the heat and mass transfer in the single pellet. R.T. wrote the second part related to the long-range gradients in packed bed reactors. All authors have read and agreed to the published version of the manuscript.

Glossary

List of Symbols

a_m	Specific surface area
A	Reactor cross section
b_w	Water adsorption equilibrium constant
c	Generic concentration
c_i	Concentration of component i
c_i°	Initial i concentration
c_b	Generic concentration of a component in the bulk
${c_i}^B$	Concentration of i in the bulk
${c_i}^P$	Concentration of i inside a catalytic particle
c_S	Generic concentration at the catalytic surface
${c_i}^S$	Concentration of i at the surface
$\mathrm{C_p}$	Average gas specific heat
${C_p}^P$	Particle specific heat
Δc	Concentration gradient
$\Delta c_{\min}$	Minimum concentration gradient
D	Reactor diameter
d_p	Particle diameter
D	Generic molecular diffusivity
D_i	Molecular diffusivity of component i
$D_{i,j}$	Mutual binary diffusion coefficient
D_{12}	Mutual binary diffusion coefficient
D_{im}	Diffusion coefficient of i in a mixture m
$\mathrm{D_{eff}}$	Effective molecular diffusivity
$(D_i)_{eff}$	Effective molecular diffusivity of component i
D_{be}	Bulk diffusion coefficient
D_{ke}	Knudsen diffusion coefficient
D_{ei}	Effective diffusivity inside particle
D_{ai}	Axial diffusivity of component i
D_{ri}	Radial diffusivity of component i
F_i	Molar flow rate of component i
F	Overall molar flow rate
G	Mass velocity
h	Film heat transfer coefficient
h_w	Wall heat transfer coefficient
ΔH	Generic reaction enthalpy
ΔH_j	Enthalpy of reaction j

Symbol	Description
J_i	Molar flux of component i
J_D, J_H	Terms for mass and heat transfer analogy
k, k_i	Generic kinetic constant
k_B	Boltzmann's constant
k_T	Generic thermal conductivity of the fluid
k_f	Thermal conductivity of the bulk
k_{eff}	Effective thermal conductivity
k_{Sol}	Thermal conductivity of the solid
K_a	Axial thermal conductivity
K_r	Radial thermal conductivity
K_e	Particle thermal conductivity
k_S	Kinetic constant
k_c	Film mass transfer coefficients (concentration gradient)
k_g	Film mass transfer coefficients (pressure gradient)
k_m	Mass transfer coefficient
L	Characteristic length
L_e	Lewis's number
m	Radial aspect ratio
m_I	Inert dilution ratio
M, M_i	Molecular weight
M_F	Average molecular weight of the mixture
M_w	Weisz modulus
N_C	Number of components
N_{re}	Number of reactions
N_r	Molar flux
N_i, N_A	Molar flux
N	Number of nodes
n	Reaction order
P	Total pressure
P_m	Methanol partial pressure
P_f	Formaldehyde partial pressure
P_w	Water partial pressure
P_{O2}	Oxygen partial pressure
P_{ma}	Axial Peclet's number for mass
P_{mr}	Radial Peclet's number for mass
P_{ha}	Axial Peclet's number for heat
P_{hr}	Radial Peclet's number for heat
P_r	Prandtl's number
Q	Rate of heat transfer
Q_v	Overall volumetric flow rate
q	Heat flux
r	Reactor radial coordinate
r_P	Particle spherical radius
R	Gas constant
R_r	Reactor radius
R_{ni}	Reaction rate at node i
R_j	Reaction rate (fluid volume)
$\overline{R_j}$	Reaction rate (catalyst mass)
r_{cj}	Intrinsic reaction rate
R_e	Reynold's number
S_v	Specific surface area
S_h	Sherwood's number
S_c	Schmidt's number

S	Selectivity
S_g	Specific surface area
T	Generic temperature
T_S	Temperature at particle surface
T_P	Temperature inside the particle
T_b	Bulk temperature
T_c	Cooling fluid temperature
ΔT_{max}	Maximum temperature difference
t	Time
u	Velocity
u_z	Velocity in z direction
U	Overall heat transfer coefficient
v_r	Reaction rate
$v_{r,i}$	Reaction rate, reaction i-th
$v_{r,j}{}^G$	Reaction rate (pellet volume)
V_{ci}	Critical volume of component i
x	Particle radial coordinate
X_i	Fractional conversion
y_i	Gas phase mole fraction component i
z	Axial reactor coordinate
Z	Reactor length
Greek Letters	
α_A	Constant in Equation (89)
α_B	Constant in Equation (89)
α_E	Reaction rate exponential parameter
α_J	Constant in Equation (38)
α_H	Constant in Equation (40)
β	Prater's number
β_J	Constant in Equation (38)
β_H	Constant in Equation (40)
γ_{dr}	Dimensionless concentration
γ_{ij}	Stoichiometric coefficient
δ	Thickness of boundary layer
ε_{dr}	Dimensionless radius
ε_B	Bed void fraction
ε_{Bs}	Bed void fraction of the solid
ε_J	Constant in Equation (38)
ε_H	Constant in Equation (40)
ε_{ij}	Interaction parameter
ε_p	Particle void fraction
η, η_j	Effectiveness factor
μ	Viscosity
θ	Porosity of the solid
ρ	Average gas density
ρ_p	Particle density
ρ_d	Intermolecular distance
σ_{ij}	Kinetic diameter
τ	Tortuosity factor
ϕ	Thiele modulus
ϕ_{LJ}	Lennard–Jones potential
y_{dr}	Dimensionless reaction rate
Ω_D	Collision integral
∇	Nabla operator

References

1. Santacesaria, E.; Tesser, R. *The Chemical Reactor from Laboratory to Industrial Plant*; Springer: Berlin, Germany, 2018.
2. Froment, G.F. The kinetics of complex catalytic reactions. *Chem. Eng. Sci.* **1987**, *42*, 1073–1087. [CrossRef]
3. Smith, J.M. *Chemical Engineering Kinetics*; McGraw-Hill: New York, NY, USA, 1981.
4. Fogler, H.S. *Elements of Chemical Reaction Engineering*; Prentice-Hall International: Upper Saddle River, NJ, USA, 1986.
5. Horak, J.; Pasek, J. *Design of Industrial Chemical Reactors from Laboratory Data*; Hayden Publishing: London, UK, 1978.
6. Satterfield, C.N.; Sherwood, T.K. *The Role of Diffusion in Catalysis*; Addison-Wesley Publishing: Boston, MA, USA, 1963.
7. Levenspiel, O. *The Chemical Reactor Omnibook*; OSU Book Store: Corvallis, OR, USA, 1984.
8. Satterfield, C.N. *Heterogeneous Catalysis in Practice*; Addison-Wesley Publishing: Boston, MA, USA, 1972.
9. Holland, C.D.; Anthony, R.G. *Fundamentals of Chemical Reaction Engineering*; Prentice-Hall: London, UK, 1979.
10. Westerterp, K.R.; van Swaaij, W.P.M.; Beenackers, A.A.C.M. *Chemical Reactor Design and Operation*; John Wiley & Sons: New York, NY, USA, 1984.
11. Davis, M.E.; Davis, R.J. *Fundamentals of Chemical Reaction Engineering*; Dover Publications, Inc.: New York, NY, USA, 2003.
12. Vogel, G.H. *Process Development Wiely—VCH*; John Wiley & Sons: Weinheim, Germany, 2005.
13. Winterbottom, J.M.; King, M.B. *Reactor Design for Chemical Engineers*; Stanley Thornes Ltd.: Cheltenham Glos, UK, 1999.
14. Wheeler, A. Reaction rates and selectivity in catalyst pores. In *Advances in Catalysis*; Academic Press: Cambridge, MA, USA, 1951; Volume 3.
15. Bird, R.B.; Stewart, W.E.; Lightfoot, E.N. *Fenomeni di Trasporto*; Casa Editrice Ambrosiana: Milan, Italy, 1970.
16. Missen, R.W.; Mims, C.A.; Saville, B.A. *Chemical Reaction Engineering and Kinetics*; John Wiley & Sons: New York, NY, USA, 1999.
17. Carberry, J.J. Physico chemical aspects of mass and heat transfer in heterogeneous catalysis. In *Catalysis*; Springer: Berlin/Heidelberg, Germany, 1987; pp. 131–171.
18. Thiele, E.W. Relation between catalytic activity and size of particle. In *Industrial and Engineering Chemistry*; ACS Publications: Washington, DC, USA, 1939; Volume 31, pp. 916–920.
19. Weisz, P.B.; Prater, C.D. Interpretation of measurements in experimental catalysis. In *Advances in Catalysis*; Academic Press: New York, NY, USA, 1954; Volume 6, pp. 143–196. [CrossRef]
20. Santacesaria, E. Catalysis and transport phenomena in heterogeneous gas-solid and gas-liquid-solid systems. In *Catalysis Today*; Elsevier Science: Amsterdam, The Netherlands, 1997; Volume 34, pp. 411–420.
21. Froment, G.F.; Bishoff, K.B. *Chemical Reaction Analysis and Design*; John Wiley & Sons: New York, NY, USA, 1990.
22. Woodside, W.W.; Messner, J.H. Thermal conductivity of porous media. *J. Appl. Phys.* **1961**, *32*, 1688. [CrossRef]
23. Hindmarsh, A.C. LSODE and LSODI, two initial value ordinary differential equation solvers. *ACM Signum* **1980**, *15*, 10–11. [CrossRef]
24. Palm, W.J. *Introduction to MATLAB for Engineers*; Mc Graw-Hill: New York, NY, USA, 2011.
25. Tesser, R.; Santacesaria, E. Catalytic oxidation of methanol to formaldehyde: An example of kinetics with transport phenomena in a packed -bed reactor. *Catal. Today* **2003**, *77*, 325–333. [CrossRef]
26. Mars, J.; Krevelen, D.W. Oxidations carried out by means of vanadium oxide catalysts. *Chem. Eng. Sci.* **1954**, *3*, 41. [CrossRef]
27. Santacesaria, E.; Morbidelli, M.; Carrà, S. Kinetics of the catalytic oxidation of methanol to formaldehyde. *Chem. Eng. Sci.* **1981**, *36*, 909–918. [CrossRef]
28. Dente, M.; Collina, A.; Pasquon, I. Verifica di un reattore tubolare per l'ossidazione del metanolo a formaldeide. *La Chimica Industria* **1966**, *48*, 581–588.
29. Riggs, J.B. *Introduction to Numerical Methods for Chemical Engineers*; Texas Tech University Press: Lubbock, TX, USA, 1988.

30. Carrà, S.; Forzatti, P. Engineering aspects of selective hydrocarbons oxidation. *Catal. Rev. Sci. Eng.* **1977**, *15*, 1–52. [CrossRef]
31. Carberry, J.J. *Chemical and Catalitic Reaction Engineering*; McGraw-Hill: New York, NY, USA, 1976.
32. Lee, H.H. *Heterogeneous Reactors Design*; Butterwoth Publisher: Oxford, UK, 1984.
33. Froment, G.F. Fixed bed catalytic reactors—Current design status. *Ind. Eng. Chem.* **1967**, *59*, 18–27. [CrossRef]

3

Estimation of the Biot Number using Genetic Algorithms: Application for the Drying Process

Krzysztof Górnicki *, Radosław Winiczenko and Agnieszka Kaleta

Department of Fundamental Engineering, Warsaw University of Life Sciences, Nowoursynowska 164 St., 02-787 Warsaw, Poland

* Correspondence: krzysztof_gornicki@sggw.pl

Abstract: The Biot number informs researchers about the controlling mechanisms employed for heat or mass transfer during the considered process. The mass transfer coefficients (and heat transfer coefficients) are usually determined experimentally based on direct measurements of mass (heat) fluxes or correlation equations. This paper presents the method of Biot number estimation. For estimation of the Biot number in the drying process, the multi-objective genetic algorithm (MOGA) was developed. The simultaneous minimization of mean absolute error (MAE) and root mean square error (RMSE) and the maximization of the coefficient of determination R^2 between the drying model and experimental data were considered. The Biot number can be calculated from the following equations: $Bi = 0.8193\exp(-6.4951T^{-1})$ (and moisture diffusion coefficient from $D/s^2 = 0.00704\exp(-2.54T^{-1})$) (RMSE = 0.0672, MAE = 0.0535, $R^2 = 0.98$) or $Bi = 1/0.1746\log(1193847T)$ ($D/s^2 = 0.0075\exp(-6T^{-1})$) (RMSE = 0.0757, MAE = 0.0604, $R^2 = 0.98$). The conducted validation gave good results.

Keywords: Biot number; genetic algorithms; drying

---

1. Introduction

The dimensionless Biot number (Bi) is present in partial differential equations in cases when the surface boundary conditions (of the third kind) are written in a dimensionless form. The Biot number informs researchers about the relationship between the internal and external fluxes [1]. As far as heat transfer is concerned, Bi expresses the ratio between the internal and external resistances [2]. Therefore, it can be stated that the heat Biot number is a measure of the temperature drop in the material with respect to the difference of the temperatures between the solid surface and the surrounding medium [3]. Assuming similarity between heat and mass transfer, the Biot number used for mass exchange is gained by equating internal and external mass fluxes at the interface [3,4] and the discussed number is defined as follows:

$$Bi = \frac{h_m L}{D}, \tag{1}$$

The analysis of Biot numbers enables researchers to answer questions regarding the controlling mechanisms employed for heat or mass transfer during the considered process [3]. Dincer [3] stated that the values of Bi for mass transfer can be divided into the following groups:

$Bi < 0.1$ (the surface resistance across the surrounding medium boundary layer is much bigger in comparison with the internal resistance to the mass diffusion within the solid body);

$0.1 < Bi < 100$ (the values of the internal and external resistance can be treated as comparable);

$Bi > 100$ (external (surface) resistance is much lower than the internal resistance).

Ruiz-López et al. [5,6] assumed that for $Bi > 40$, it can be accepted that the internal resistance to mass exchange is the only mechanism controlling the rate of the drying process. In such a case, the moisture content of the solid body surface reaches its equilibrium value at once. Wu and

Irudayaraj [7] experimentally stated that drying can only be treated as an isothermal process for very low Biot numbers.

The mass transfer coefficients (and heat transfer coefficients) are usually determined experimentally based on direct measurements of mass (heat) fluxes or correlation equations. The mass transfer coefficient can also be calculated from the dimensionless Sherwood number (*Sh*). The *Sh* can be expressed as a function of the Reynolds number (*Re*) and the Schmidt number (*Sc*) (forced convection), as a function of the Grashof number (mass) (Gr_m) and the *Sc* (natural convection), and as a function of the Archimedes number (*Ar*) and the *Sc* (vacuum-microwave drying) [8].

The importance of the heat and mass Biot number has been shown in several publications. Rovedo et al. [4] analysed the drying process of shrinking potato slabs. The numerical solving of the drying model using various forms of *Bi* gave a deeper insight into the process. Huang and Yeh [9] considered an inverse problem in simultaneously estimating the heat and mass Biot numbers during the drying of a porous material. Dincer and Hussain [10] developed a new Biot number and lag factor correlation, which gave good agreement between the predicted and measured values of moisture content. Chen and Peng [11] analysed the values of modified Biot numbers during hot air drying of small, moist, and porous objects. Giner et al. [1] considered the variableness of heat and mass *Bi* during the drying of wheat and apple-leather, whereas Xie et al. [12] modelled the pulsed vacuum drying of rhizoma dioscoreae slices using the Dincer model [13], which describes the moisture ratio with a correlation between *Bi* and the lag factor.

An accurate estimation of the Biot number is essential for an efficient heat and mass transfer analysis, leading to optimum operating conditions and an efficient process.

The aim of the multi-objective optimization task is to optimize the several objective functions simultaneously with many criteria. These issues have long been of interest to researchers using traditional optimization and search techniques. In the case of multi-criteria optimization, the concept of the optimal solution is not as obvious as in the case of one criterion. If we do not agree in advance to compare the values of different criteria, then we must propose a definition of optimality that respects the integrity of each of them. This approach is called optimality in the Pareto sense. It is convenient to classify possible solutions of multi-criteria optimization tasks as dominated and non-dominated (*Pareto-optimal*) solutions.

Multi-criteria optimization of drying technology is used for conveyer-belt dryer design [14,15], fluidized bed dryers [16,17], control of a drying process [18], batch drying of rice [19], and drying [20] and rehydrated apple issue [21]. Non-preference multi-criterion optimization methods with the Pareto-optimal set were used in the papers [14–17]. The researchers developed a mathematical model of the fluidized bed dryer and determined the colour deterioration laws for potato slices. A multi-criterion optimization of the thermal processing was conducted by [22] and [23]. The authors developed an intelligent hybrid method for identifying the optimal processing conditions. The complex method applied to different shapes was subjected to the processing boundary conditions to find the best process temperature and to maximize the retention of thiamine. Winiczenko et al. predicted the quality indicators of the drying [20] and rehydrated [21] apple issue using a non-sorting genetic algorithm.

The aim of the present study is to determine the Biot number using the genetics algorithm (GA). The method is applied in the process of drying.

2. Materials and Methods

2.1. Material

The research material was parsley roots of the Berlińska variety. The parsley roots were purchased at a local market in Warsaw. The material was cut into 6 mm thick slices, which were dried in natural convection conditions at the temperatures of 40, 50, 60, and 70 °C.

A detailed description of the equipment used, measurements performed, and their accuracy may be found in [24].

2.2. Moisture Transfer Analysis

It can be assumed that the moisture movement inside the dried solid body is only a diffusion movement in the convection drying of food products. The unsteady-state, one-dimensional mass exchange equation within a slice (treated as an infinite plane) of a thickness of 2 s can be expressed in the following form [25]:

$$\frac{\partial M}{\partial t} = D\frac{\partial^2 M}{\partial x^2}\ (t > 0;\ -s < x < s). \tag{2}$$

The following common assumptions were taken in Equation (2):

- Shape and volume of the slice do not change during drying;
- Mass diffusion coefficient is constant;
- Equation (2) is subjected to the following conditions:

Moisture content at any point of the slice is the same at the beginning of drying (initial condition):

$$M(x,0) = M_0 = \text{const}, \tag{3}$$

The mass flux from the surface of the slice is expressed in terms of the moisture content difference between the surface and the equilibrium moisture content (boundary conditions of the third kind):

$$\pm D\frac{\partial M(\pm s,t)}{\partial x} = k[M(\pm s,t) - M_e], \tag{4}$$

An analytical solution of Equation (2) at the initial condition in the form of Equation (3) and at the boundary conditions given by Equation (4) with respect to the mean moisture content as function of time can be written in the following form [25]:

$$MR = \frac{M - M_e}{M_0 - M_e} = \sum_{i=1}^{\infty} B_i \exp\left[-\mu_i{}^2\frac{D}{s^2}t\right], \tag{5}$$

where

$$B_i = \frac{2Bi^2}{\mu_i{}^2 Bi^2 + Bi + \mu_i{}^2}, \tag{6}$$

$$ctg\mu_i = \frac{1}{Bi}\mu_i, \tag{7}$$

The Biot number in Equations (6) and (7) was assumed to be dependent on the temperature using the following formulas:

$$Bi = a_b \exp\left(-\frac{b_b}{T}\right), \tag{8a}$$

$$Bi = a_b \ln(b_b T), \tag{8b}$$

$$Bi = a_b \log(b_b T), \tag{8c}$$

$$Bi = a_b \ln(T) + b_b, \tag{8d}$$

$$Bi = a_b T^{-b_b}, \tag{8e}$$

$$Bi = \frac{1}{a_b \ln(b_b T)}, \tag{8f}$$

$$Bi = \frac{1}{a_b \log(b_b T)}, \tag{8g}$$

$$Bi = \frac{1}{a_b \exp\left(-\frac{b_b}{T}\right)}, \tag{8h}$$

Dependencies (8a)–(8d) indicate the increase of the Biot number with the increase of the temperature, whereas dependencies (8e)–(8h) indicate the decrease of Bi with the increase of the temperature (i.e., [26]).

The moisture diffusion coefficient in Equation (5) was assumed to be dependent on the temperature using the following formula (i.e., [27,28]):

$$\frac{D}{s^2} = a_d \exp\left(-\frac{b_d}{T}\right) \tag{9}$$

An increase of the temperature according to Equation (9) results in a increase of the moisture diffusion coefficient.

2.3. Optimization

The optimization problem has been divided into two tasks. The first multi-objective optimization (MOO) task was to determine a Biot number and mass diffusion coefficient from the moisture ratio (MR) model (5) for each drying temperature: 40, 50, 60, and 70 °C. The second optimization task was to find the constants in equations for calculation of the Biot number (8a)–(8h) and mass diffusion coefficient (9).

The mean absolute error (MAE) and root mean square error (RMSE) were minimized, whereas the coefficient of determination R^2 was maximized for the difference between the data and objective function for both optimized tasks. The algorithm randomly selects a set of models by minimizing the error between the proposed model and experimental data. Then, it evolves them to create the best fit. Therefore, the MOO problem was expressed as follows:

$$\text{Min} = \begin{cases} \text{Min RMSE} = \sqrt{\frac{1}{N}\sum_{i=1}^{N}\left(MR_{\text{pred}} - MR_{\text{exp}}\right)^2} \\ \text{Min MAE} = \frac{1}{N}\sum_{i=1}^{N}\left|MR_{\text{pred}} - MR_{\text{exp}}\right| \\ \text{Max } R^2 = 1 - \frac{\sum_{i=1}^{N}\left(MR_{\text{pred}} - MR_{\text{exp}}\right)^2}{\sum_{i=1}^{N}\left(MR_{\text{pred}} - \overline{MR}_{\text{exp}}\right)^2} \end{cases} . \tag{10}$$

The multi-objective optimization was carried out using an elitist, non-dominated sorting genetic algorithm (NSGA II). The Pareto front for this problem was created using Optimization Toolbox™ realized in MATLAB R2018. The multi-objective genetic algorithm (MOGA) options are shown in Table 1.

Table 1. The multi-objective genetic algorithm (MOGA) settings.

Parameter	Value
Crossover coefficient	0.8
Crossover function	Intermediate
Mutation coefficient	0.2
Mutation function	Uniform
Number generations	500·number of variables
Population size	60·number of variables
Selection function	Turnament size = 2

3. Results

3.1. Case 1. Optimizaton of the Biot Number

The results of optimization described by Equation (10) are presented in Table 2 and Figures 1 and 2. The Pareto set of the Biot number for the T of 40 °C indicates that in the case of searching for the smallest values of RMSE, the best solution is ID 40_1 (0.04441). For ID 40_2, the value of R^2 is the greatest (0.9937). The solution of MAE minimization is ID 40_5 (0.03652). The set of the best solutions for 40 °C (marked in Figure 1) indicates the greatest dispersion of solutions compared with the best solutions for other drying temperatures examined. The Biot number determined for the drying temperature of 50 °C has the smallest RMSE for ID 50_1 (0.03731). The value of R^2 is the greatest for ID 50_2 (0.9955). The solution of MAE minimization is ID 50_3 (0.03228). The dispersion of the best solutions is smaller than for the temperature of 40 °C (Figure 1), and the solutions are characterised by the best statistics. The Biot number determined for the drying temperature of 60 °C has the smallest RMSE for ID 60_1 and 60_2 (0.04072–0.04073, difference 0.00001). For ID 60_4, the value of R^2 is the greatest (0.9948). The solution of MAE minimization is ID 60_4 i 60_3 (0.03640 and 0.03641, difference 0.00001). The set of best solutions for 60 °C (Figure 1) is characterised by small (the smallest) dispersion. The Biot number determined for the drying temperature of 70 °C has the smallest RMSE and the greatest R^2 for ID 70_1–70_3 (0.03888–0.03889 and 0.9943, respectively). The solution of MAE minimization is ID 70_4 (0.03420), with the solutions for ID 70_5, 70_2, 70_3, and 70_1 being slightly worse (0.03424, 0.03427, 0.03428, and 0.03429, difference 0.00009). The set of best solutions for 70 °C (Figure 1) is only characterised by a very small amount of dispersion when the MAE minimization criterion is not considered.

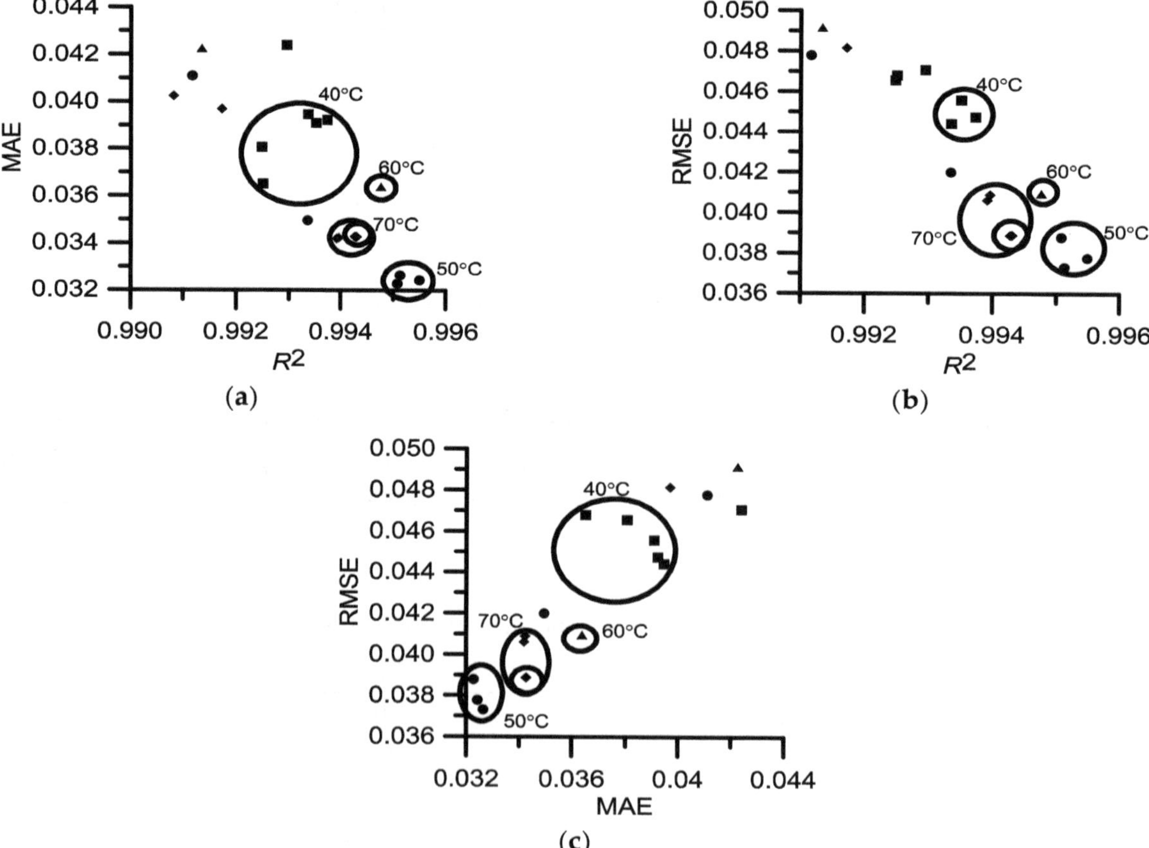

Figure 1. Results of the statistical analysis of Pareto optimal sets for the Biot number: (■) T = 40°C, (●) T = 50°C, (▲) T = 60°C, and (♦) T = 70°C.

Table 2. Pareto optimal set for the Biot number (*Bi*) and D/s^2 and results of the statistical analysis.

Temperature, °C	ID	*Bi*	D/s^2 (min^{-1})	R^2	RMSE	MAE
40	40_1	0.7233	0.00610	0.9934	0.04441	0.03947
	40_2	0.7243	0.00601	0.9937	0.04474	0.03924
	40_3	0.7016	0.00625	0.9935	0.04558	0.03910
	40_4	0.7548	0.00442	0.9925	0.04656	0.03807
	40_5	0.7450	0.00454	0.9925	0.04680	0.03652
	40_6	0.7555	0.00601	0.9930	0.04707	0.04240
	40_7	0.7749	0.00532	0.9858	0.07744	0.06666
	40_8	0.6745	0.00883	0.9820	0.07770	0.06498
	40_9	0.6901	0.00883	0.9804	0.08622	0.07129
	40_10	0.6555	0.00990	0.9748	0.10327	0.08541
50	50_1	0.7245	0.00615	0.9951	0.03731	0.03264
	50_2	0.7231	0.00605	0.9955	0.03777	0.03243
	50_3	0.7012	0.00635	0.9951	0.03879	0.03228
	50_4	0.6967	0.00698	0.9934	0.04200	0.03496
	50_5	0.6967	0.00747	0.9912	0.04779	0.04110
	50_6	0.7283	0.00734	0.9897	0.06001	0.05115
	50_7	0.6506	0.00884	0.9867	0.06764	0.05661
	50_8	0.7040	0.00824	0.9858	0.07501	0.06318
	50_9	0.6540	0.00912	0.9836	0.07945	0.06742
	50_10	0.7040	0.00873	0.9828	0.08853	0.07409
60	60_1	0.7017	0.00788	0.9944	0.04072	0.03646
	60_2	0.7017	0.00791	0.9943	0.04073	0.03654
	60_3	0.7086	0.00780	0.9946	0.04085	0.03641
	60_4	0.7001	0.00775	0.9948	0.04096	0.03640
	60_5	0.7095	0.00872	0.9913	0.04916	0.04225
	60_6	0.7337	0.00838	0.9917	0.05097	0.04363
	60_7	0.6507	0.01012	0.9889	0.05853	0.04684
	60_8	0.6513	0.01016	0.9888	0.05896	0.04720
	60_9	0.7152	0.00958	0.9875	0.06637	0.05393
	60_10	0.6791	0.01046	0.9845	0.07567	0.06103
70	70_1	0.7003	0.01012	0.9943	0.03888	0.03429
	70_2	0.7005	0.01011	0.9943	0.03888	0.03427
	70_3	0.7014	0.01011	0.9943	0.03889	0.03428
	70_4	0.6789	0.01062	0.9939	0.04061	0.03420
	70_5	0.6898	0.01060	0.9940	0.04089	0.03424
	70_6	0.7043	0.01120	0.9917	0.04815	0.03969
	70_7	0.6525	0.01260	0.9908	0.05378	0.04025
	70_8	0.6705	0.01324	0.9878	0.06406	0.05103
	70_9	0.6914	0.01286	0.9878	0.06602	0.05290
	70_10	0.6636	0.01376	0.9860	0.07202	0.05885

The values of the Biot numbers obtained as a result of optimization for the temperature of 40 °C (ID 40_1–40_5) are within the range of 0.723–0.755; for 50 °C (ID 50_1–50_3), are within the range of 0.701–0.725; for 60 °C (ID 60_1–60_4), are within the range of 0.700–0.709; for 70 °C (ID 70_1–70_3), are within the range of 0.700–0.701; and including the MAE criterion (ID 70_1–70_5), are within the range of 0.679–0.701. The smallest RMSE (0.0373) and MAE (0.0323) were obtained for $T = 50$ °C, whereas the greatest errors for were obtained for $T = 40$ °C. The results of optimization indicate that, initially, the Biot number does not change (or changes slightly) with the increase of temperature (40–50 °C), followed by the decrease in its value (50–70 °C). The obtained results of optimization also indicate that the increase of T results in the increase of the moisture diffusion coefficient (i.e., [26,28]) (Table 2).

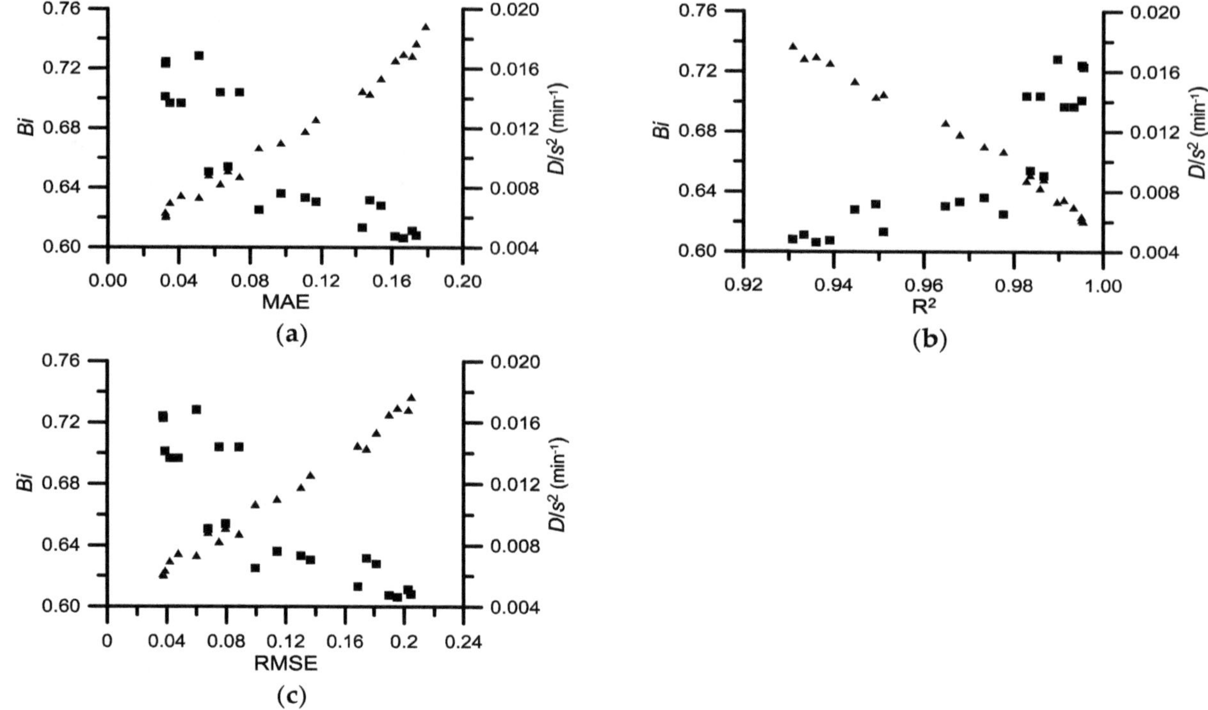

Figure 2. Pareto optimal set for the Biot number (■) and D/s^2 (▲) for T = 50 °C.

In accordance with Equations (5)–(7), the value of the calculated reduced moisture content MR depends not only on the Biot number, but also, assuming the accuracy of the description by the model, on the moisture diffusion coefficient D. Figure 2 shows an example of the relation between Bi and the moisture diffusion coefficient. The increase of the Biot number requires the value of D to be reduced (Figure 2), with better fitting of the model (5) to experimental data being obtained for greater Bi values. For smaller values of Bi, the errors (RMSE and MAE) are greater, and R^2 is smaller, even after the increase of D (Figure 2).

3.2. Case 2a. Optimization of Parameters of the Function for Determining the Biot Number (Equation (8a))

The optimization task described by Equation (10) considers the Biot number from Equation (8a) in MR_{pred} (Equations (5)–(7)). Therefore, the parameters a_b and b_b of the equation (Equation (8a)) were sought in this strategy.

The results of optimization are presented in Table 3, Figure 3, and Figure 4. The best solution is ID 2a_1 (the smallest MAE = 0.05346 and the greatest R^2 = 0.9822) and ID 2a_2 (the smallest RMSE = 0.06695 and slightly worse MAE and R^2 compared with ID 2a_1).

Table 3. Pareto optimal set for constants in Equations (8a) and (9) and results of the statistical analysis.

ID	a_b	b_b	a_d	b_d	R^2	RMSE	MAE
2a_1	0.81932	6.49505	0.00704	2.54000	0.9822	0.06724	0.05346
2a_2	0.81976	6.53818	0.00720	2.54000	0.9817	0.06695	0.05358
2a_3	0.81935	6.48207	0.00717	2.28576	0.9815	0.06717	0.05374
2a_4	0.81959	6.43306	0.00722	2.06814	0.9812	0.06744	0.05405
2a_5	0.82018	6.47067	0.00700	1.57058	0.9814	0.06783	0.05405
2a_6	0.82042	6.47042	0.00705	1.57079	0.9813	0.06779	0.05411
2a_7	0.81959	6.43306	0.00722	1.56814	0.9807	0.06804	0.05455
2a_8	0.81964	6.52297	0.00732	1.91671	0.9807	0.06794	0.05457
2a_9	0.81955	6.46056	0.00735	1.58362	0.9803	0.06864	0.05514
2a_10	0.81958	6.45445	0.00750	1.87225	0.9800	0.06911	0.05557

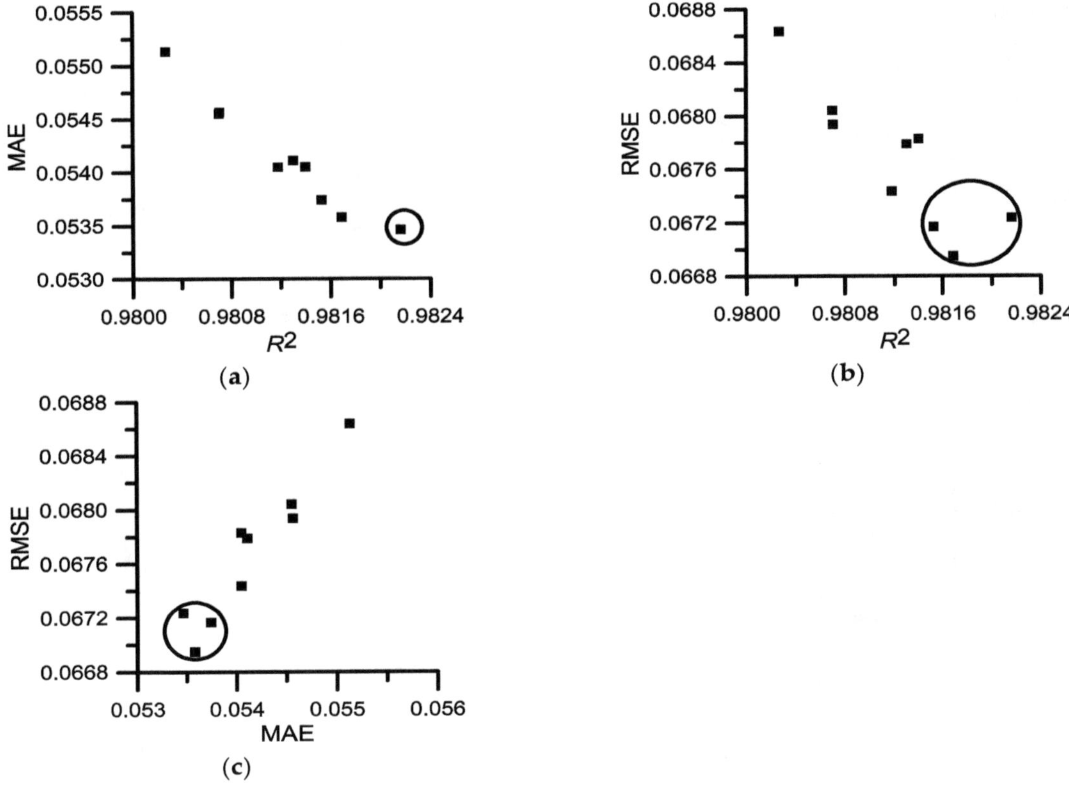

Figure 3. Pareto optimal sets for constants of Equation (8a).

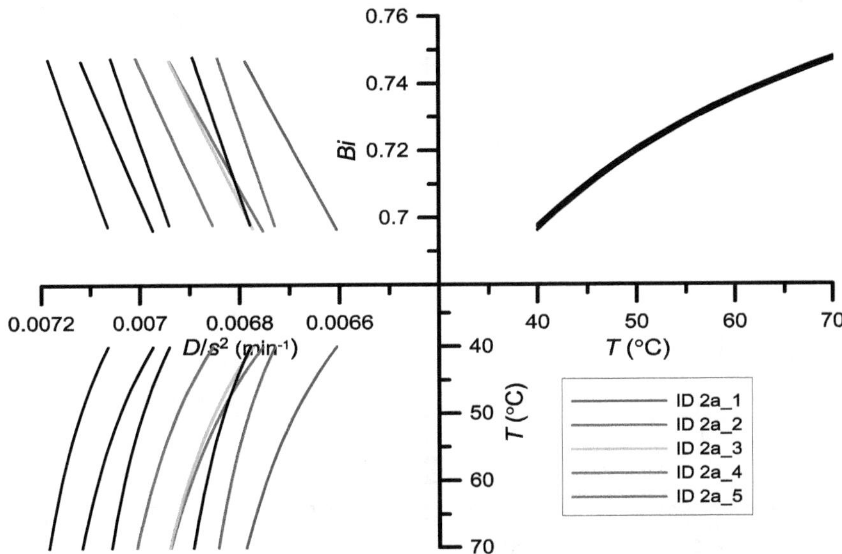

Figure 4. Pareto optimal set for the function of the Biot number and D/s^2: Equations (8a) and (9).

The courses of the function (Equation (8a)) with parameters a_b and b_b obtained as a result of optimization are very similar (Figure 5) (overlap almost entirely–similar values of parameters (Table 3)), and the values of the Biot number obtained from them for the temperatures of 40–70°C are within the range of 0.696523–0.74672 for ID 2a_1 and 0.696144–0.746659 for ID 2a_2. The courses of the function (Equation (9)) with parameters a_d and b_d obtained as a result of optimization differ from each other (Figure 4) and the best solution of the optimization task was obtained for smaller values of the moisture diffusion coefficient. The values of D/s^2 for the temperatures of 40–70°C are within the range of 0.006606–0.006788 for ID 2a_1 and 0.006753–0.00694 for ID 2a_2.

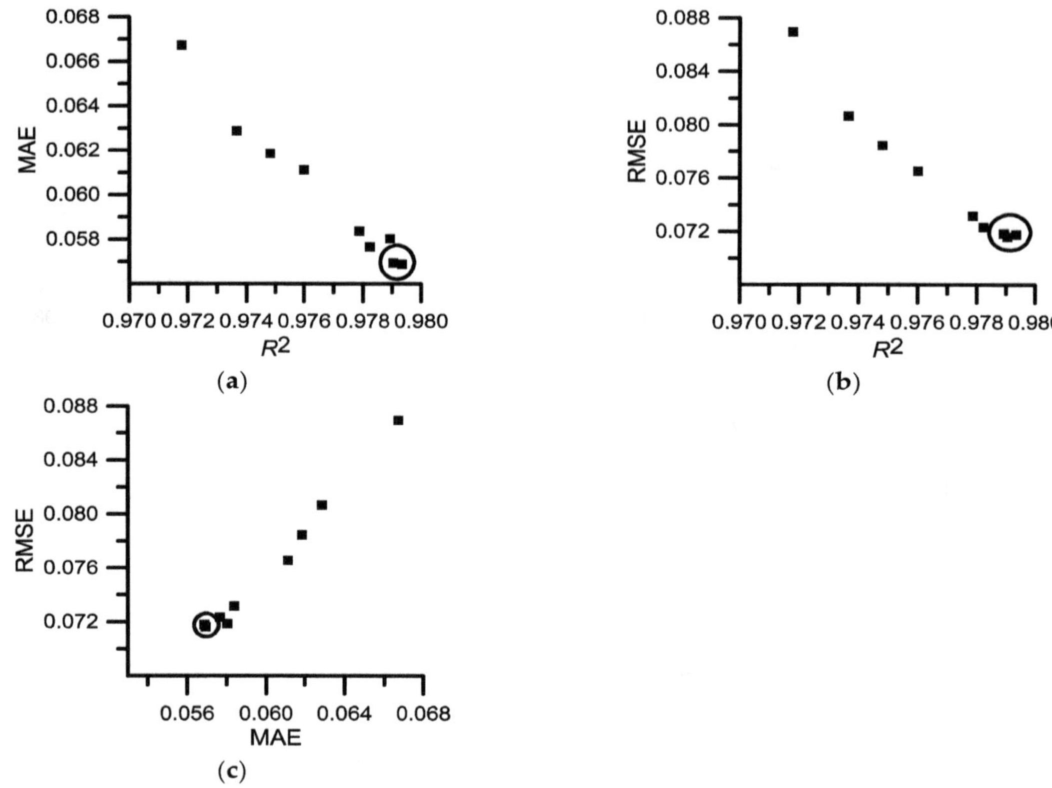

Figure 5. Pareto optimal sets for constants of Equation (8b).

3.3. *Case 2b. Optimization of Parameters of the Function for Determining the Biot number (Equation (8b))*

The results of optimization are presented in Table 4 and Figures 5 and 6. The best solutions are ID 2b_1 (the smallest MAE = 0.05688 and the greatest $R^2 = 0.9794$) and ID 2b_2 (the smallest RMSE = 0.07160 and slightly smaller R^2 (difference 0.0004)).

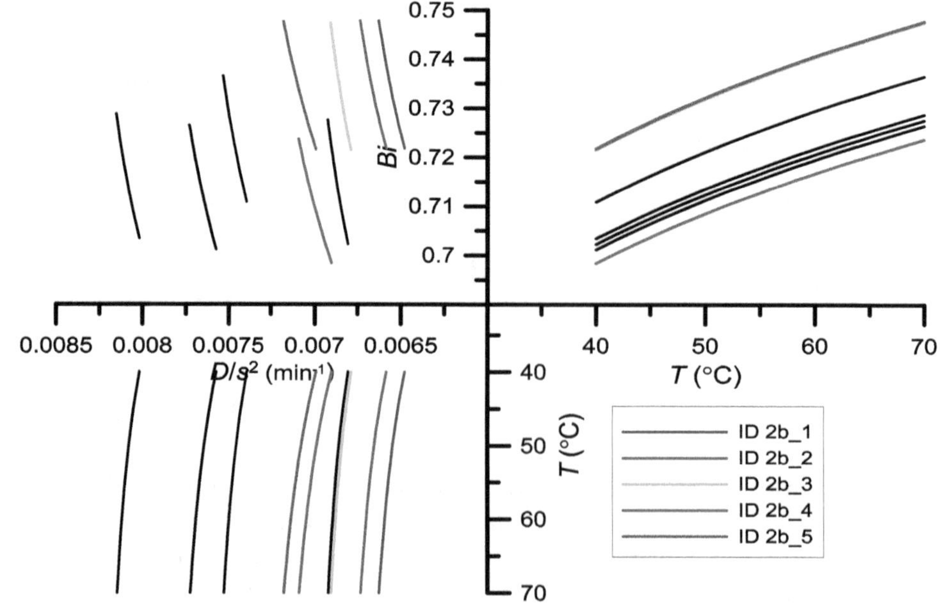

Figure 6. Pareto optimal set for the function of the Biot number and D/s^2: Equations (8b) and (9).

The courses of the function (Equation (8b)) with parameters a_b and b_b obtained as a result of optimization are very similar (Figure 6) (ID 2b_1, 2b_2, and 2b_3 overlap almost entirely - similar values of parameters (Table 4)), and the values of the Biot numbers obtained from the function for the temperatures of 40–70 °C are within the range of 0.721854–0.747842 both for ID 2b_1 and ID 2b_2, and they are the greatest compared with the other solutions. The courses of the function (Equation (9)) with parameters a_d and b_d obtained as a result of optimization differ from each other (Figure 6) and the best solution of the optimization task was obtained for smaller values of the moisture diffusion coefficient. The values of D/s^2 for the temperatures of 40–70°C are within the range of 0.006477–0.006624 for ID 2b_1 and 0.006582–0.006731 for ID 2b_2.

Table 4. Pareto optimal set for constants in Equations (8b) and (9) and results of the statistical analysis.

ID	a_b	b_b	a_d	b_d	R^2	RMSE	MAE
2b_1	0.04644	140795.3	0.006825	2.094928	0.9794	0.07178	0.05688
2b_2	0.04644	140795.5	0.006935	2.091899	0.9790	0.07160	0.05694
2b_3	0.046449	139323	0.007073	1.661161	0.9782	0.07233	0.05766
2b_4	0.045152	130655.1	0.007347	2.498029	0.9789	0.07185	0.05803
2b_5	0.046592	133294.6	0.007432	2.446309	0.9779	0.07316	0.05837
2b_6	0.045209	139490.1	0.007085	1.60367	0.9760	0.07654	0.06113
2b_7	0.045793	138027.1	0.007722	1.747985	0.9748	0.07845	0.06186
2a_8	0.045224	135305.8	0.007933	1.8519	0.9737	0.08067	0.06288
2a_9	0.045296	138802.9	0.008329	1.525632	0.9718	0.08710	0.06673
2b_10	0.046629	139913.5	0.008114	1.605228	0.9718	0.08986	0.06883

3.4. *Case 2c. Optimization of Parameters of the Function for Determining the Biot Number (Equation (8c))*

The results of optimization are presented in Table 5 and Figures 7 and 8. The best solutions are ID 2c_1 (the smallest MAE = 0.05693 and the greatest R^2 = 0.9793) and ID 2c_2 (the smallest RMSE = 0.07168843 and slightly smaller R^2 (difference 0.0003)).

Table 5. Pareto optimal set for constants in Equations (8c) and (9) and results of the statistical analysis.

ID	a_b	b_b	a_d	b_d	R^2	RMSE	MAE
2c_1	0.105914	161892.9	0.006818	1.999354	0.9793	0.07184	0.05693
2c_2	0.105920	161892.9	0.006939	1.999354	0.9790	0.07168	0.05701
2c_3	0.105956	161896.9	0.007083	1.836184	0.9784	0.07212	0.05751
2c_4	0.105896	162115	0.007115	1.717303	0.9782	0.07242	0.05773
2c_5	0.105920	161892.9	0.007259	1.999354	0.9780	0.07274	0.05803
2c_6	0.102556	162170.2	0.007465	1.623216	0.9777	0.07301	0.05877
2c_7	0.102552	162127.3	0.007589	1.80125	0.9775	0.07331	0.05887
2c_8	0.102543	162188.4	0.007620	1.608689	0.9772	0.07384	0.05919
2c_9	0.102548	162077.8	0.007761	1.823051	0.9770	0.07446	0.05951
2c_10	0.103676	162170.9	0.007290	1.999114	0.9765	0.07493	0.05968

The courses of the function (Equation (8c)) with parameters a_b and b_b obtained as a result of optimization are very similar (Figure 8) (overlap almost entirely–similar values of parameters (Table 5)), and the values of the Biot numbers obtained from the function for the temperatures of 40–70°C are within the range of 0.721413–0.747151 for ID 2c_1 and 0.721450–0.747192 for ID 2c_2, and they are the greatest compared with other solutions. The courses of the function (Equation (9)) with parameters a_d and b_d obtained as a result of optimization differ from each other (Figure 8) and the best solution of the optimization task was obtained for smaller values of the moisture diffusion coefficient. The values of D/s^2 for the temperatures of 40–70°C are within the range of 0.00649–0.00663 for ID 2c_1 and 0.0066–0.00674 for ID 2c_2.

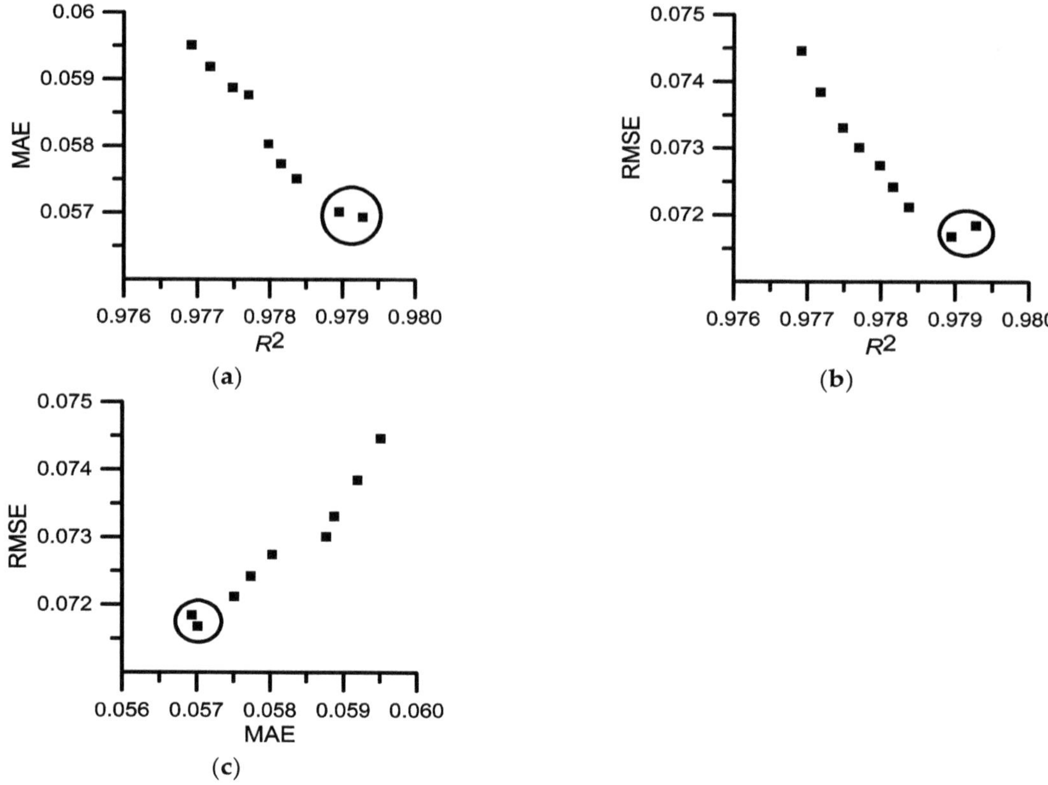

Figure 7. Pareto optimal sets for constants of Equation (8c).

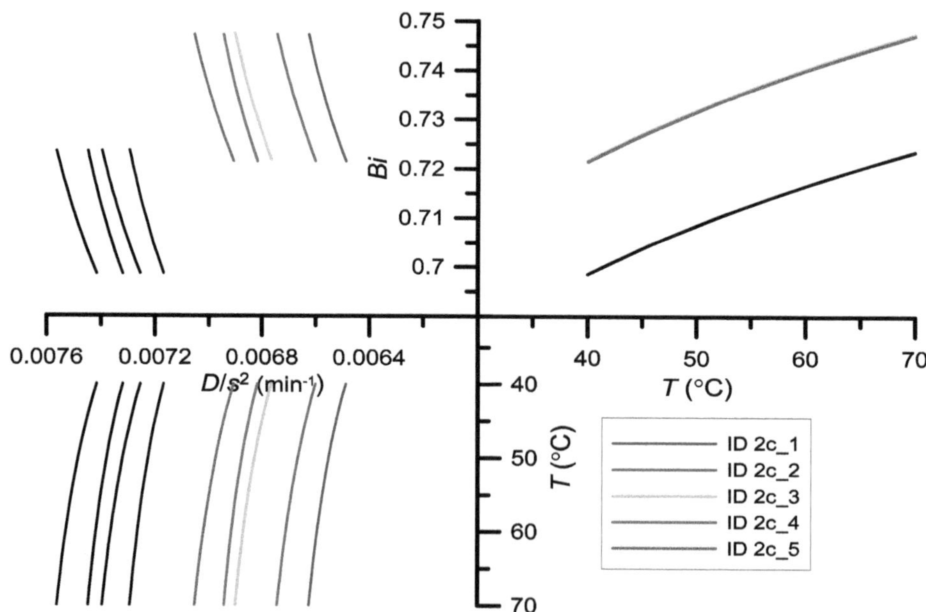

Figure 8. Pareto optimal set for the function of the Biot number and D/s^2: Equations (8c) and (9).

3.5. Case 2d. Optimization of Parameters of the Function for Determining the Biot Number (Equation (8d))

The results of optimization are presented in Table 6 and Figures 9 and 10. The performed optimization also allowed us to obtain constants for Equation (8d) and Equation (9) and then to calculate the Biot number and the moisture diffusion coefficient (Table 6). The best solutions are ID 2d_1 (the smallest MAE = 0.05378 and the greatest R^2 = 0.9818) and ID 2d_2 (the smallest RMSE = 0.06742 and only slightly worse R^2 (difference 0.0004)).

Table 6. Pareto optimal set for constants in Equations (8d) and (9) and results of the statistical analysis.

ID	a_b	b_b	a_d	b_d	R^2	RMSE	MAE
2d_1	0.087671	0.374624	0.006984	1.988010	0.9818	0.06763	0.05378
2d_2	0.087777	0.374868	0.007130	1.988010	0.9814	0.06742	0.05390
2d_3	0.099520	0.329786	0.007546	2.461728	0.9804	0.06837	0.05507
2d_4	0.091440	0.359263	0.007757	2.217054	0.9794	0.07085	0.05688
2d_5	0.100428	0.303184	0.008240	2.052006	0.9783	0.07329	0.05852
2d_6	0.091342	0.356587	0.007770	2.413837	0.9778	0.07300	0.05883
2d_7	0.078979	0.400324	0.007770	1.930612	0.9773	0.07414	0.05956
2d_8	0.072832	0.412350	0.007200	1.973931	0.9772	0.07705	0.06194
2d_9	0.077064	0.381598	0.008456	2.366783	0.9758	0.07764	0.06241
2d_10	0.029274	0.581316	0.008014	1.660092	0.9746	0.07869	0.06313

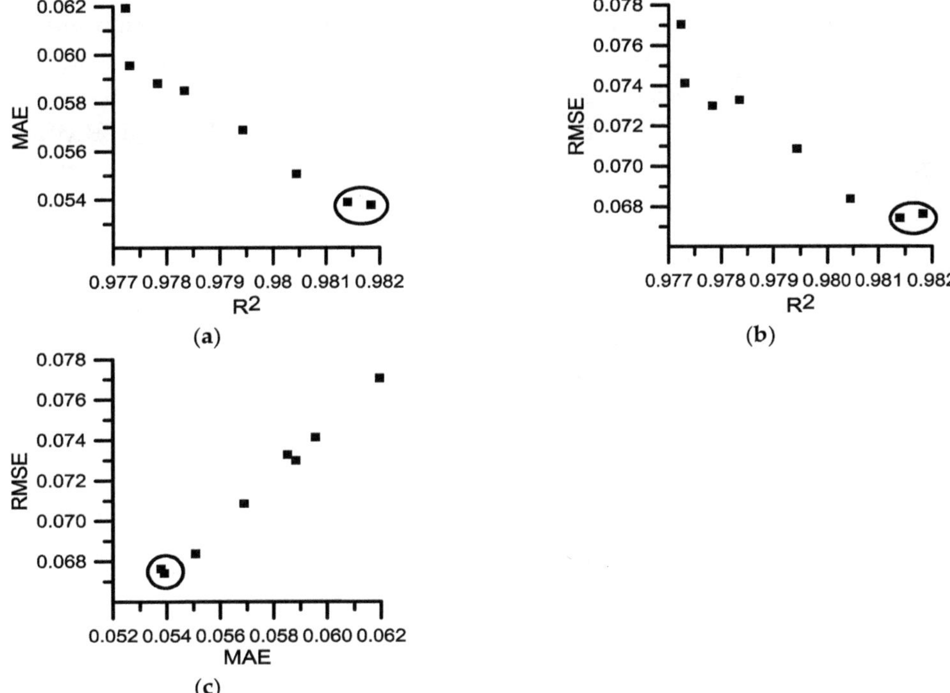

Figure 9. Pareto optimal sets for constants of Equation (8d).

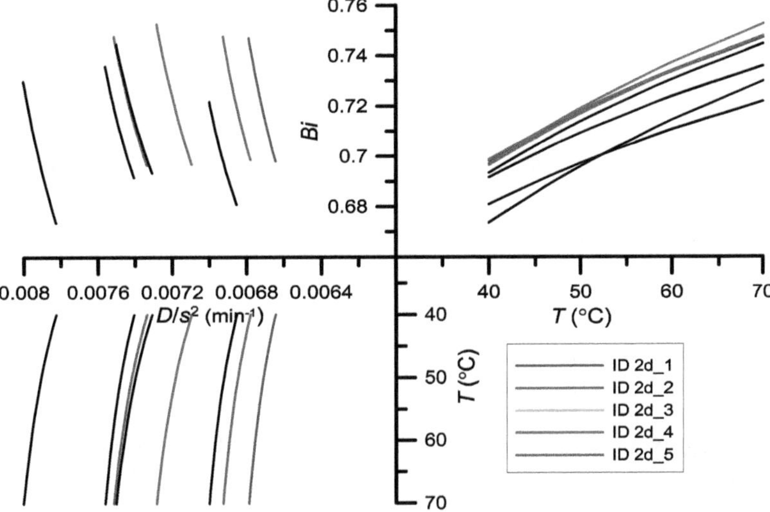

Figure 10. Pareto optimal set for the function of the Biot number and D/s^2: Equations (8d) and (9).

The courses of the function (Equation (8d)) with parameters a_b and b_b obtained as a result of optimization are very similar (Figure 10) (overlap almost entirely–similar values of parameters (Table 6)), and the values of the Biot numbers obtained from the function for the temperatures of 40–70 °C are within the range of 0.698032–0.747095 for ID 2d_1 and 0.698668–0.747789 for ID 2d_2, and are the greatest compared with other solutions. The courses of the function (Equation 9) with parameters a_d and b_d obtained as a result of optimization differ from each other (Figure 10) and the best solution of the optimization task was obtained for smaller values of the moisture diffusion coefficient. The values of D/s^2 for the temperatures of 40–70°C are within the range of 0.00665–0.00679 for ID 2d_1 and 0.00678–0.00693 for ID 2d_2.

3.6. Case 2e. Optimization of Parameters of the Function for Determining the Biot Number (Equation (8e))

The results of optimization are presented in Table 7 and Figures 11 and 12. The optimization also allowed us to obtain constants Equation (8e) and Equation (9) and then to calculate the Biot number and the moisture diffusion coefficient (Table 7). The set of best solutions is made up of ID 2e_1–ID 2e_5, with ID 2e_1 being characterised by the smallest value of MAE = 0.06633, ID 2e_4 by the smallest value of RMSE = 0.08183, and ID 2e_5 by the greatest R^2 = 0.9771. For ID 2e_1, RMSE is greater than for ID 2e_4 by only 0.0002 and R^2 is smaller than for ID 2e_5 by only 0.0003. For ID 2e_4, MAE is greater than for ID 2e_1 by only 0.0002 and R^2 is only 0.0003 smaller than for ID 2e_5. For ID 2e_5, MAE is greater than for ID 2e_1 by only 0.0002 and RMSE is greater than for ID 2e_4 by only 0.00006. For ID 2e_2, ID 2e_3, and ID 2e_6, R^2 is smaller than the greatest one (ID 2e_5) by 0.0002, 0.00007, and 0.0001, respectively; RMSE is greater than the smallest one (ID 2e_4) by 0.00013, 0.00014, and 0.000005, respectively; and MAE is greater than the smallest one (ID 2e_1) by 0.00001, 0.0009, and 0.00025, respectively.

Table 7. Pareto optimal set for constants in Equations (8e) and (9) and results of the statistical analysis.

ID	a_b	b_b	a_d	b_d	R^2	RMSE	MAE
2e_1	1.298971	0.124544	0.006152	4.960183	0.9768	0.08207	0.06633
2e_2	1.449848	0.153382	0.006185	4.996656	0.9769	0.08196	0.06634
2e_3	1.290019	0.131396	0.006165	4.989554	0.9770	0.08197	0.06642
2e_4	1.289663	0.129626	0.006287	4.989283	0.9768	0.08183	0.06651
2e_5	1.456449	0.165094	0.006188	4.984513	0.9771	0.08189	0.06652
2e_6	1.466073	0.166745	0.006246	4.961883	0.9770	0.08184	0.06658
2a_7	1.557768	0.172773	0.006043	3.882209	0.9762	0.08287	0.06694
2e_8	1.562403	0.184398	0.006047	3.886745	0.9765	0.08278	0.06715
2e_9	1.56531	0.183702	0.006119	3.888001	0.9763	0.08269	0.06719
2e_10	1.276498	0.120256	0.005896	2.895483	0.9756	0.08381	0.06743

The courses of the function (Equation (8e)) with parameters a_b and b_b obtained as a result of optimization differ from each other (Figure 12), and the values of *Bi* obtained from them are the greatest. The values of *Bi* obtained for the temperatures of 40–70 °C are within the range of 0.823371–0.721929: 0.820488–0.76525 for ID 2e_1, 0.823371–0.755645 for ID 2e_2, 0.794496–0.738172 for ID 2e_3, 0.79948–0.743539 for ID 2e_4, 0.792145–0.722239 for ID 2e_5, and 0.792538–0.721929 for ID 2e_6. The courses of the function (Equation (9)) with parameters a_d and b_d obtained as a result of optimization are similar (Figure 12) and the best solution of the optimization task was obtained for smaller values of the moisture diffusion coefficient (except for ID 2e_4). The values of D/s^2 obtained for the temperatures of 40–70 °C are within the range of 0.00543–0.00573 for ID 2e_1, 0.00546–0.00576 for ID 2e_2, 0.00544–0.00574 for ID 2e_3, 0.00555–0.00585 for ID 2e_4, 0.00546–0.00576 for ID 2e_5, and 0.00552–0.00582 for ID 2e_6.

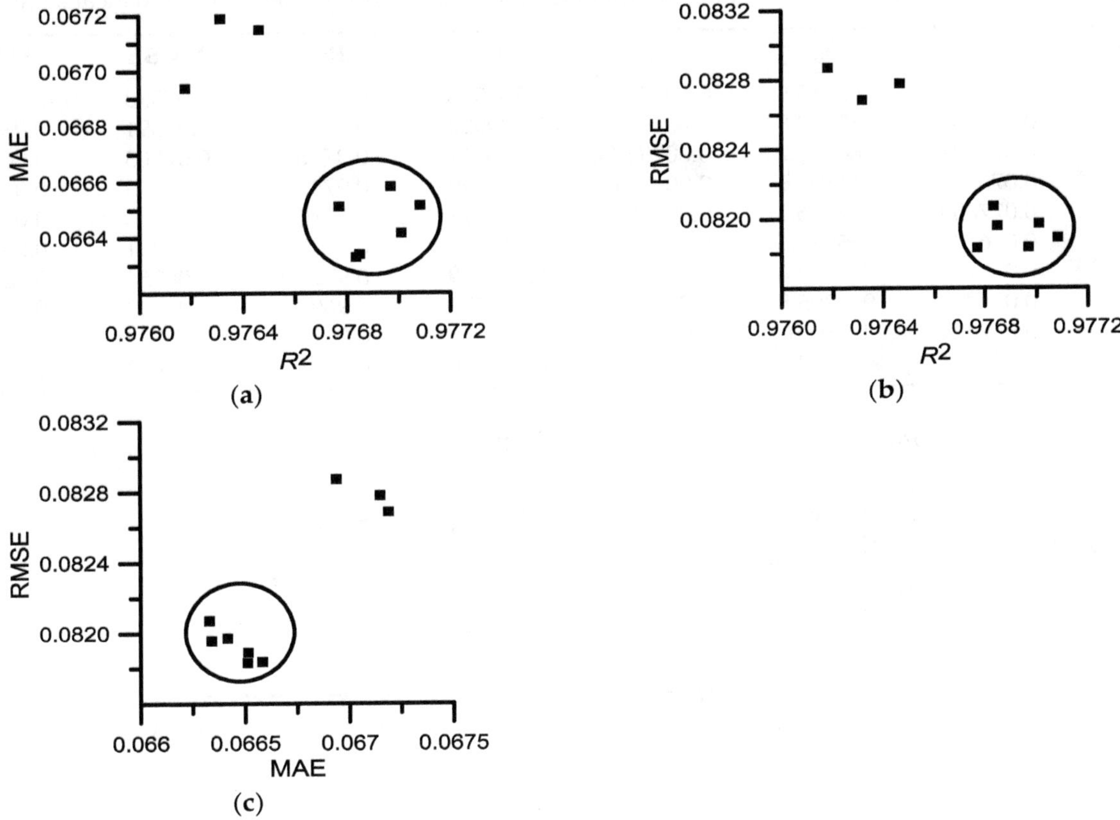

Figure 11. Pareto optimal sets for constants of Equation (8e).

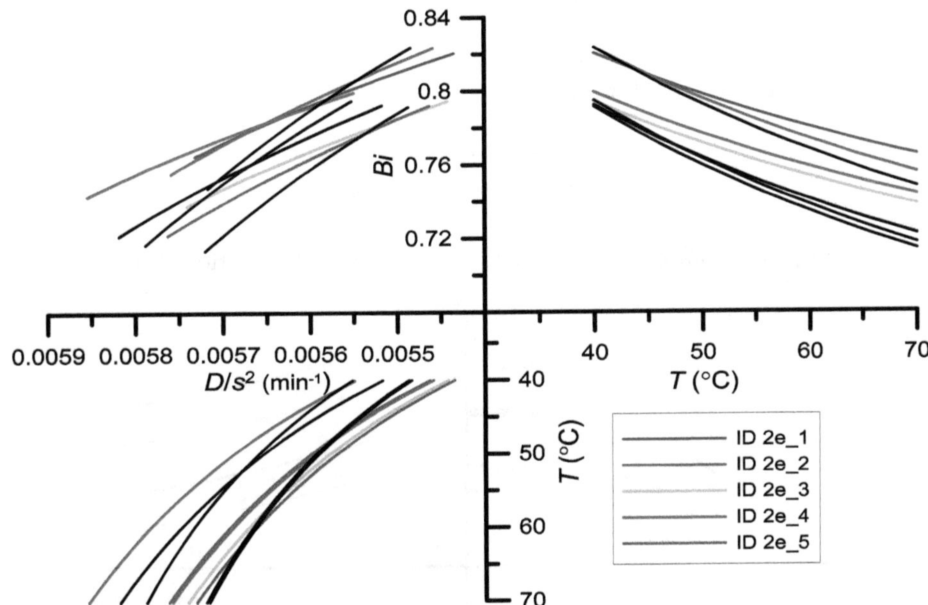

Figure 12. Pareto optimal set for the function of the Biot number and D/s^2: Equations (8e) and (9).

3.7. Case 2f. Optimization of Parameters of the Function for Determining the Biot Number (Equation (8f))

The results of optimization are presented in Table 8 and Figures 13 and 14. The best solutions are ID 2f_1 (the smallest MAE = 0.06056 and the greatest R^2 = 0.9764) and ID 2f_2 (the smallest RMSE = 0.07591, and compared with 2f_1, slightly worse R^2 (smaller by 0.0002) and MAE (greater by 0.00002)).

Table 8. Pareto optimal set for constants in Equations (8f) and (9) and results of the statistical analysis.

ID	a_b	b_b	a_d	b_d	R^2	RMSE	MAE
2f_1	0.077244	862213.6	0.007368	5.739963	0.9764	0.07598	0.06056
2f_2	0.077236	862220.8	0.007449	5.729922	0.9762	0.07591	0.06058
2f_3	0.077207	864646.7	0.007868	5.355261	0.9748	0.07803	0.06199
2f_4	0.073875	1059418	0.006741	2.720529	0.9745	0.07933	0.06235
2a_5	0.077859	862396.3	0.007587	5.708511	0.9730	0.08043	0.06483
2a_6	0.076727	865525.6	0.007648	4.417204	0.9713	0.08331	0.06520
2f_7	0.078282	862213.6	0.007368	5.741489	0.9727	0.08254	0.06684
2f_8	0.076328	960205.5	0.007648	2.969758	0.9698	0.08642	0.06705
2f_9	0.072421	1059416	0.006582	1.131156	0.9690	0.08750	0.06884
2f_10	0.070248	1137211	0.006003	0.042246	0.9696	0.08812	0.06893

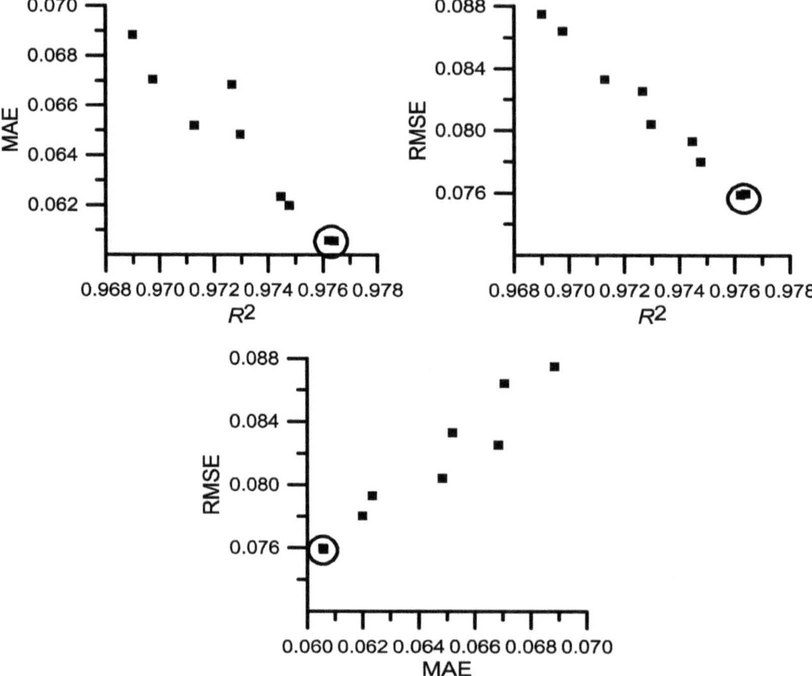

Figure 13. Pareto optimal sets for constants of Equation (8f).

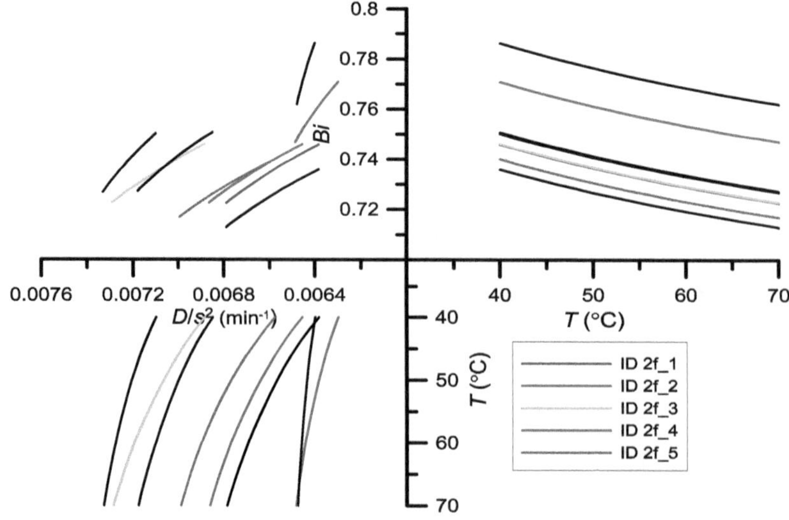

Figure 14. Pareto optimal set for the function of the Biot number and D/s^2: Equations (8f) and (9).

The courses of the function (Equation (8f)) with parameters a_b and b_b obtained as a result of optimization differ from each other (Figure 14). The values of the Biot number obtained for the temperatures of 40–70 °C are within the range of 0.745901–0.722602 for ID 2f_1 and 0.745978–0.722677 for ID 2f_2. The courses of the function (Equation 9) with parameters a_d and b_d obtained as a result of optimization differ from each other (Figure 14). The values of D/s^2 obtained for the temperatures of 40–70 °C are within the range of 0.006383–0.006788 for ID 2f_1 and 0.006454–0.006863 for ID 2f_2.

3.8. Case 2g. Optimization of Parameters of the Function for Determining the Biot Number (Equation (8g))

The results of optimization are presented in Table 9 and Figure 15–16. The best solution is ID 2g_1 (the smallest both MAE = 0.06041, RMSE = 0.07566 and the greatest R^2 = 0.9764).

Table 9. Pareto optimal set for constants in Equations (8g) and (9) and results of the statistical analysis.

ID	a_b	b_b	a_d	b_d	R^2	RMSE	MAE
2g_1	0.17462	1193847	0.007492	5.999963	0.9764	0.07566	0.06041
2g_2	0.17462	1193847	0.007492	4.984338	0.9754	0.07677	0.06124
2g_3	0.20101	109835.1	0.007664	7.999996	0.9748	0.07827	0.06191
2g_4	0.20101	109834.6	0.007698	7.999992	0.9747	0.07826	0.06192
2g_5	0.20097	109790.8	0.007727	7.968262	0.9746	0.07830	0.06195
2g_6	0.19994	114067.1	0.007438	6.988878	0.9743	0.07932	0.06265
2g_7	0.18938	116127.4	0.007119	8.999996	0.9760	0.07926	0.06310
2g_8	0.18938	116103.4	0.007247	8.999909	0.9758	0.07911	0.06312
2g_9	0.18944	115792.8	0.007267	8.882749	0.9749	0.07991	0.06341
2g_10	0.19830	114642.6	0.007446	5.755039	0.9731	0.08041	0.06346

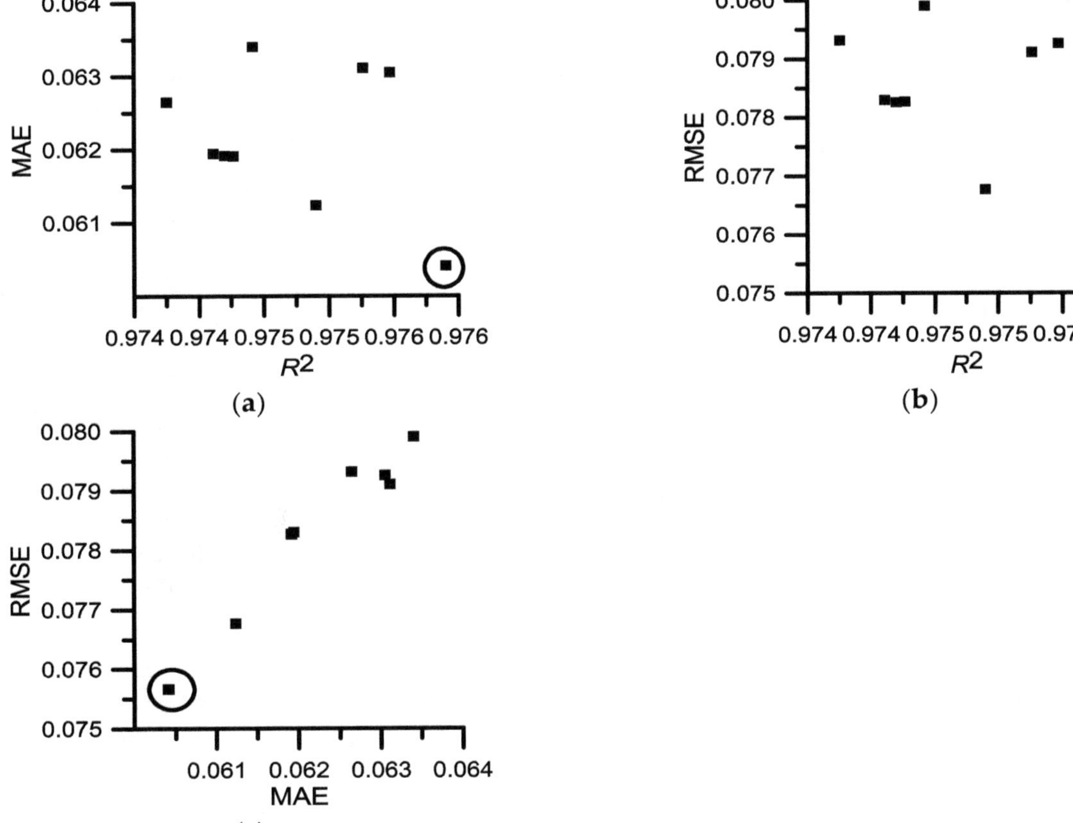

Figure 15. Pareto optimal sets for constants of Equation (8g).

The courses of the function (Equation (8g)) with parameters a_b and b_b and the function (Equation (9)) with parameters a_d and b_d obtained as a result of optimization differ from each other (Figure 16). The values of the Biot number and the values of D/s^2 obtained for the temperatures of 40–70 °C for ID 2g_1 change slightly and are within the range of 0.745762–0.722883 and 0.006448–0.006877, respectively.

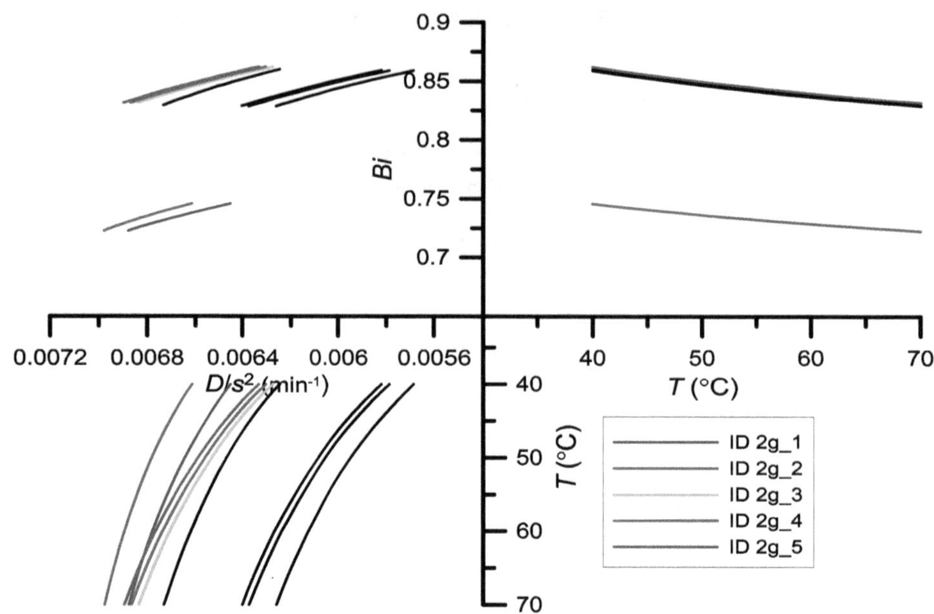

Figure 16. Pareto optimal set for the function of the Biot number and D/s^2: Equations (8g) and (9).

3.9. *Case 2h. Optimization of Parameters of the Function for Determining the Biot Number (Equation (8h))*

The results of optimization are presented in Table 10 and Figures 17 and 18. The best solutions are ID 2h_1 (the smallest MAE = 0.06293 and the greatest R^2 = 0.9740) and ID 2h_2 (the smallest RMSE = 0.07925, with MAE being slightly worse than for ID 2h_1 (the difference 0.00003)).

Table 10. Pareto optimal set for constants in Equations (8h) and (9) and results of the statistical analysis.

ID	a_b	b_b	a_d	b_d	R^2	RMSE	MAE
2h_1	1.434697	2.507106	0.006923	2.14	0.9740	0.07927	0.06293
2h_2	1.434897	2.509596	0.006967	2.14	0.9739	0.07925	0.06296
2h_3	1.431146	2.528872	0.007108	2.13075	0.9736	0.07946	0.06315
2h_4	1.431491	2.57301	0.007147	2.047283	0.9734	0.07970	0.06330
2h_5	1.429736	2.524792	0.007201	2.118279	0.9733	0.07981	0.06337
2h_6	1.429389	2.530306	0.007283	2.102477	0.9731	0.08025	0.06364
2h_7	1.432067	2.515218	0.007349	2.128414	0.9729	0.08062	0.06385
2h_8	1.430119	2.532143	0.007376	2.103931	0.9728	0.08085	0.06400
2h_9	1.42891	2.540204	0.007397	2.086483	0.9727	0.08104	0.06412
2h_10	1.435259	2.545177	0.007423	2.112372	0.9727	0.08118	0.06418

The courses of the function (Equation (8h)) with parameters a_b and b_b obtained as a result of optimization differ slightly from each other (Figure 18). The values of the Biot number obtained for the temperatures of 40–70 °C are within the range of 0.742097–0.722428 for ID 2h_1 and within the range of 0.742039–0.722353 for ID 2h_2. The courses of the function (Equation (9)) with parameters a_d and b_d obtained as a result of optimization are similar (Figure 18) and the best solution of the optimization task was obtained for smaller values of the moisture diffusion coefficient. The values of D/s^2 obtained for the temperatures of 40–70 °C are within the range of 0.006562–0.006714 for ID 2h_1 and 0.006604–0.006757 for ID 2h_2.

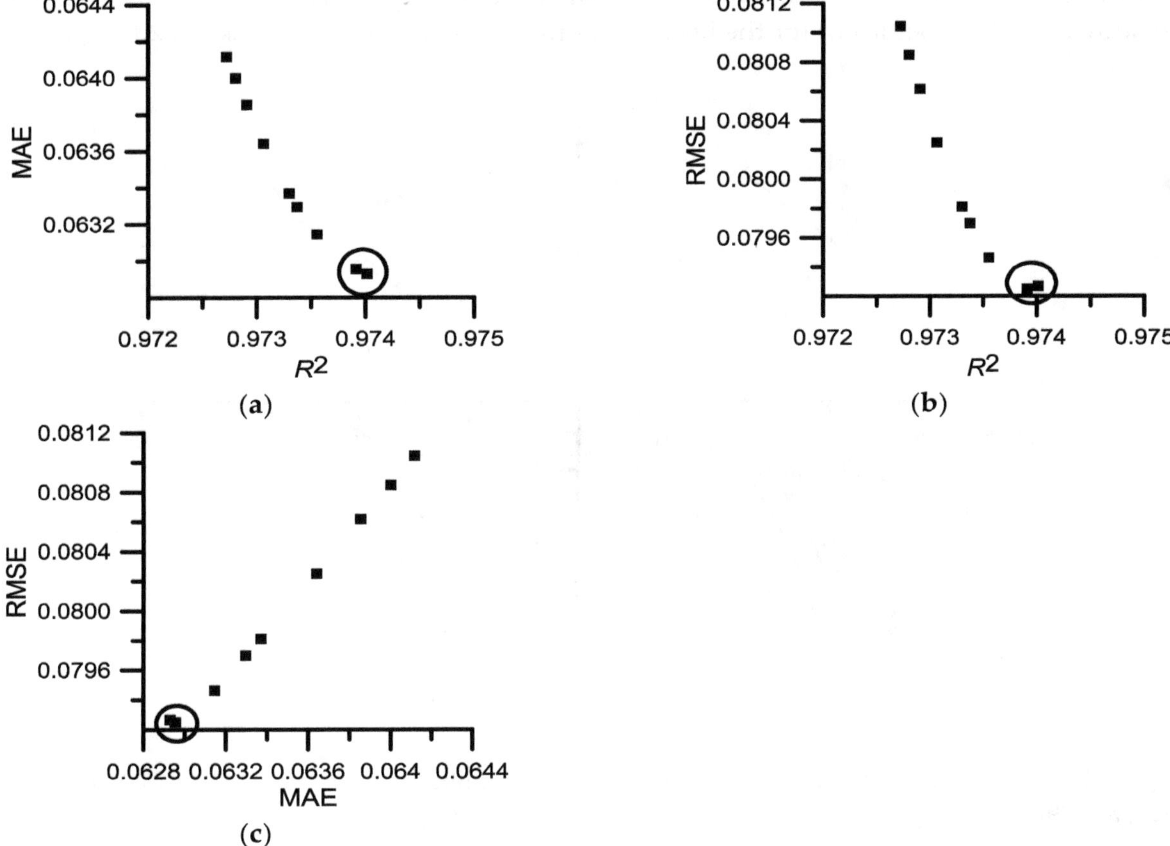

Figure 17. Pareto optimal sets for constants of Equation (8h).

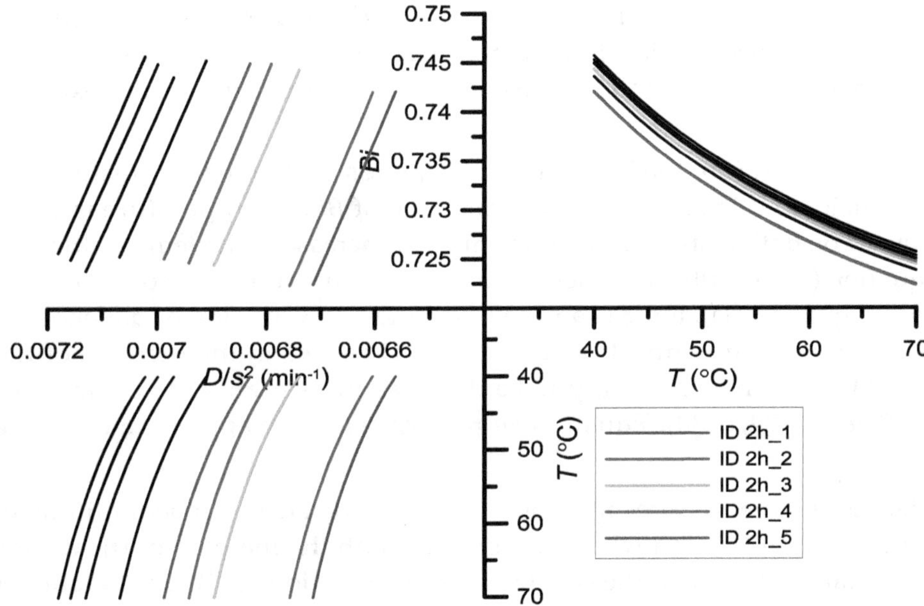

Figure 18. Pareto optimal set for the function of the Biot number and D/s^2: Equations (8h) and (9).

Among the functions (Equation (8a))–(Equation (8d)) indicating the increase of the Biot number with the increase of the temperature, the best one was the function described by the equation (Equation (8a)), and the second best function (Equation (8d)) was only slightly worse. Among the functions (Equation (8e))–(Equation (8h)) indicating the decrease of the Biot number with the increase of the temperature, the best one was the function described by the equation (Equation (8g)), and the second best function (Equation (8h)) was only slightly worse. Figure 19 shows the Biot number and the moisture diffusion coefficient for the best solutions of the optimization task 2a, 2d, 2g, and 2h.

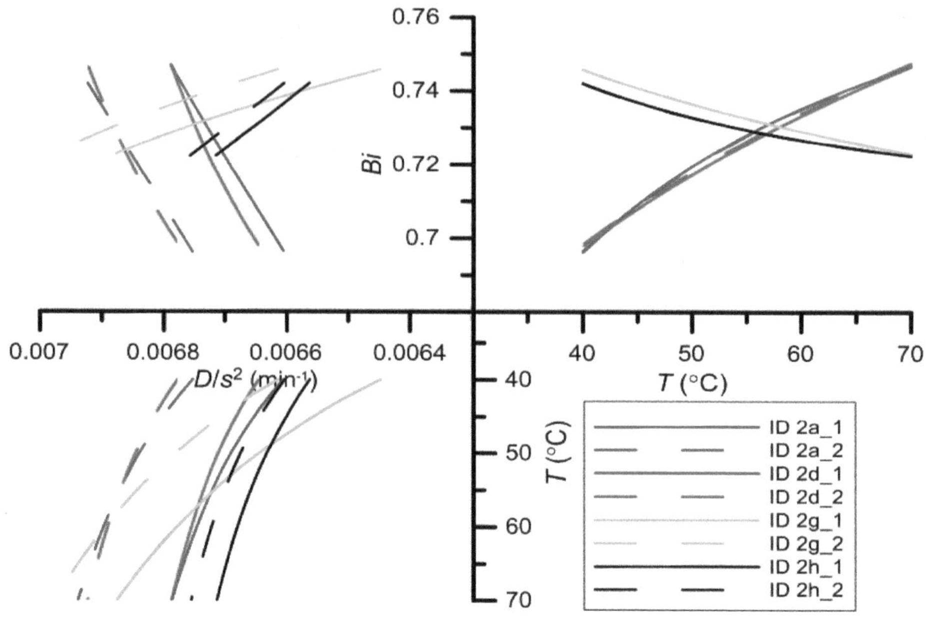

Figure 19. Pareto optimal set for the function of the Biot number and D/s^2: Equations (8a), (8d), (8g), (8h), and (9).

The range of variability of the Biot number for ID 2a_1, 2a_2, and 2d_1 is identical (the courses of the functions are almost identical), and it ranges between 0.70 and 0.75, whereas for ID 2g_1, 2g_2, 2h_1, and 2h_2, the range is smaller, between 0.72 and 0.75. D/s^2 calculated using (Equation (9)) changes most for ID 2g_1 (0.00645–0.00688). For ID 2g_2, the changes are smaller (0.00661–0.00698), and for ID 2a_2 and 2d_2, the changes are the smallest (the courses of the functions are almost identical), ranging between 0.00675 and 0.00685.

According to (Equation (1)), the Biot number depends on the mass transfer coefficient h_m and mass diffusion coefficient D. Therefore, the dependence of Bi on temperature is determined by the temperature influence on both mentioned coefficients. An increase of the temperature (according to the literature (i.e., Equation (9))) results in an increase of the moisture diffusion coefficient. In the definition of the Biot number (Equation (1)), there is a ratio h_m/D and the Biot number depends precisely on this relation. A greater temperature impact on h_m than on D results in the increase of the Biot number with temperature. Many authors have applied a dimensionless number to estimate the value of the mass transfer coefficient [8,24,29,30]. Equation $Sh = f(Gr,Sc)$ is often applied under natural convection conditions [31].

The dependences obtained as a result of optimization indicate that a better fitting of the model (5) to the experimental data is obtained when Bi increases with the increase of the drying temperature according to the (Equation (8a)) with the constants of the equation for ID 2a_1 (Table 3):

$$Bi = 0.8193\exp\left(-6.4951T^{-1}\right), \tag{11}$$

Simultaneously, D/s^2 is calculated based on (Equation (9)) with the constants of the equation for ID 2a_1 (Table 3):

$$D/s^2 = 0.00704\exp\left(-2.54T^{-1}\right), \tag{12}$$

However, only slightly worse fitting is obtained when Bi decreases with the increase of the drying temperature according to (Equation (8g)) with the constants of the equation for ID 2g_1 (Table 9):

$$Bi = 1/0.1746\log(119347T) \tag{13}$$

Simultaneously, D/s^2 is calculated based on (Equation (9)) with the constants of the equation for ID 2g_1 (Table 9):

$$D/s^2 = 0.0075\exp\left(-6T^{-1}\right), \tag{14}$$

The values of D/s^2 obtained from Equation (12) (for 40–70 °C) are the following: 0.00665–0.00679, and they are in the range of 0.00645–0.00688 obtained from Equation (14). The D values lie within the general range of 10^{13} to 10^6 m^2s^{-1} for food materials [32,33]. However the range of the Biot number calculated from Equation (13) (0.7729–0.7458) is wider than that obtained from Equation (11) (0.6965–0.6747).

The acceptance in the drying model (Equation (5)) of the Biot number and D/s^2 determined from Equations (11) and (12) or from Equations (13) and (14) results in various fits of the discussed model at individual considered temperatures. The MAE values for the model (Equation (5)) (Bi and D/s^2 from Equations (11) and (12)) are the following: 0.0690, 0.0340, 0.0426, and 0.0725 (for 40, 50, 60, and 70 °C, respectively), whereas for Bi and D/s^2 determined from Equations (13) and (14), the MAE values are much higher for extreme temperatures, namely 40 and 70 °C, and they are the following: 0.0815, 0.0377, 0.0.0425, and 0.0856 (for the mentioned temperatures, respectively).

The results of the validation of Biot number estimation are the following. The validation was done using the experimental data of drying kinetics of parsley root (Berlińska variety, 6 mm thick slices) dried in natural convection conditions at 55 °C. Biot number values were the following: 0.7281 and 0.7326 (Equations (11) and (13), respectively), whereas D/s^2 equalled 0.0067 (Equations (12) and (14)). Regardless of whether Equations (11) and (12) or (13) and (14) were used, R^2 = 0.9955, RMSE = 0.0376, and MAE = 0.0336, so the validation can be treated as satisfactory.

4. Conclusions

The analysis of Biot numbers enables questions about the controlling mechanisms employed for heat or mass transfer during the considered process to be answered.

This paper used a multi-objective optimization method (based on GA and Pareto optimization) for determination of the Biot number. A MOO GA method with a consideration of the simultaneous maximization of R^2 and minimization of RMSE and MAE between experimental data and the drying model was successfully applied.

The optimum values of the Biot number and constants of function used to determine the Biot number, gained by the MOO GA, were found. Eight types of equations for determining the Bi were tested. The Biot number can be calculated from the following equations: $Bi = 0.8193\exp(-6.4951T^{-1})$ (and moisture diffusion coefficient from D/s^2=0.00704exp($-2.54T^{-1}$)) or Bi=1/0.1746log(1193847T) ($D/s^2 = 0.0075\exp(-6T^{-1})$). The results of statistical analysis were as follows: RMSE = 0.0672, MAE = 0.0535, and R^2 = 0.98 for the first equation for Bi and D/s^2 and RMSE = 0.0757, MAE = 0.0604, and R^2 = 0.98 for the second one. The conducted validation gave good results: RMSE = 0.0376, MAE = 0.0336, and R^2 = 0.9955.

Author Contributions: K.G.: Proposal of the research topic, experiments, modelling, and writing of manuscript; R.W.: Optimization and writing of manuscript; A.K.: Writing of manuscript and critical revision of manuscript.

Nomenclature

a_b, b_b	constants in Equations (8a)–(8h)
a_d, b_d	constants in Equation (9)
$Bi = h_mL/D$	Biot number for mass transfer

c	concentration
D	mass diffusion coefficient (m^2s^{-1})
D_{AB}	moisture diffusivity in the gaseous phase (m^2s^{-1})
g	acceleration of gravity (ms^{-2})
$Gr = gL^3\beta\Delta c/\nu^2$	Grashof number
h_m	mass transfer coefficient (ms^{-1})
L	characteristic dimension (m)
M	moisture content (kg H_2O kg^{-1} d.m.)
M_0	initial moisture content (kg H_2O kg^{-1} d.m.)
M_e	equilibrium moisture content (kg H_2O kg^{-1} d.m.)
MR	moisture ratio
MR_{pred}	predicted moisture ratio
MR_{exp}	moisture ratio from experiment
$\overline{MR}_{exp}$	average value of moisture ratio from experiment
n	number of data
s	half of plane thickness (m)
$Sc = \nu/D_{AB}$	Schmidt number
$Sh = h_mL/D_{AB}$	Sherwood number
T	drying air temperature (°C)
t	time (s)
x	coordinate (m)
Greek symbols	
β	coefficient of expansion
v	kinematic viscosity (m^2s^{-1})

References

1. Giner, S.A.; Irigoyen, R.M.T.; Cicuttín, S.; Fiorentini, C. The variable nature of Biot numbers in food drying. *J. Food Eng.* **2010**, *101*, 214–222. [CrossRef]
2. Cuesta, F.J.; Lamúa, M.; Alique, R. A new exact numerical series for the determination of the Biot number: Application for the inverse estimation of the surface heat transfer coefficient in food processing. *Int. J. Heat Mass Transf.* **2012**, *55*, 4053–4062. [CrossRef]
3. Dincer, I. Moisture transfer analysis during drying of slab woods. *Heat Mass Transf.* **1998**, *34*, 317–320. [CrossRef]
4. Rovedo, C.O.; Suarez, C.; Viollaz, P. Analysis of moisture profiles, mass Biot number and driving forces during drying of potato slabs. *J. Food Eng.* **1998**, *36*, 211–231. [CrossRef]
5. Ruiz-López, I.I.; Ruiz-Espinosa, H.; Arellanes-Lozada, P.; Bárcenas-Pozos, M.E.; García-Alvarado, M.A. Analytical model for variable moisture diffusivity estimation and drying simulation of shrinkable food products. *J. Food Eng.* **2012**, *108*, 427–435. [CrossRef]
6. Ruiz-López, I.I.; Ruiz-Espinosa, H.; Luna-Guevara, M.L.; García-Alvarado, M.A. Modeling and simulation of heat and mass transfer during drying of solids with hemispherical shell geometry. *Comput. Chem. Eng.* **2011**, *35*, 191–199. [CrossRef]
7. Wu, Y.; Irudayaraj, J. Analysis of heat, mass and pressure transfer in starch based food systems. *J. Food Eng.* **1996**, *29*, 399–414. [CrossRef]
8. Górnicki, K.; Kaleta, A. Some problems related to mathematical modelling of mass transfer exemplified of convection drying of biological materials. In *Heat and Mass Transfer*; Hossain, M., Ed.; IntechOpen: Rijeka, Croatia, 2011.
9. Huang, C.-H.; Yeh, C.-Y. An inverse problem in simultaneous estimating the Biot numbers of heat and moisture transfer for a porous material. *Int. J. Heat Mass Transf.* **2002**, *45*, 4643–4653. [CrossRef]
10. Dincer, I.; Hussain, M.M. Development of a new Biot number and lag factor correlation for drying applications. *Int. J. Heat Mass Transf.* **2004**, *47*, 653–658. [CrossRef]
11. Chen, X.D.; Peng, X. Modified Biot number in the context of air drying of small moist porous objects. *Dry. Technol.* **2005**, *23*, 83–103. [CrossRef]

12. Xie, Y.; Gao, Z.; Liu, Y.; Xiao, H. Pulsed vacuum drying of rhizoma dioscoreae slices. *LWT* **2017**, *80*, 237–249. [CrossRef]
13. Dincer, I.; Dost, S. A modelling study for moisture diffusivities and moisture transfer coefficients in drying of solid objects. *Int. J. Energy Res.* **1996**, *20*, 531–539. [CrossRef]
14. Kiranoudis, C.T.; Markatos, N.C. Pareto design of conveyor-belt dryers. *J. Food Eng.* **2000**, *46*, 145–155. [CrossRef]
15. Kiranoudis, C.T.; Maroulis, Z.B.; Marinos-Kouris, D. Product quality multi-objective dryer design. *Dry. Technol.* **1999**, *17*, 2251–2270. [CrossRef]
16. Krokida, M.K.; Kiranoudis, C.T. Pareto design of fluidized bed dryers. *Chem. Eng. J.* **2000**, *79*, 1–12. [CrossRef]
17. Krokida, M.K.; Kiranoudis, C.T. Product quality multi-objective optimization of fluidized bed dryers. *Dry. Technol.* **2000**, *18*, 143–163. [CrossRef]
18. Quirijns, E.J. Modelling and Dynamic Optimisation of Quality Indicator Profiles during Drying. PhD Thesis, Wageningen University, Wageningen, The Netherland, April 2006.
19. Olmos, A.; Trelea, I.C.; Courtois, F.; Bonazzi, C.; Trystram, G. Dynamic optimal control of batch rice drying process. *Dry. Technol.* **2002**, *20*, 1319–1345. [CrossRef]
20. Winiczenko, R.; Górnicki, K.; Kaleta, A.; Martynenko, A.; Janaszek-Mańkowska, M.; Trajer, J. Multi-objective optimization of convective drying of apple cubes. *Comput. Electron. Agric.* **2018**, *145*, 341–348. [CrossRef]
21. Winiczenko Górnicki, K.; Kaleta, A.; Janaszek-Mankowska, M.; Trajer, J. Multi-objective optimization of the apple drying and rehydration processes parameters. *Emir. J. Food Agric. EJFA* **2018**, *30*, 1–9.
22. Chen, C.R.; Ramaswamy, H.S. Modeling and optimization of variable retort temperature (VRT) thermal processing using coupled neural networks and genetic algorithms. *J. Food Eng.* **2002**, *53*, 209–220. [CrossRef]
23. Erdoğdu, F.; Balaban, M.O. Complex method for nonlinear constrained multi-criteria (multi-objective function) optimization of thermal processing. *J. Food Process Eng.* **2003**, *26*, 357–375. [CrossRef]
24. Górnicki, K.; Kaleta, A. Modelling convection drying of blanched parsley root slices. *Biosyst. Eng.* **2007**, *97*, 51–59. [CrossRef]
25. Luikov, A.V. *Analytical Heat Diffusion Theory*; Academic Press Inc.: New York, NY, USA, 1970.
26. Zielinska, M.; Markowski, M. Drying behavior of carrots dried in a spout–fluidized bed dryer. *Dry. Technol.* **2007**, *25*, 261–270. [CrossRef]
27. Celma, A.R.; Cuadros, F.; López-Rodríguez, F. Convective drying characteristics of sludge from treatment plants in tomato processing industries. *Food Bioprod. Process.* **2012**, *90*, 224–234. [CrossRef]
28. Nguyen, T.H.; Lanoisellé, J.L.; Allaf, T.; Allaf, K. Experimental and fundamental critical analysis of diffusion model of airflow drying. *Dry. Technol.* **2016**, *34*, 1884–1899. [CrossRef]
29. Górnicki, K.; Kaleta, A. Drying curve modelling of blanched carrot cubes under natural convection condition. *J. Food Eng.* **2007**, *82*, 160–170. [CrossRef]
30. Chen, X.D.; Lin, S.X.Q.; Chen, G. On the ratio of heat to mass transfer coefficient for water evaporation and its impact upon drying modeling. *Int. J. Heat Mass Transf.* **2002**, *45*, 4369–4372. [CrossRef]
31. Bird, R.B.; Stewart, W.E.; Lightfoot, E.N. *Transport Phenomena*; Rev. 2nd ed.; Wiley: New York, NY, USA, 2007; ISBN 978-0-470-11539-8.
32. Zogzas, N.P.; Maroulis, Z.B.; Marinos-Kouris, D. Moisture diffusivity data compilation in foodstuffs. *Dry. Technol.* **1996**, *14*, 2225–2253. [CrossRef]
33. Mujumdar, A.S. *Handbook of Industrial Drying*, 4th ed.; CRC Press, Taylor & Francis Group: Boca Raton, FL, USA, 2015; ISBN 978-1-4665-9665-8.

4

Experimental Study on a Thermoelectric Generator for Industrial Waste Heat Recovery Based on a Hexagonal Heat Exchanger

Rui Quan [1,2,*], Tao Li [1], Yousheng Yue [1], Yufang Chang [2] and Baohua Tan [3]

1 Hubei Key Laboratory for High-efficiency Utilization of Solar Energy and Operation Control of Energy Storage System, Hubei University of Technology, Wuhan 430068, China; z2283902662@163.com (T.L.); David7689@163.com (Y.Y.)

2 Hubei Collaborative Innovation Center for High-efficiency Utilization of Solar Energy, Hubei University of Technology, Wuhan 430068, China; changyf@hbut.edu.cn

3 School of Science, Hubei University of Technology, Wuhan 430068, China; tan_bh@126.com

* Correspondence: quan_rui@126.com

Abstract: To study on the thermoelectric power generation for industrial waste heat recovery applied in a hot-air blower, an experimental thermoelectric generator (TEG) bench with the hexagonal heat exchanger and commercially available Bi_2Te_3 thermoelectric modules (TEMs) was established, and its performance was analyzed. The influences of several important influencing factors such as heat exchanger material, inlet gas temperature, backpressure, coolant temperature, clamping pressure and external load current on the output power and voltage of the TEG were comparatively tested. Experimental results show that the heat exchanger material, inlet gas temperature, clamping pressure and hot gas backpressure significantly affect the temperature distribution of the hexagonal heat exchanger, the brass hexagonal heat exchanger with lower backpressure and coolant temperature using ice water mixture enhance the temperature difference of TEMs and the overall output performance of TEG. Furthermore, compared with the flat-plate heat exchanger, the designed hexagonal heat exchanger has obvious advantages in temperature uniformity and low backpressure. When the maximum inlet gas temperature is 360 °C, the maximum hot side temperature of TEMs is 269.2 °C, the maximum clamping pressure of TEMs is 360 kg/m^2, the generated maximum output power of TEG is approximately 11.5 W and the corresponding system efficiency is close to 1.0%. The meaningful results provide a good guide for the system optimization of low backpressure and temperature-uniform TEG, and especially demonstrate the promising potential of using brass hexagonal heat exchanger in the automotive exhaust heat recovery without degrading the original performance of internal combustion engine.

Keywords: industrial waste heat recovery; thermoelectric generator; hexagonal heat exchanger; temperature distribution; output performance

---

1. Introduction

The continuously updated development of green energy techniques is a good alternative to resolve the global energy crisis and increase environmental protection. Owing to several advantages, such as little vibration, high reliability and durability and no moving parts, thermoelectric modules (TEMs) have been widely developed in photovoltaic, automotive, military, aerospace, wearable devices, wireless sensor networks and microelectronic applications over the past years [1–9]. Jaziri et al. [1] described and concluded the exploitation of thermoelectric generators (TEGs) in various fields starting from low-power applications (medical and wearable devices, IoT: internet of things, and WSN: wireless sensor network) to high-power applications (industrial electronics, automotive engines and aerospace). Liu et al. [2] developed a TEG based on concentric filament architecture for low power aerospace microelectronic devices, which was able to produce an electrical voltage of 83.5 mVand an electrical power of 32.1 μW with a planar heat source and temperature of 398.15 K. Willars-Rodríguez et al. [3] created and studied a solar hybrid system including photovoltaic (PV) module, concentrating Fresnel lens, thermo-electric generator (TEG) and running water heat extracting unit. Demir et al. [4] proposed a novel system for recovering waste heat of the automobile by a system based thermoelectric generator, and presented the variations of material properties, efficiency and generated power with respect to temperature and position. Meng et al. [5] proposed a technical solution recycling exhaust gas sensible heat based on thermoelectric power generation, which can produce about 1.47 kW electrical energy per square meter and achieve a conversion efficiency of 4.5% for exhaust gas at 350 degrees. Proto et al. [6] analyzed the results of measurements on thermal energy harvesting through a wearable thermoelectric generator (TEG) placed on the arms and legs whose generated power values were in the range from 5 to 50 μW. Kim et al. [7] demonstrated a self-powered wireless sensor node (WSN) driven by a flexible thermoelectric generator (f-TEG) with a significantly improved degree of practicality, and developed a large-area f-TEG that can be wrapped around heat pipes with various diameters, improving their usability and scalability. Leonov et al. [8] studied a thermoelectric energy harvesting on people that generated power in a range of 5–0.5 mW at ambient temperatures of 15 °C–27 °C, respectively. Holgate et al. [9] conceptualized and modeled a newly designed enhanced multi-mission radioisotope thermoelectric generator (MMRTG) that utilized the more efficient skutterudite-based thermoelectric materials, and presented a discussion of the motivations, modeling results and key design factors. With the rapid development of the economy and society, there is considerable unused waste heat dissipated in industrial heat-generating processes such as power industrial boilers, steelmaking, central air-conditioning, heating equipment and so on.

Thermoelectric generator (TEG) is a promising energy source that includes a hot source (heat exchanger) and a cold source (cooling unit); it can convert the heat into electricity based on the temperature difference of the installed TEMs that are sandwiched between the heat exchanger and cooling unit. In the case of TEG for waste heat recovery, many studies have focused on the application and optimization of the flat-plate heat exchanger to obtain higher power output or evaluate its performance [10–14]. For instance, regarding automotive applications, Zhang et al. [10] developed an automotive exhaust thermoelectric generator (AETEG) with 300 TEGs and water cooling for recovering diesel engine exhaust heat, and obtained 1002.6 W power and 2.1% efficiency when the average exhaust temperature was 823 K and the mass flow rate was 480 g/s. Szybist et al. [11] conducted an experiment to investigate 72 TEGs installed on the downstream of three-way catalytic converter and cooled by engine coolant based on a medium-sized gasoline passenger sedan, and 50 W generation power could be generated in the three typical cycles such as FTP (Federal Test Protocol), HWFET (Highway Fuel Economy Test) and US06 cycles. Kim et al. [12] put forward a novel AETEG with heat pipes and measured a maximum power of 350W with 443 K at the evaporator surface of heat pipes. Merkisz et al. [13] placed 24 TEGs at a distance of about 1.5 m from the end of three-way catalytic converter for a gasoline engine and generated the maximum power (189.3W) and maximum efficiency (1.3%) of AETEG, corresponding to the engine speed of 2600 rpm and 2200 rpm, respectively. Fernandez-Yanez et al. [14,15] developed a diesel engine and a gasoline engine AETEGs with engine coolant cooling, and concluded that the maximum produced power was approximately 270 W for both engines if a bypass was not included. However, two non-negligible problems existing in the flat-plate heat exchanger are the heat uniformity and the unwanted backpressure. The former plays an important role in the average temperature difference and thermoelectric efficiency of TEGs, as it dominates the energy conversion efficiency of heat to electricity. The latter reduces the efficiency of the engine for it seriously affects the original fuel economy and emission performance, or even worse, the lost performance of the engine cannot be compensated by the small generated power of TEG. Therefore, an ideal heat exchanger used in a TEG especially in an AETEG, should provide sufficient surface area to install TEMs as much as possible, increase its surface temperature uniformity, make the backpressure as low as possible and ensure the waste heat gas flow at a high value.

In this paper, we expanded the previous study to develop a low-cost and symmetrical TEG based on a brass hexagonal heat exchanger and the commercially available Bi_2Te_3 TEMs to recover the waste heat from a hot-air blower, which can simulate the internal combustion engine used in the AETEG. This study aimed to validate the promising potential and advantages of using the hexagonal heat exchanger in waste heat recovery, and analyze the influences of several important influencing factors on the output performance of the TEG. This work can provide an application guideline for industrial heat-generating processes and automotive exhaust heat recovery.

2. Experimental Setup of a TEG System

A detailed schematic diagram of the designed TEG system is shown in Figure 1; it consists of a hot-air blower, a hexagonal heat exchanger, a pump, a radiator, a water tank, a pressure regulator, temperature and pressure sensors (P_1 denotes pressure sensor, T_1 denotes inlet temperature sensor, T_2 denotes outlet temperature sensor) and six groups of TEMs and cooling units. The hot-air blower provides waste heat to the hexagonal heat exchanger, which can also simulate different driving cycles of vehicles by adjusting the gas temperature, flow and pressure. When the hot-air blower works, exhaust gas flows into the hexagonal heat exchanger to provide the hot side temperature, and the cooling water (i.e., coolant) stored in the water tank is circulated with the pump among the cooling units to form the cold side. Therefore, electricity is generated due to the temperature difference between the hot side and cold side of each TEM. To obtain higher temperature difference and better performance, the rotate speed of radiator can be controlled to precool the inlet coolant in cooling units.

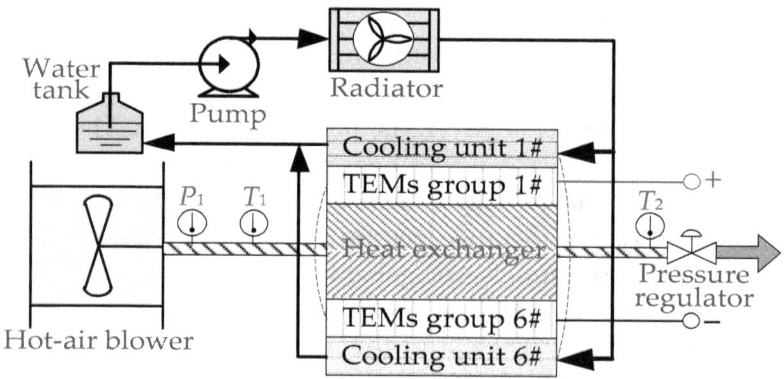

Figure 1. Schematic of an experimental setup of a TEG System. TEM: thermoelectric module.

The specific architecture of TEG and the dimensions of the hexagonal heat exchanger are shown in Figure 2. Each group of TEMs is sandwiched between the surface of the hexagonal heat exchanger and a cooling unit, and they are clamped with an adjustable force using five bolt and screw combinations. In total, there are 30 TEMs of Bi_2Te_3-based materials arranged on the six surfaces of the heat exchanger in five columns (i.e., five TEMs in each column are fixed with a common cooling box), and all the TEMs stalled above the surface of the hexagonal heat exchanger are electrically connected in series. Furthermore, to guarantee uniform cold side temperature of TEMs, all the cooling boxes are thermally connected in parallel (i.e., all the inlets of six cooling boxes are connected with the outlet of radiator, all the outlets of six cooling boxes are connected with the inlet of water tank). To evaluate the hot side temperatures distribution of the 30 TEMs, the corresponding *K*-type thermocouples below each TEM are pasted above each surface of the hexagonal heat exchanger. For the TEG, the Bi_2Te_3 TEMs (Model Name: TEHP1-1264-0.8) are manufactured by Thermonamic Electronics (Jiangxi, China) Corp. Ltd., whose high conductivity graphite paper is used as the thermally conductive interface material. The specific parameters of the principal components of the TEG system are provided in Table 1.

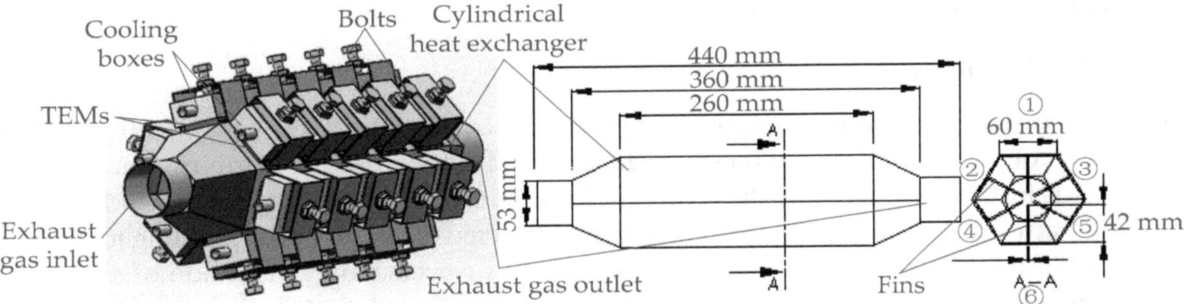

Figure 2. TEG architecture.

Table 1. Parameters of the principal components of a TEG system.

Parameters	Value
Dimension of TEM	40 mm × 40 mm × 40 mm
Maximum operation temperature of TEM	400 °C
Rated operation temperature of TEM	330 °C
Maximum conversion efficiency of TEM	6.5%
Dimension of cooling box	250 mm × 50 mm × 20 mm
Material of cooling box	Aluminum
Thickness of cooling box	1 mm
Thickness of heat exchanger	2 mm
Maximum power of pump	40 W
Maximum flow of pump	5000 L/H
Rated power of radiator	100 W (24V DC)
Rated power of hot-air blower	2000 W

The real experimental setup of a TEG system is shown in Figure 3; the output of TEG is connected with an adjustable electronic load, and the experiments described in the next section were carried out to evaluate several important influencing factors such as heat exchanger material, inlet gas temperature, backpressure, coolant temperature and external load current on its temperature distribution and output performance.

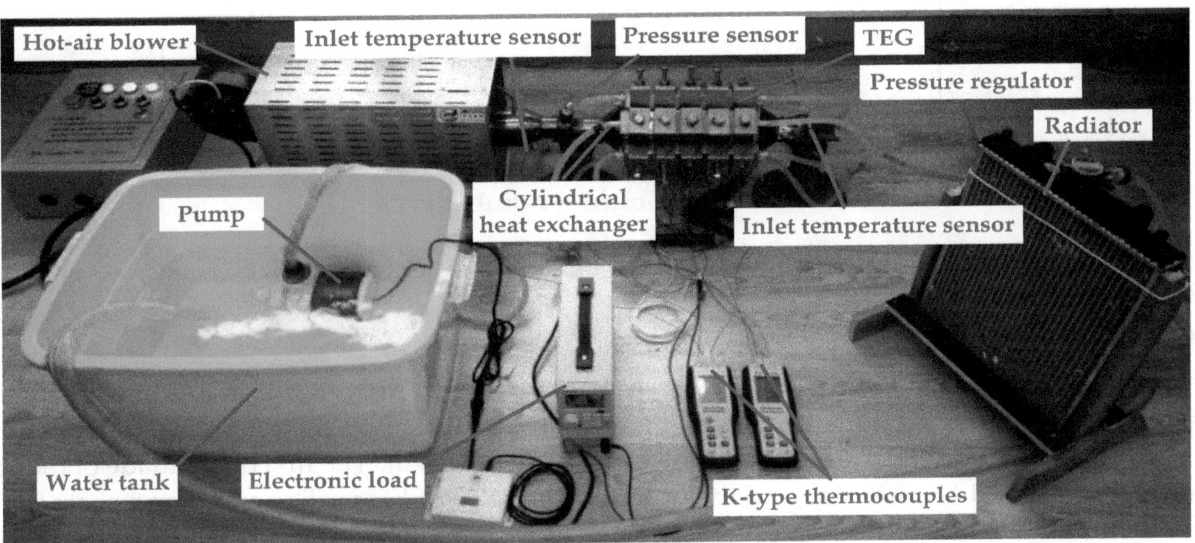

Figure 3. Real experimental setup of a TEG system.

3. Experimental Results and Discussions

3.1. Temperature Distribution

To analyze the temperature uniformity of the hexagonal heat exchanger, 30 independent K-type thermocouples pasted above its six surfaces are divided into five columns, and each K-type thermocouple is installed right below the central hot side of the corresponding single TEM. The specific surface temperature distribution of the 30 temperature detected locations corresponding to the 30 TEMs is shown in Figure 4. On this occasion, the hexagonal heat exchanger is made of stainless steel (Model Name: 304), its inlet temperature is 260 °C, the pressure regulator is fully open and the effective transfer size of its interior fins is 260 mm in length and 42 mm in width (1.5 mm in thickness). For the interior cavity of heat exchange is much larger than the inlet tube, the gas flows quickly across the inlet section. Furthermore, the gas flow decreases accordingly from column 1 to column 5 because of the pressure caused by the interior fins, which enhances the heat transfer between the hot gas and heat exchanger. Thus, it can be concluded that the section closer to the inlet has the lowest surface temperature, the temperatures of the detected locations marked with blue near the exhaust gas outlet is higher than those in the central area and closer to the exhaust gas inlet because of the gas eddy caused by the sudden narrow outlet tube and the temperatures of the six detected locations in the same column are almost the same (the maximum temperature is below 2 °C). Thus, compared with our previously designed chaos-shaped flat-plate heat exchanger [16,17], the temperature uniformity of the hexagonal heat exchanger is better, which is in good agreement with our expected results.

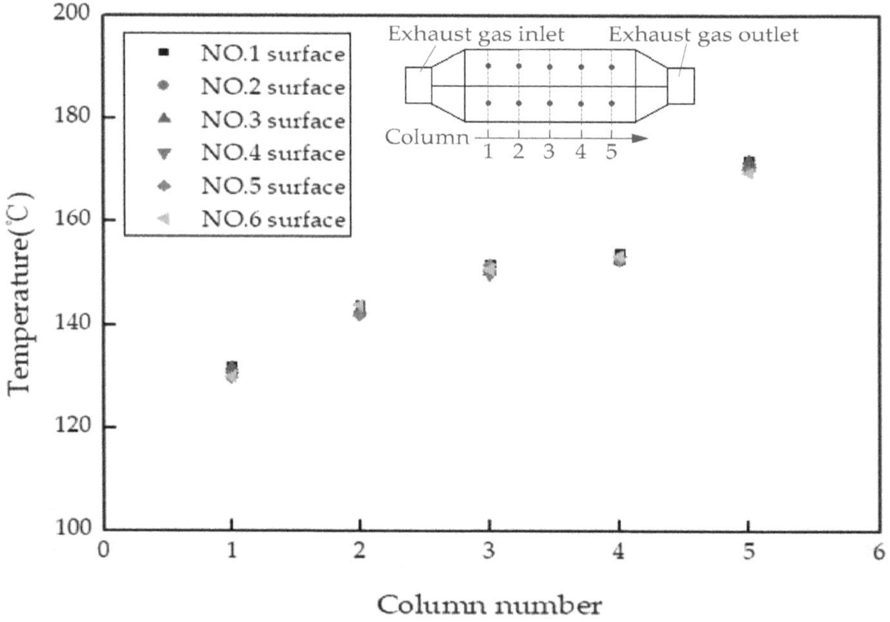

Figure 4. The surface temperature distribution of the hexagonal heat exchanger.

3.2. Influence of Heat Exchanger Material

For the above designed hexagonal heat exchanger shown in Figure 2, two kinds of different metal materials are adopted without changing its dimension to analyze their influences on the surface temperature distribution and the output performance of TEG system. The first one is made of the above stainless steel (Model Name: 304), and the other one is made of brass. The compared characteristic of TEG with the two different kinds of hexagonal heat exchangers is shown in Figure 5. In this case, the clamping pressure of TEMs above each surface increases is 120 kg/m^2, the inlet gas pressure is 70 Pa, the coolant is pumped without radiator, the coolant flow is 5000 L/h and its original temperature is equal to the ambient temperature (27 °C).

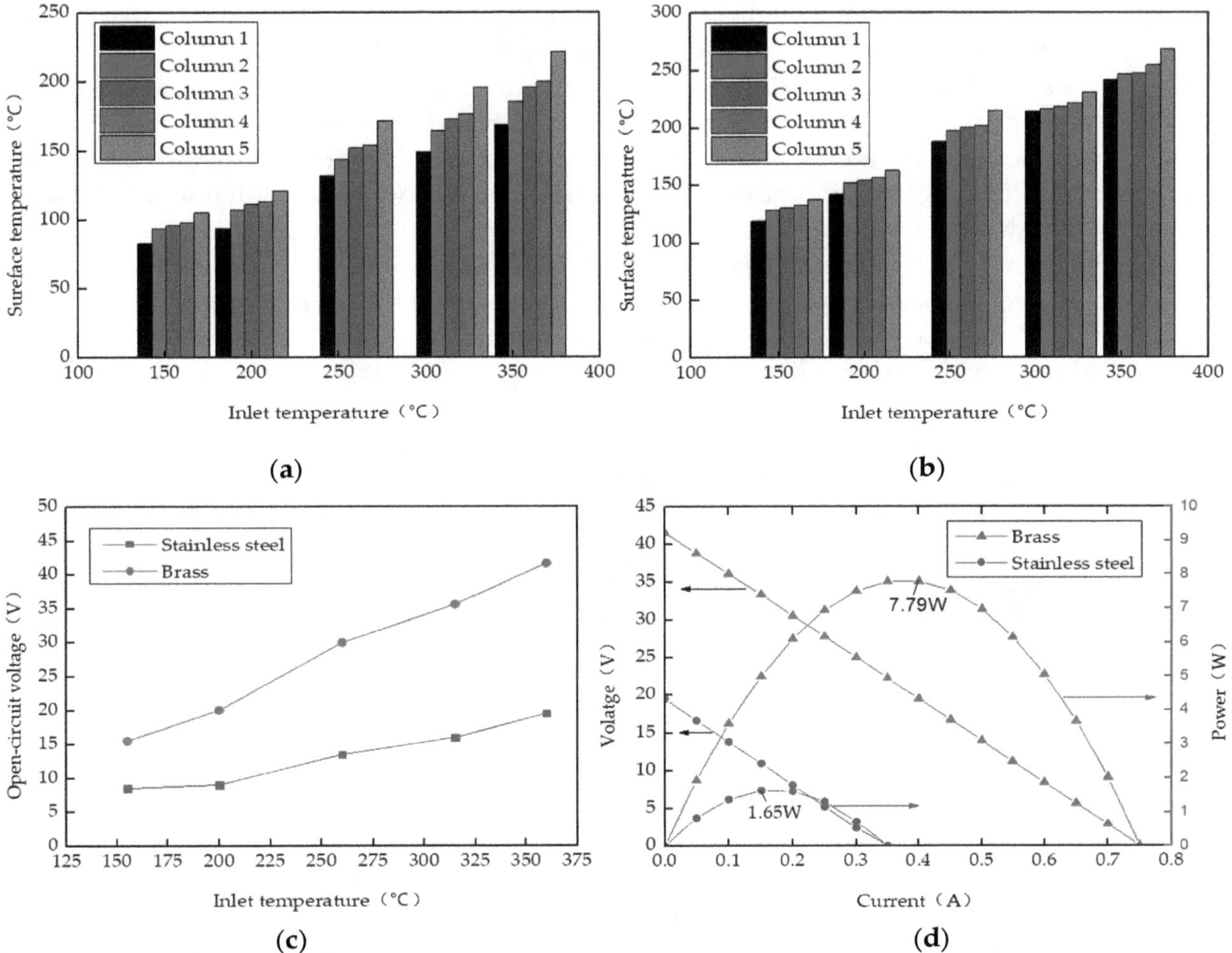

Figure 5. The compared characteristic performance of TEG with different heat exchanger materials. (**a**) Surface temperature distribution of the stainless steel heat exchanger. (**b**) Surface temperature distribution of the brass heat exchanger. (**c**) Open-circuit voltage of TEG with inlet gas temperatures. (**d**). Voltage and power versus current when the maximum inlet temperature is 360 °C.

Figure 5a,b demonstrate the surface temperature distribution of the two heat exchangers with different inlet temperatures of 155 °C, 200 °C, 260 °C, 315 °C and 360 °C, respectively. Considering the above uniform temperature distribution in each column shown in Figure 4, only the temperature detected locations in the NO.1 surface are selected as the average temperatures of the hexagonal heat exchangers (i.e., the hot side temperatures of TEMs) in each column. It is obvious that the average temperatures of both the stainless steel and brass hexagonal heat exchangers are in direct proportion to the inlet gas temperature. According to the Fourier equation, the absorbed heat (denoted Q) of heat exchanger can be calculated as follows:

$$Q = KA\Delta T/d \tag{1}$$

where K is the heat conductivity coefficient, A is the heat transfer area, ΔT is the temperature variation and d is the heat transfer distance. For the average heat conductivity of stainless steel 304 is about 16.2 W/m.K, while the one of brass is about 120 W/m.K, which is almost seven times of the former. The detected locations temperatures in the same column validated that the brass hexagonal heat exchanger is much better than the stainless steel hexagonal heat exchanger in heat transfer with the same inlet gas temperature. Properly speaking, when the maximum inlet gas temperature is 360 °C, the maximum temperature of the stainless steel hexagonal heat exchanger (in column 5) is 222.1 °C, while the one of the brass hexagonal heat exchanger (in column 5) is 269.2 °C, which is an increase of

21.2%. According to Equation (1), it can be calculated that the absorbed heat of brass heat exchanger is increased by about 930% because of its higher heat conductivity coefficient.

Figure 5c shows the corresponding open-circuit voltage of TEG with the above two kinds of hexagonal heat exchangers, the open-circuit voltage of TEG based on the two different hexagonal heat exchangers increases with the augment of its inlet gas temperature. Figure 5d provides the measured characteristics curves of voltage and power versus current of TEG with two different hexagonal heat exchangers when the maximum inlet gas temperature is 360 °C. It can be seen that the open-circuit voltage of TEG with the stainless steel hexagonal heat exchanger is only 19.5 V, while the one with the brass hexagonal heat exchanger is 41.6 V, which is an increase of 113.3%. In addition, the maximum power of TEG with the stainless steel hexagonal heat exchanger is only 1.65 W when the current is 0.15 A, while the one with the brass hexagonal heat exchanger is 7.79 W, which is an increase of nearly four times.

For the above coolant of ambient temperature is adopted, the cold side temperatures of TEMs in the same column of both the two kinds of different hexagonal heat exchangers can be seen similar. Furthermore, as shown in Figure 5b, the TEMs installed in the same column above the brass hexagonal heat exchanger have much higher hot side temperatures with the same inlet gas temperatures, it can be concluded that the temperature difference of TEMs with the brass hexagonal heat exchanger is much larger than that with the stainless steel hexagonal heat exchanger for more hot gas heat is absorbed on the same occasion because of the higher heat transfer coefficient of brass. Thus, the brass hexagonal heat exchanger has overwhelming heat-conducting property advantage over the stainless steel hexagonal heat exchanger despite its high cost, and the TEG system with the brass hexagonal heat exchanger has a better output performance.

3.3. *Influence of Backpressure*

Considering the advantage of the above brass hexagonal heat exchanger used in TEG, the influence of inlet gas backpressure on TEG based on the brass hexagonal heat exchanger is shown in Figure 6. On this occasion, the clamping pressure of TEMs above each surface increases is 120 kg/m^2, the coolant is pumped without radiator, the coolant flow is 5000 L/h and its original temperature is equal to the ambient temperature (27 °C), the inlet gas backpressure is adjusted by the regulator opening. For the detected temperature locations in the 5th column of the brass hexagonal heat exchanger have the highest temperature values, Figure 6a,b show the maximum surface temperatures of the brass hexagonal heat exchanger and the open-circuit voltage of TEG corresponding to different inlet gas temperatures with different inlet gas backpressures of 70 Pa, 180 Pa and 220 Pa, respectively. Figure 6c,d show the voltage and output power versus current characteristics with the above different inlet gas backpressures when the maximum inlet gas temperature is 360 °C.

Larger inlet gas backpressure means lower maximum surface temperature and open-circuit voltage with the same inlet gas temperature; the reason of this is that large inlet gas backpressure will decrease the rotation speed of hot-air blower's fan and the corresponding flow of hot gas. In this case, the heat transfer between hot gas and heat exchanger is reduced, which leads to a lower temperature difference of TEMs based on the same cooling condition. When the maximum inlet gas temperature is 360 °C, the maximum surface temperature of the brass hexagonal heat exchanger is 268.2 °C (70 Pa), 250.1 °C (180 Pa) and 241.3 °C (220 Pa); the open-circuit of TEG is 41.6 V (70 Pa), 35.1 V (180 Pa) and 29.9 V (220 Pa). Furthermore, the open-circuit voltage and output power of TEG are in inverse proportion to the inlet gas backpressures, decreased both output voltage and power with the same output current are accompanied with the augment of inlet gas backpressures. Since the cold side temperature of TEMs in each column with different inlet gas backpressures can be regarded as the same because of the ambient temperature coolant, it can be concluded that the lower inlet gas backpressure is, the larger inlet gas flow will be. Lower inlet gas backpressure contributes to higher temperature differences and better output performance of TEMs, since it ensures higher rotating speed of hot-air blower, and efficient heat conduction between the gas and brass hexagonal heat exchanger on the same occasion.

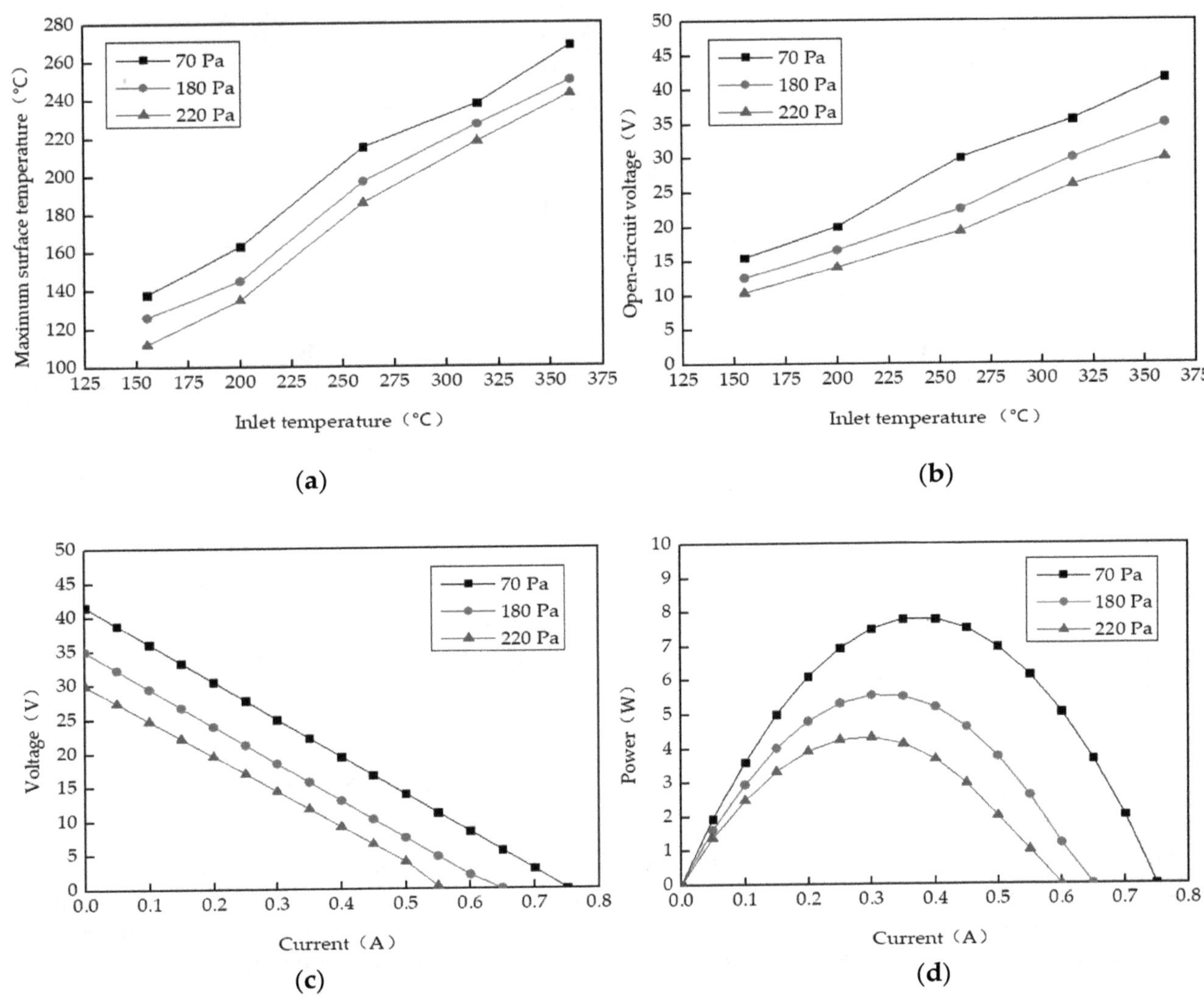

Figure 6. The compared characteristic performance of TEG based on the brass hexagonal heat exchanger with different inlet gas backpressures. (**a**) Maximum surface temperatures of the brass heat exchanger with inlet gas temperatures. (**b**) Open-circuit voltage of TEG with inlet gas temperatures. (**c**) Voltage versus current when the maximum inlet temperature is 360 °C. (**d**) Power versus current when the maximum inlet temperature is 360 °C.

3.4. *Influence of Coolant Temperatures*

To ensure lower cold side temperatures and higher temperature differences of TEMs with the same hot side temperatures, three different kinds of coolant were used in the test as follows: case 1: pumped coolant of ambient temperature without radiator; case 2: pumped coolant of ambient temperature with radiator of 100% speed; case 3: pumped coolant of ice water mixture (0 °C) without radiator. Figure 7 shows the characteristic performance of TEG based on the brass hexagonal heat exchanger in the above three cases when the clamping pressure of TEMs above each surface increases is 120 kg/m^2, the coolant flow is 5000 L/h, the ambient temperature is 27 °C and the inlet gas backpressure is maintained at 70 Pa.

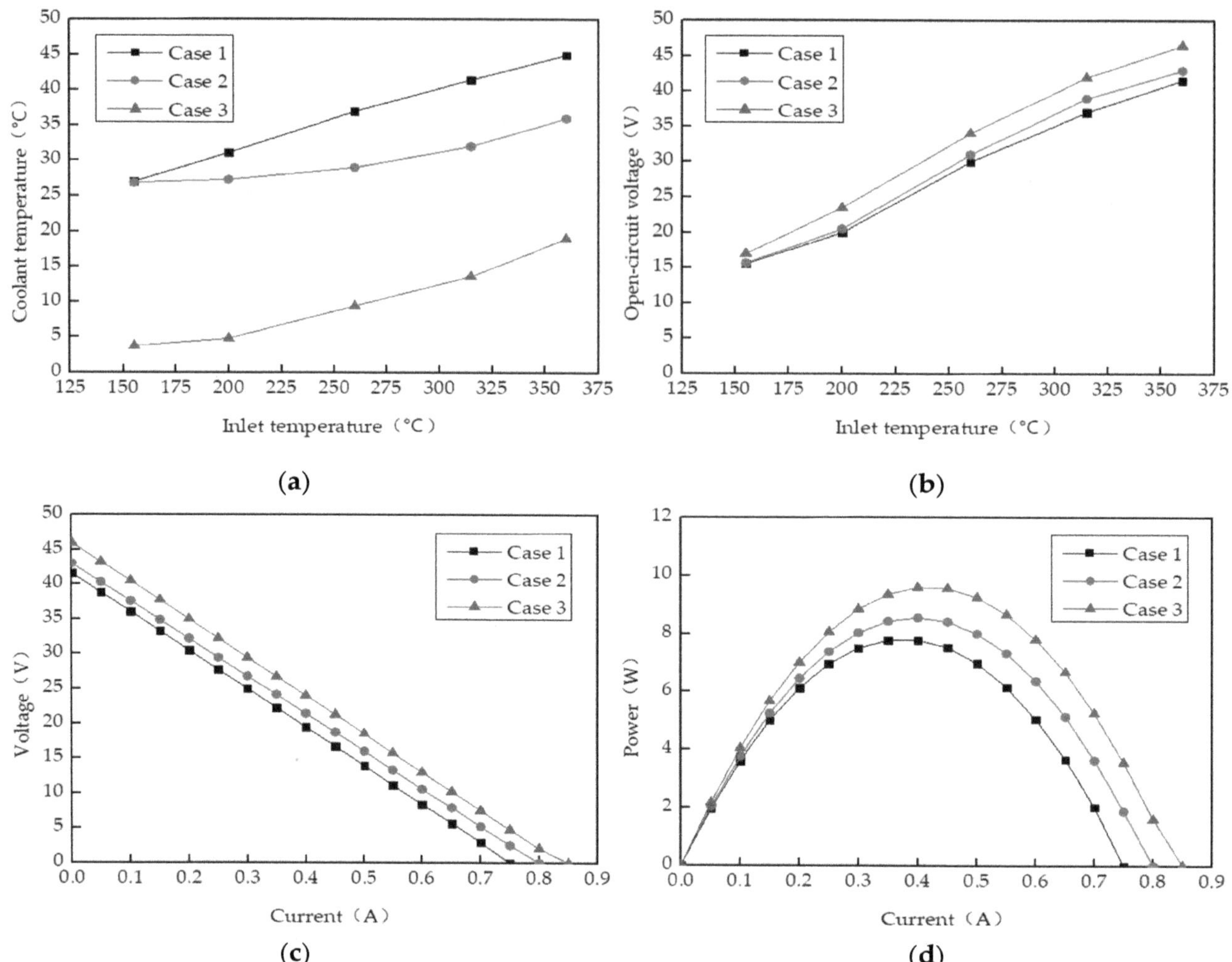

Figure 7. The compared characteristic performance of TEG based on the brass hexagonal heat exchanger in different coolant cases. (**a**) Coolant temperatures of TEG with inlet gas temperatures. (**b**) Open-circuit voltage of TEG with inlet gas temperatures. (**c**) Voltage versus current when the maximum inlet temperature is 360 °C. (**d**) Power versus current when the maximum inlet temperature is 360 °C.

For Figure 7a, the inlet gas temperatures are changed every five minutes to ensure the steady heat conduction among TEMs, heat exchanger and cooling boxes. The coolant temperatures rise as the inlet gas temperatures increase, the coolant temperature in case 1 is the highest, the coolant temperature in case 2 takes second place, while the coolant temperature in case 3 is the lowest. When the maximum inlet gas temperature is 360 °C, the maximum coolant temperature in case 1 is 45.9 °C, the one in case 2 is 36.6 °C and the one in case 3 is 20.2 °C. Figure 7b shows the open-circuit voltage of TEG with different inlet gas temperatures based on the above three different kinds of coolant. It is obvious that the open-circuit voltage of TEG in case 3 is the largest, the one in in case 2 takes second place, while the one in case 1 is the lowest with the same inlet gas temperature. Figure 7c,d show the performance characteristics of TEG in the above three cases with different load currents when the maximum inlet gas temperature is 360 °C. It can be seen that both the output voltage and power of TEG corresponding to the same load current increase evidently with lower coolant temperatures, the maximum power of TEG in case 1, 2 and 3 is 7.79 W, 8.56 W and 9.65 W, respectively. Additionally, the coolant temperature will increase evidently without a radiator because of the large thermal conductivity of TEMs, and the coolant temperature with radiator can be maintained in a relatively stable range. To ensure larger output power and higher output voltage, a coolant of ice water mixture is recommended in the cooling unit of TEG, as it can evidently enlarge the temperature difference of TEM without consuming extra radiator power.

3.5. Influence of Clamping Pressure

Considering the thermal contact resistance caused by the clamping pressure may affect the hot side and cold side temperatures of TEG, the effect of different clamping pressures on the overall performance of TEG is shown in Figure 8. On this occasion, the coolant of ice water mixture (0 °C) is pumped without radiator, the coolant flow is 5000 L/h, the ambient temperature is 27 °C, the inlet gas backpressure is maintained at 70 Pa and the maximum inlet temperature of heat exchanger is 360 °C. The maximum output power of TEG is 9.65 W, 10.32 W and 11.49 W with different clamping pressures of 120 kg/m^2, 240 kg/m^2 and 360 kg/m^2, respectively. The corresponding open-circuit voltage of TEG is 46.2 V, 49.1 V and 53.3 V, respectively.

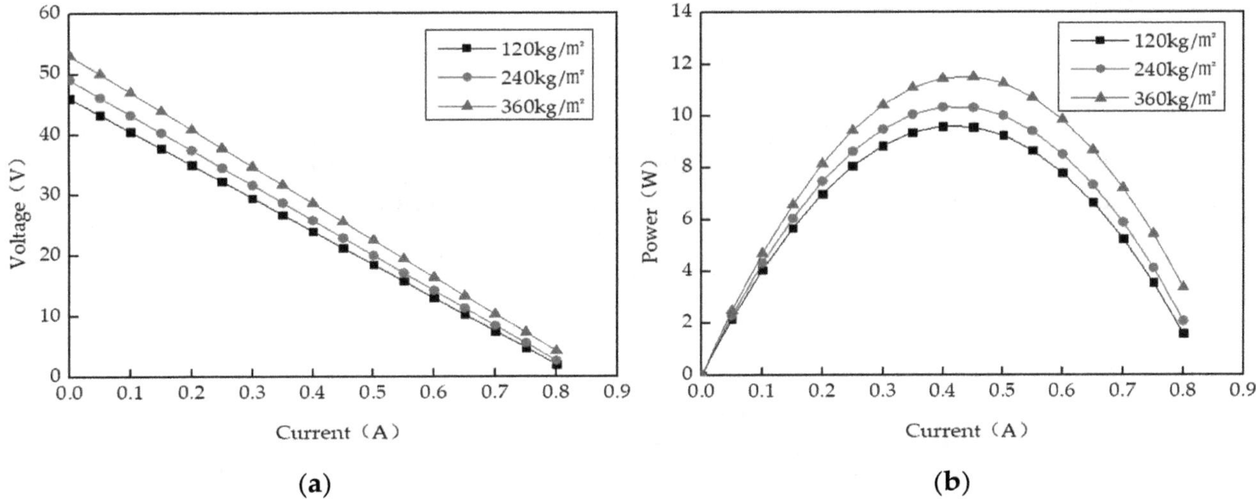

Figure 8. TEG performance with clamping pressure of 120 kg/m^2, 240 kg/m^2 and 360 kg/m^2. (**a**) Voltage versus current curves, (**b**) power versus current curves.

From the separate characteristic curves shown in Figure 8a, the slope of voltage versus current decreases with the increased clamping pressure, which means that the inner resistance of TEG increased accordingly. Combined with the above presented results, it demonstrates that the thermal contact resistance of TEG can be reduced for the empty air gap is decreased with large clamping pressure. Thus, much more inlet gas heat from the brass heat exchanger can be absorbed by the hot sides of TEMs, and increasing heat from the cold sides of TEMs can be brought off with large clamping pressure. The larger clamping pressure is, the lower thermal insulator is, which contributes to the enhanced output performance becaused of the lower thermal contact resistance. Therefore, to ensure larger output power and high efficiency, AETEG should be clamped as tight as possible within the allowable pressure of each TEM.

3.6. System Efficiency

For the 30 TEMs of Bi_2Te_3 based materials used in TEG, the output of TEG can be expressed as follows [1,18,19]:

$$V_{\mathrm{TEG}} = \sum_{\mathrm{i}=1}^{240} n\alpha_{\mathrm{PNi}}\Delta T_{\mathrm{i}} = n\left(\alpha_{\mathrm{pi}} - \alpha_{\mathrm{ni}}\right)\left(T_{\mathrm{Hi}} - T_{\mathrm{Li}}\right) \tag{2}$$

$$R_{\mathrm{TEG}} = \sum_{\mathrm{i}=1}^{240} R_{\mathrm{i}} = 240(nl_{\mathrm{p}}/(\sigma_{\mathrm{p}}A_{\mathrm{p}}) + nl_{\mathrm{n}}/(\sigma_{\mathrm{n}}A_{\mathrm{n}})) \tag{3}$$

$$\alpha_{\mathrm{PNi}} = \left(22224 + 930.6 \times 0.5 \times \Delta T_{\mathrm{i}} - 0.9905 \times (0.5 \times \Delta T_{\mathrm{i}})^2\right) \times 10^{-9} \tag{4}$$

where n is the p-type and the n-type semiconductor galvanic arms number in each TEM and V_{TEG} and R_{TEG} are the open circuit voltage and internal resistance of TEG, respectively. α_{PNi} is the relative Seebeck coefficient (V/K), α_{Pi} and α_{ni} are the Seebeck coefficients of the p-type and the n-type semiconductor galvanic arms, respectively. T_{Hi} and T_{Li} are the hot side and cold side temperature (K), respectively. l_p, σ_P and A_p are the leg length (m), electricity resistivity (Ωm) and cross-sectional area (m^2) of a p-type semiconductor galvanic arm, respectively, while l_n, σ_n and A_n are the leg length, electricity resistivity and cross-sectional area of an n-type semiconductor galvanic arm, respectively.

As shown in Figures 5d–8b the output power of the TEG reaches its maximum value (denoted P_{TEG}) when the external load resistance (denoted R_m) is equal to its internal resistance, and is expressed as:

$$P_{TEG} = U_{TEG}^2/(4R_m) \quad (5)$$

The heat input to the hexagonal heat exchanger is obtained as [20]:

$$Q_h = G_h\rho_h C_h \Delta T_h = G_h \rho_h C_h (T_{hi} - T_{ho}) \quad (6)$$

where G_h is the volume flow rate of inlet gas, ρ_h is the density of inlet gas, C_h is the heat capacity of inlet gas at constant pressure, while T_{hi} and T_{ho} are the inlet gas temperature and outlet gas temperature of hot gas, respectively. In this case, the maximum TEG system efficiency η can be calculated as follows:

$$\eta = P_{TEG}/Q_h \quad (7)$$

Figure 9 shows the temperature drop and outlet temperature of hot gas along the brass hexagonal heat exchanger when the inlet gas temperature changes from 155 °C to 360 °C. On this occasion, the clamping pressure of TEMs above each surface increases to 360 kg/m^2, the pumped coolant of ice water mixture (0 °C) without radiator is adopted, the ambient temperature is 27 °C and the inlet gas backpressure is 70 Pa (the corresponding flow rate is 0.18 m^3/h). It can be seen that both the temperature drops and outlet temperature increases with increasing inlet gas temperatures; the maximum temperature drop between the inlet and outlet gas is within 50 °C.

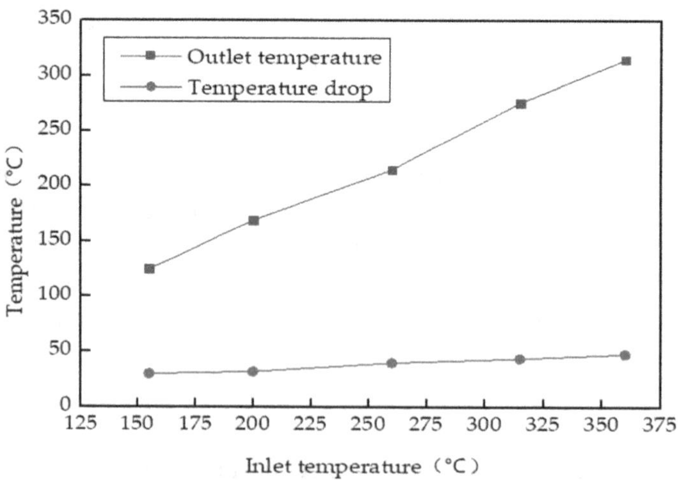

Figure 9. The decrease in temperature and the outlet gas temperature for different inlet gas temperatures.

Experimental results of both the maximum output power and corresponding system efficiency of TEG are shown in Figure 10 with respect to the inlet gas temperatures. As shown in Figure 10, for fixed inlet gas flow rates and backpressures, the maximum output power and system efficiency of TEG based on the coolant of ice water mixture increase with increasing inlet gas temperature. When the maximum inlet gas temperature is 360 °C, the outlet gas temperature is 315.1 °C, the generated maximum output power of TEG is 11.49 W, and the corresponding system efficiency is 0.96%. For the maximum temperature limitation (360 °C) of hot-air blower outlet, the maximum surface temperature of the brass hexagonal heat exchanger shown in Figure 6 is only 269.2 °C, which is much lower than the operated operation temperature of TEM (330 °C). Thus, if the maximum surface temperature of the brass hexagonal heat exchanger can be raised to 330 °C, it can be deduced that the maximum output power of TEG will approach 20 W, and the corresponding system efficiency will be above 1.5%.

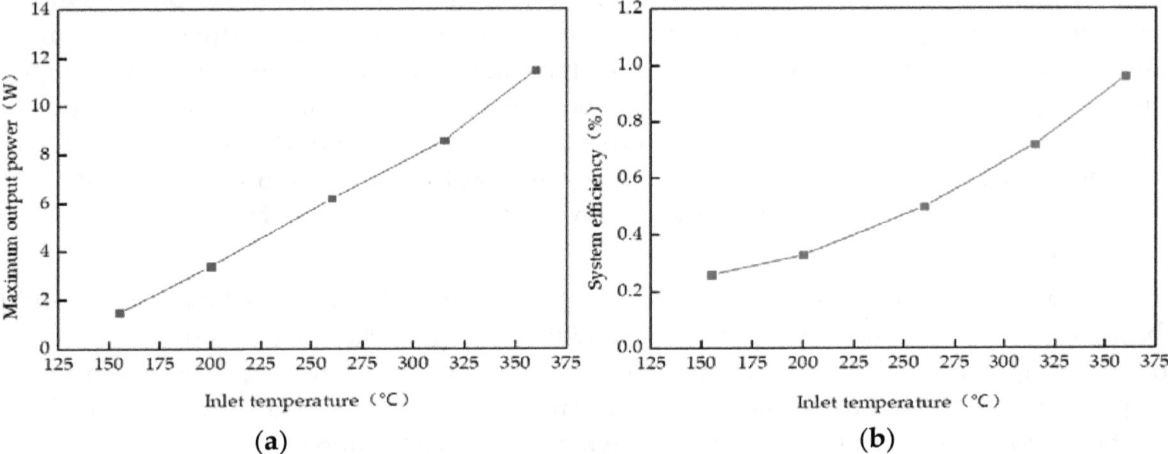

Figure 10. The overall output performance of TEG with respect to the inlet gas temperature. (**a**) Maximum output power; (**b**) system efficiency.

Furthermore, to validate the above numerical model of TEG, Figure 11 shows the comparison between the predicted and measured maximum power and corresponding system efficiency with different matched load resistance, the clamping pressure of TEMs above each surface increases is 360 kg/m^2, the pumped coolant of ice water mixture (0 °C) without radiator is adopted, the coolant flow is 5000 L/h, the ambient temperature is 27 °C and the inlet gas backpressure is 70 Pa (the corresponding flow rate is 0.18 m^3/h). It can be seen that the above numerical model can predict the performances of TEG when the inlet gas temperature changes from 125 °C to 360 °C. At low temperatures, there is a good agreement between the experimental performances and the predicted results. However, at high temperatures, the discrepancy between the experimental performances and predicted values increases evidently and is also acceptable. The main reason is that the heat losses from the hot sides to cold sides of TEMs are not considered in the numerical model, which plays a more important role in the heat transfer at high temperatures. Overall, to estimate the performance and establish the precise model of TEG, the properties of the thermoelectric materials used in TEMs should be assumed to be variables, the heat losses from the uncovered surface, TEMs gap and hot sides to cold sides should be taken into account.

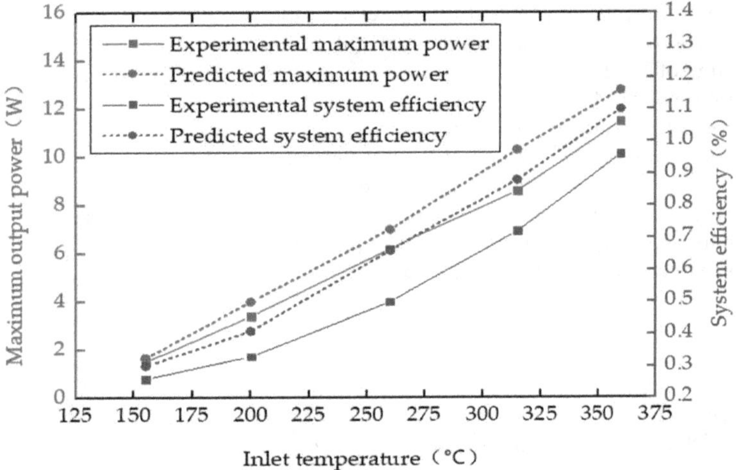

Figure 11. Comparison between the predicted and measured maximum power and corresponding system efficiency of TEG.

4. Conclusions

Waste heat recovery based on TEMs presents a promising research focus worldwide, but enhancing the output performance and system efficiency of TEG remains a significant challenge. In this study, a TEG system with the low-temperature and common commercially available Bi_2Te_3 TEMs and a hexagonal heat exchanger is designed for the waste heat recovery of an industrial hot-air blower. The influences of different operating conditions such as material, backpressure, inlet gas temperature, clamping pressure and coolant temperature on the temperature distribution, open-circuit voltage, the maximum output power and the system efficiency of TEG are reported from the experimental measurements. The experimental results demonstrate that the surface temperature distribution of hexagonal heat exchanger is uniform in each column, and is greatly affected by the adopted material of heat exchanger and the backpressure of inlet gas.

The comparisons indicate that the brass hexagonal heat exchanger has better heat conduction, lower backpressure can enhance the gas flow rate and the average surface temperature of heat exchanger (i.e., hot side temperature of TEMs) and the coolant of ice water mixture contributes to lower cold side temperature of TEMs. Furthermore, the maximum output power and system efficiency of TEG is proportional to the practical temperature differences of TEMs caused by inlet gas temperatures, backpressures, clamping pressures and coolant temperature. The designed brass hexagonal heat exchanger has low pressure drop, and it is very suitable for the automotive exhaust thermoelectric generator. Further experiments are planned to optimize the TEG and apply it in the automotive exhaust heat recovery.

Author Contributions: T.L. and Y.Y. performed the experiments and data analysis. R.Q. designed the experiments and wrote the manuscript. Y.C. and B.T. offered valuable discussions in analyses and revised the manuscript. All authors have read and agreed to the published version of the manuscript.

Acknowledgments: "This work was supported by the National Natural Science Foundation of China (No. 51977061, 61903129, 51407063)"and "This work was supported by the Doctor Scientific Research Foundation of Hubei University of Technology (No. BSQD13064)".

List of Notations

P_1 inlet gas temperature sensor
T_1 inlet gas temperature sensor
T_2 outlet gas temperature sensor
n p-type and the n-type semiconductor galvanic arms number
V_{TEG} open circuit voltage of TEG
R_{TEG} internal resistance of TEG
α_{PNi} relative Seebeck coefficient
α_{pi} Seebeck coefficients of the p-type semiconductor galvanic arms
α_{ni} Seebeck coefficients of the n-type semiconductor galvanic arms
T_{Hi} hot side temperature of TEM
T_{Li} cold side temperature of TEM
l_p leg length of a p-type semiconductor galvanic arm
σ_p electricity resistivity of a p-type semiconductor galvanic arm
A_p cross-sectional area of a p-type semiconductor galvanic arm
l_n leg length of a n-type semiconductor galvanic arm
σ_n electricity resistivity of a n-type semiconductor galvanic arm
A_n cross-sectional area of a p-type semiconductor galvanic arm
P_{TEG} maximum output power of TEG
R_m external load resistance
Q_h heat input to the cylindrical heat exchanger
G_h volume flow rate of inlet gas
ρ_h density of inlet gas
C_h heat capacity of inlet gas at constant pressure
T_{hi} inlet gas temperature
T_{ho} outlet gas temperature
η maximum TEG system efficiency

References

1. Jaziri, N.; Boughamoura, A.; Müller, J.; Mezghani, B.; Tounsi, F.; Ismail, M. A comprehensive review of thermoelectric generators: Technologies and common applications. *Energy Rep.* **2019**. [CrossRef]
2. Liu, K.; Tang, X.; Liu, Y.; Yuan, Z.; Li, J.; Xu, Z.; Zhang, Z.; Chen, W. High performance and integrated design of thermoelectric generator based on concentric filament architecture. *J. Power Sources* **2018**, *393*, 161–168. [CrossRef]
3. Willars-Rodríguez, F.J.; Chávez-Urbiola, E.A.; Vorobiev, P.; Vorobiev, Y.V. Investigation of solar hybrid system with concentrating Fresnel lens, photovoltaic and thermoelectric generators. *Int. J. Energy Res.* **2017**, *41*, 377–388. [CrossRef]
4. Demir, M.E.; Dincer, I. Performance assessment of a thermoelectric generator applied to exhaust waste heat recovery. *Appl. Therm. Eng.* **2017**, *120*, 694–707. [CrossRef]
5. Meng, F.K.; Chen, L.G.; Feng, Y.L.; Xiong, B. Thermoelectric generator for industrial gas phase waste heat recovery. *Energy* **2017**, *135*, 83–90. [CrossRef]
6. Proto, A.; Bibbo, D.; Cerny, M.; Vala, D.; Kasik, V.; Peter, L.; Conforto, S.; Schmid, M.; Penhaker, M. Thermal energy harvesting on the bodily surfaces of arms and legs through a wearable thermo-electric generator. *Sensors* **2018**, *18*, 1927. [CrossRef] [PubMed]
7. Kim, Y.; Gu, H.M.; Kim, C.; Choi, H.; Lee, G.; Kim, S.; Yi, K.; Lee, S.; Cho, B. High-performance self-powered wireless sensor node driven by a flexible thermoelectric generator. *Energy* **2018**, *162*, 526–533. [CrossRef]
8. Leonov, V. Thermoelectric energy harvesting of human body heat for wearable sensors. *IEEE Sens. J.* **2013**, *13*, 2284–2291. [CrossRef]
9. Holgate, T.-C.; Bennett, R.; Hammel, T.; Caillat, T.; Keyser, S.; Sievers, B. Increasing the efficiency of the multi-mission radioisotope thermoelectric generator. *J. Electron. Mater.* **2015**, *44*, 1814–1821. [CrossRef]
10. Zhang, Y.L.; Cleary, M.; Wang, X.W.; Kempf, N.; Schoensee, L.; Yang, J.; Joshi, G.; Meda, L. High-temperature and high-power-density nanostructured thermoelectric generator for automotive waste heat recovery. *Energy Convers. Manag.* **2015**, *105*, 946–950. [CrossRef]
11. Szybist, J.; Davis, S.; Thomas, J.F.; Kaul, B.C. Performance of a Half-Heusler thermoelectric generator for automotive application. *SAE Tech. Pap.* **2018**, 1. [CrossRef]
12. Kim, S.K.; Won, B.C.; Rhi, S.H.; Kim, S.H.; Yoo, J.H.; Jang, J.C. Thermoelectric power generation system for future hybrid vehicles using hot exhaust gas. *J. Electron. Mater.* **2011**, *40*, 778–783. [CrossRef]
13. Merkisz, J.; Fuc, P.; Lijewski, P.; Ziolkowski, A.; Galant, M.; Siedlecki, M. Analysis of an increase in the efficiency of a spark ignition engine through the application of an automotive thermoelectric generator. *J. Electron. Mater.* **2016**, *45*, 4028–4037. [CrossRef]
14. Fernandez-Yanez, P.; Armas, O.; Capetillo, A.; Martinez-Martinez, S. Thermal analysis of a thermoelectric generator for light-duty diesel engines. *Appl. Energy* **2018**, *226*, 690–702. [CrossRef]
15. Fernandez-Yanez, P.; Armas, O.; Kiwan, R.; Stefanopoulou, A.G.; Boehman, A.L. A thermoelectric generator in exhaust systems of spark-ignition and compression-ignition engines. A comparison with an electric turbo-generator. *Appl. Energy* **2018**, *229*, 80–87. [CrossRef]
16. Quan, R.; Zhou, W.; Yang, G.Y.; Quan, S.H. A hybrid maximum power point tracking method for automobile exhaust thermoelectric generator. *J. Electron. Mater.* **2017**, *46*, 2676–2683. [CrossRef]
17. Quan, R.; Liu, G.Y.; Wang, C.J.; Zhou, W.; Huang, L.; Deng, Y.D. Performance investigation of an exhaust thermoelectric generator for military SUV application. *Coatings* **2018**, *8*, 45. [CrossRef]
18. Fraisse, G.; Ramousse, J.; Sgorlon, D.; Goupil, C. Comparison of different modeling approaches for thermoelectric elements. *Energy Convers. Manag.* **2013**, *65*, 351–366. [CrossRef]
19. Quan, R.; Wang, C.J.; Wu, F.; Chang, Y.F.; Deng, Y.D. Parameter matching and optimization of an isg mild hybrid powertrain based on an automobile exhaust thermoelectric generator. *J. Electron. Mater.* **2019**. [CrossRef]
20. Niu, X.; Yu, J.L.; Wang, S.Z. Experimental study on low-temperature waste heat thermoelectric generator. *J. Power Sources* **2009**, *188*, 621–626. [CrossRef]

5

Thermal Conductivity of Korean Compacted Bentonite Buffer Materials for a Nuclear Waste Repository

Seok Yoon *, WanHyoung Cho, Changsoo Lee and Geon-Young Kim

Division of Radioactive Waste Disposal Research, Korea Atomic Energy Research Institute (KAERI), 989-111, Daedeok-daero, Yuseong-gu, Daejeon 34057, Republic of Korea; cho0714@kaeri.re.kr (W.C.); leecs@kaeri.re.kr (C.L.); kimgy@kaeri.re.kr (G.-Y.K.)

* Correspondence: busybeeyoon@kaist.ac.kr

Abstract: Engineered barrier system (EBS) has been proposed for the disposal of high-level waste (HLW). An EBS is composed of a disposal canister with spent fuel, a buffer material, backfill material, and a near field rock mass. The buffer material is especially essential to guarantee the safe disposal of HLW, and plays the very important role of protecting the waste and canister against any external mechanical impact. The buffer material should also possess high thermal conductivity, to release as much decay heat as possible from the spent fuel. Its thermal conductivity is a crucial property since it determines the temperature retained from the decay heat of the spent fuel. Many studies have investigated the thermal conductivity of bentonite buffer materials and many types of soils. However, there has been little research or overall evaluation of the thermal conductivity of Korean Ca-type bentonite buffer materials. This paper investigated and analyzed the thermal conductivity of Korean Ca-type bentonite buffer materials produced in Gyeongju, and compared the results with various characteristics of Na-type bentonites, such as MX80 and Kunigel. Additionally, this paper suggests various predictive models to predict the thermal conductivity of Korean bentonite buffer materials considering various influential independent variables, and compared these with results for MX80 and Kunigel.

Keywords: bentonite buffer material; Ca-type bentonite; thermal conductivity; predictive models

1. Introduction

Spent fuels from nuclear energy sources release decay heat and harmful radiation for extended periods, creating longstanding issues with high-level waste (HLW). Among the various types of disposal systems, deep geological repositories based on the concept of engineered barrier system (Figure 1), which safely isolates HLWs from human society using a surrounding buffer, backfill, and near-field rock, are preferred in most countries, owing to their safety and reliability [1,2]. In EBS system, canisters packed with spent fuel are sealed with buffer and gap materials. The buffer is an important component of the repository. By filling the void between the near-field rock and the canister it minimizes groundwater inflow from intact rock while protecting the disposed HLW from any mechanical impact. For this reason, buffers must possess low hydraulic conductivity, to minimize the inflow of water from surrounding rocks saturated with groundwater. Furthermore, the buffer plays an important role in dissipating decay heat, and for this reason buffers must have high thermal conductivity to release as much as decay heat as possible from the spent fuel [3,4].

In order to satisfy these buffer material criteria, researches have been conducted to determine the most adequate candidate materials. The studies found that bentonite is the most suitable material [5,6]. Bentonite belongs to the smectite group, which contains large amounts of montmorillonite.

Bentonite forms 2:1 layer platy structures, and consists of two silica tetrahedral layers and an aluminum hydroxide octahedral layer. Anions generated from the isomorphous substitution in the montmorillonite readily absorb cations (Na^+, Mg^{2+}, Ca^{2+}, etc.) between the layers to become electrostatically neutral. Bentonite can be largely classified as Na-type bentonite or Ca-type bentonite depending on the exchangeable cations absorbed to become neutral. In Korea, Ca-type bentonite has been produced in the Gyeongju region by CLARIANT KOREA, and since Ca-type bentonite is known to satisfy the appropriate performance criteria, it can be considered one of the candidate buffer materials for HLW repository facilities in Korea [7]. Ca-type bentonite produced before 2015 in Korea is called KJ-I, and after 2015, it is known as KJ-II.

Many studies have been conducted to evaluate the complex thermal-hydro-mechanical (THM) behaviors of such buffers. The thermal conductivity of bentonite buffer materials considering the temperature limit is one of the most important design parameters to guarantee the entire safety performance in a disposal system [8]. However, even though many studies have investigated Na-type bentonite thermal conductivity [4,9–12], relatively few have investigated the thermal conductivity behavior models of the Ca-type bentonite produced in Korea. Therefore, this study measured the thermal conductivity of the Ca-type Korean bentonite and compared results with Na-type bentonite. Furthermore, this study suggested various predictive models for thermal conductivity considering dry density and degree of saturation, which are the main governing factors used to describe the thermal conductivity behavior of the buffer.

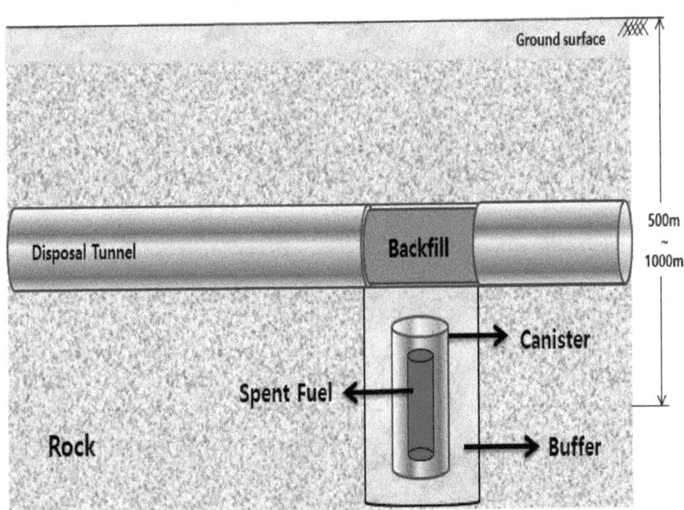

Figure 1. Engineered barrier system.

2. Laboratory Experiment

2.1. Mineral Composition of Bentonite

Ca-type bentonites (KJ-I and KJ-II) produced from Gyeongju contains montmorillonite (70%), feldspar (29%), and small amounts of quartz (~1%). Figure 2 shows XRD analysis results for KJ-I and KJ-II, and Table 1 presents a quantitative analysis of the thermal conductivity of the constitutive minerals. The quantitative analyses were conducted three times. The amount of montmorillonite was a little higher in KJ-I, but there was not a big difference in mineral composition between KJ-I and KJ-II.

Unlike the Ca-type bentonite in this study, for Na-type bentonite the percent of montmorillonite varies. According to the previous researches conducted by Villar [12] and Tang et al. [11], MX80, which is a very well-known commercial Na-type bentonite, contains montmorillonite (92%) and quartz (3%). Another Na-type bentonite considered a potential buffer material in Japan, called Kunigel, has a relatively smaller portion of montmorillonite. It mainly contains montmorillonite (46~49%) and quartz (29~38%) [3,11]. Table 2 shows the clay properties considering the Atterberg limit [13,14] Every bentonite is classified as CH with very high plasticity based on the unified soil classification system (USCS) [15]. On the whole, the Atterberg limit values of KJ-I and KJ-II were much higher than Na-type bentonites, such as MX-80 and Kunigel. It is thought that Na-type bentonite has higher compressibility and a little less tendency to be mechanically stable than Ca-type bentonite. Furthermore, MX80 showed the highest Atterberg limit, which means that MX80 has much more expansive characteristics.

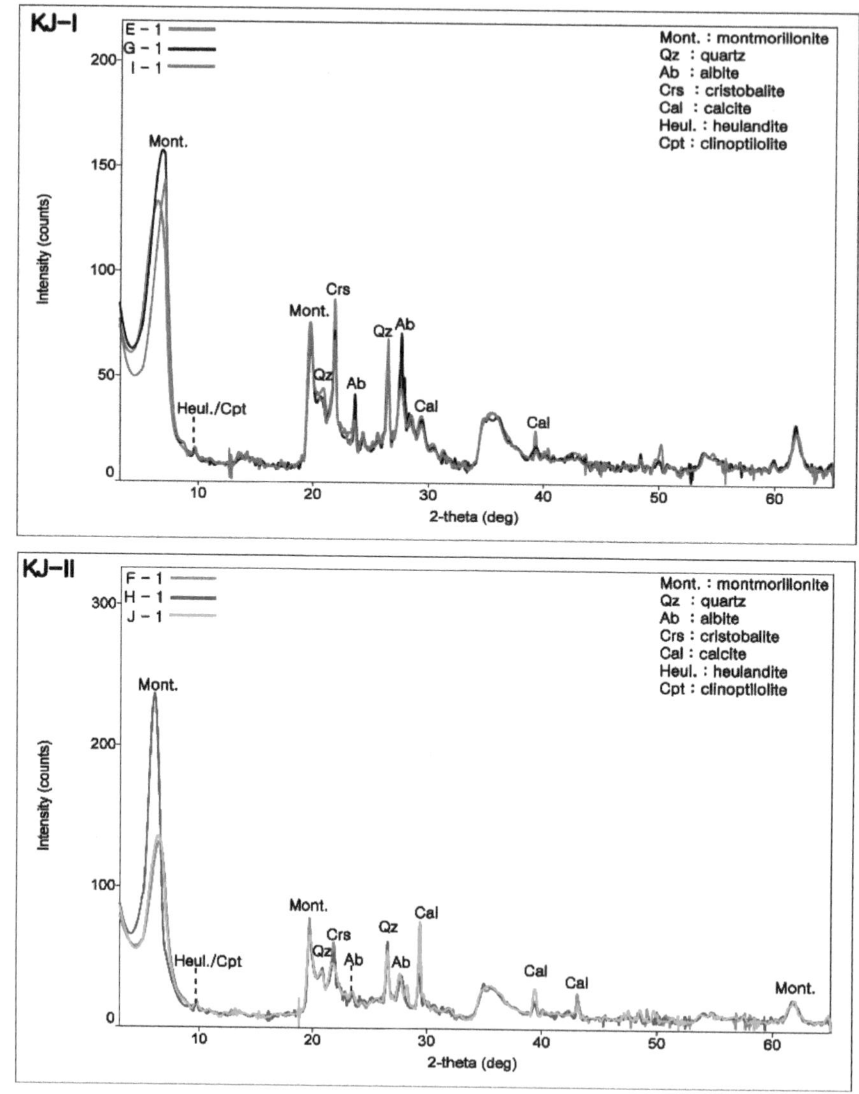

Figure 2. X-ray diffraction pattern of the KJ-I and KJ-II powders.

Table 1. Quantitative XRD analysis for mineral constituents of KJ-I and KJ-II powders.

Bentonite Type		**KJ-I**				**KJ-II**		
Sample No.	**1**	**2**	**3**	**Avg.**	**4**	**5**	**6**	**Avg.**
Montmorillonite	60.0	67.4	62.1	63.2	63.4	61.7	60.5	61.9
Albite (λ = 1.96 W/mK)	27.2	22.2	27.5	25.6	19.4	22.8	20.4	20.9
Quartz (λ = 7.69 W/mK)	5.0	4.8	5.0	4.9	5.8	4.9	5.3	5.3
Cristobalite (λ = 6.15 W/mK)	3.6	1.8	3.5	3.0	4.0	4.5	3.7	4.1
Calcite (λ = 3.59 W/mK)	2.4	2.0	2.0	2.1	4.3	3.3	6.8	4.8
Heulandite (λ = 1.09 W/mK)	1.8	1.7	Miner	1.8	3.0	2.7	3.3	3.0

Table 2. Geotechnical properties of bentonites.

	Specific Gravity	Liquid Limit (%)	Plastic Limit (%)	Plastic Index (%)	USCS
KJ-I	2.74	244.5	46.1	198.4	CH
KJ-II	2.71	146.7	28.4	118.3	CH
MX-80 [11]	2.76	520	42	478	CH
Kunigel [11]	2.79	415	32	383	CH

2.2. Equipment for Measuring Thermal Conductivity

The thermal conductivity of the compacted bentonite buffer materials were measured using QTM-500 (Kyoto Electronics Manufacturing Company, Kyoto, Japan), based on the transient line source method [16]. In this approach, an impulse of thermal flow is supplied by hot wire into the specimen, and temperature rise is measured within a certain time. As the temperature rises, the thermal conductivity is measured using Equation (1):

$$\lambda = \frac{Q}{4\pi\Delta T}\ln\left(\frac{t_2}{t_1}\right) \tag{1}$$

where λ is the thermal conductivity ($W{\cdot}m{\cdot}K^{-1}$), Q is the heat capacity per unit length ($W{\cdot}m^{-1}$), T is the temperature (K), and t is the time (s). The bentonite powders were compressed using the floating die method, and the sample size was 100 mm × 50 mm × 20 mm.

3. Experimental Results

Thermal Conductivity

It is known that the thermal conductivity of the compacted bentonite buffer materials mainly depends on the degree of saturation and dry density [3,5,17,18]. Thus, thermal conductivity of the KJ-II bentonite was measured with various water contents and dry densities. This paper collected 142 datasets for KJ-I [5,18], and 34 datasets for KJ-II was obtained using the QTM-500 equipment. Table 3 provides a summary of the statistical quantities used for the analysis, and Figure 3 depicts the thermal conductivity in proportion to dry density and degree of saturation. On the whole, the thermal conductivity of KJ-II was slightly higher than that of KJ-I because KJ-II has more minerals with high thermal conductivity, including quartz, cristobalite, and calcite, than KJ-I. TANG et al. [11] also explained that the thermal conductivity of the constitutive minerals can affect the total thermal conductivity of the bentonite buffer materials. In addition, the thermal conductivity of KJ-II was measured by drying path, while that of KJ-I was measured by wetting path. It is known that thermal conductivity is higher when measured by drying than by wetting [11,19]. However, there was no great difference in thermal conductivity values between KJ-I and KJ-II except for the low saturation. In comparison, the thermal conductivities of KJ-I and KJ-II were slightly higher than that of MX80 because of mineral composition and the high degree of saturation.

Table 3. Summary of descriptive statistical quantities.

		N	Minimum	Maximum	Average	Standard Deviation	Skewness	Kurtosis
KJ-I	Dry density (Mg/m^3)	142	1.200	1.800	1.510	0.154	−0.223	0.095
	Degree of saturation (%)		0.000	1.000	0.469	0.244	0.159	−0.022
	Thermal conductivity (W/(m·K))		0.301	1.445	0.722	0.248	0.686	−0.203
KJ-II	Dry density (Mg/m^3)	34	1.572	1.803	1.693	0.068	−0.024	−0.963
	Degree of saturation (%)		0.000	0.678	0.177	0.233	1.185	−0.184
	Thermal conductivity (W/(m·K))		0.627	1.046	0.805	0.116	0.761	−0.360

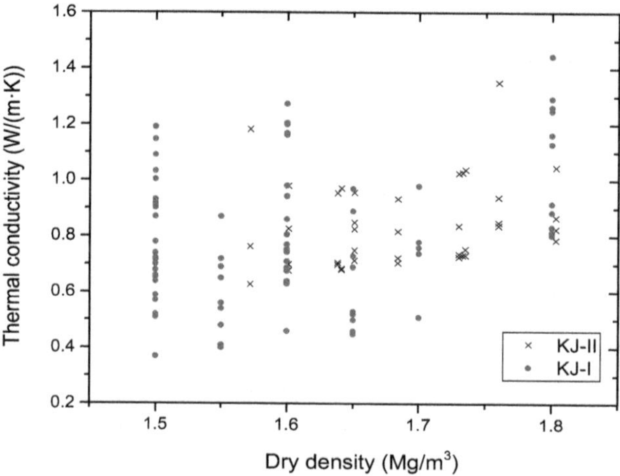

(**a**) Thermal conductivity vs. dry density

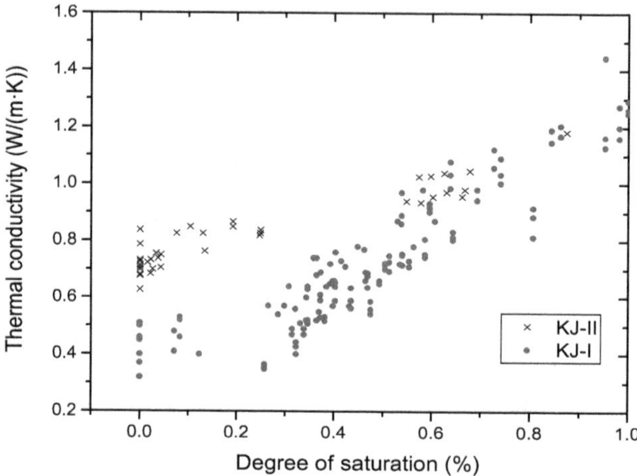

(**b**) Thermal conductivity vs. degree of saturation

Figure 3. Thermal conductivity variation KJ bentonites.

4. Thermal Conductivity Models for KJ-I and KJ-II

4.1. Models with 1.6 Mg/m^3 of Dry Density

To satisfy the required functional criteria, the suggested dry density of compacted bentonite buffer for Korea disposal systems is more than 1.6 Mg/m^3 [20]. Accordingly, the thermal conductivity model was derived for a bentonite buffer dry density of 1.6 g/cm^3. Since there were only three datasets for KJ-II, as shown in Table 4, both the KJ-I and the KJ-II datasets were used, making a total of 23. Many predictive models have been proposed to predict the thermal conductivity of compacted bentonite buffer materials, and this paper mainly used three models. To begin with, a linear regression analysis was applied. Equation (2) represents the linear regression model:

$$\lambda = 0.6683S + 0.4977 \quad (2)$$

Here, λ is thermal conductivity (W/(m·K)), and S means the degree of saturation. The R^2 value was around 0.77. Wilson et al. [21] and Lee et al. [22] also used the following empirical formula, which is well known to be adequate for predicting the thermal conductivity of compacted bentonite buffer materials:

$$\lambda = \frac{A_1 - A_2}{1 + \exp((S - S_{av})/B)} + A_2 \quad (3)$$

A_1 represents the value of λ when $S = 0$, A_2 means λ when $S = 1$. S_{av} is the degree of saturation when the thermal conductivity is the average of the two extreme values, and B is a fitting parameter. Furthermore, with the A_1 and A_2 values, thermal conductivity can also be predicted, as in Equation (4) [2]:

$$\lambda = {A_1}^{1-S}{A_2}^S \tag{4}$$

Figure 4 shows fitting curves, and Table 5 represents the summary of the predictive models and the fitting parameters, especially for Equations (3) and (4). The parameters for Equations (3) and (4) were derived using the curve fitting tool from the MATLAB program. This tool has a function for deriving the optimum equation that yields the highest R^2 value.

Table 4. Summary of descriptive statistical quantities for KJ-I and KJ-II with 1.6 Mg/m^3 of dry density.

		N	Minimum	Maximum	Average	Standard Deviation	Skewness	Kurtosis
KJ-I	Degree of saturation (%)	20	0	0.982	0.625	0.252	−0.4555	0.480
	Thermal conductivity (W/(m·K))		0.460	1.274	0.889	0.238	0.177	−1.153
KJ-I + KJ-II	Degree of saturation (%)	23	0	0.982	0.556	0.301	−0.393	−0.470
	Thermal conductivity (W/(m·K))		0.460	1.274	0.869	0.229	0.391	−0.971

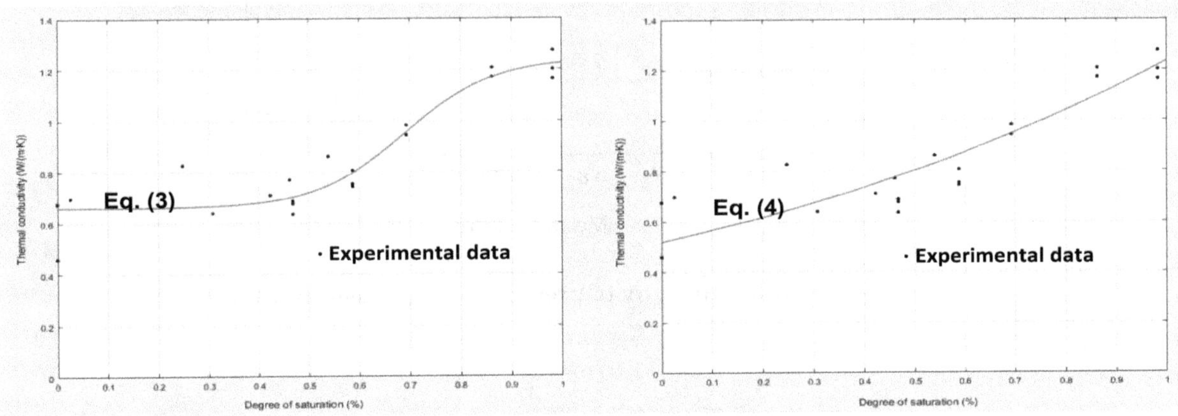

Figure 4. Fitting curves for Equations (3) and (4).

Table 5. Summary of predictive models.

	Equation (3)				Equation (4)	
Parameters	A_1	A_2	B	S_{av}	A_1	A_2
	0.6608	1.2410	0.0878	0.6906	0.5205	1.234
R^2	0.9082				0.8329	

4.2. Models Considering Various Dry Density and Degree of Saturation

Since Equations (2)–(4) only consider the degree of saturation as an independent variable, a multiple regression analysis to predict thermal conductivity was conducted considering dry density and degree of saturation as independent variables. 176 datasets of KJ-I and KJ-II bentonites were used in the regression analysis, and Equation (5) was suggested:

$$\lambda = 0.641\gamma_d + 0.624S - 0.510 \tag{5}$$

Figure 5 shows the plot of predictive values from Equation (5) and measured values. Table 6 presents the results of the regression analyses, and the R^2 value was 0.739. Based on the t analyses from Table 6 the P-value of the independent variable coefficients were lower than 0.05, which means that the two independent variables can be used statistically to predict the dependent variables [23,24].

The variance inflation factor (VIF) was lower than 10, which means there was no multicollinearity among the independent variables [24]. Table 7 shows the ANOVA analysis. Since the P-values were less than 0.01, there is a statistical significance between the independent and dependent variables[30]. A residual analysis was also conducted. The skewness was 0.210, and kurtosis was = −0.441. Since the absolute value of skewness and kurtosis was less than 2, it can be assumed that the residuals are normally distributed [24,25]. Figure 6 draws the homoscedasticity plot of the residuals, and it can be assumed that the residuals followed the homoscedasticity condition since they do not exhibit a specific pattern [25].

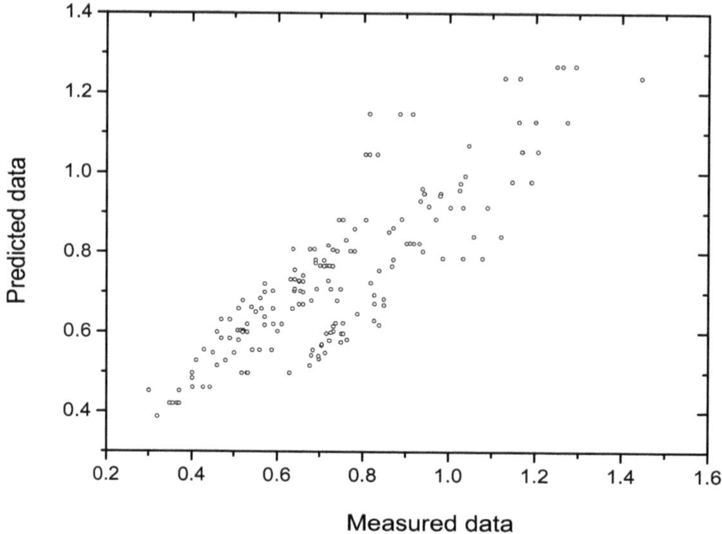

Figure 5. Thermal conductivity of predictive and measured values.

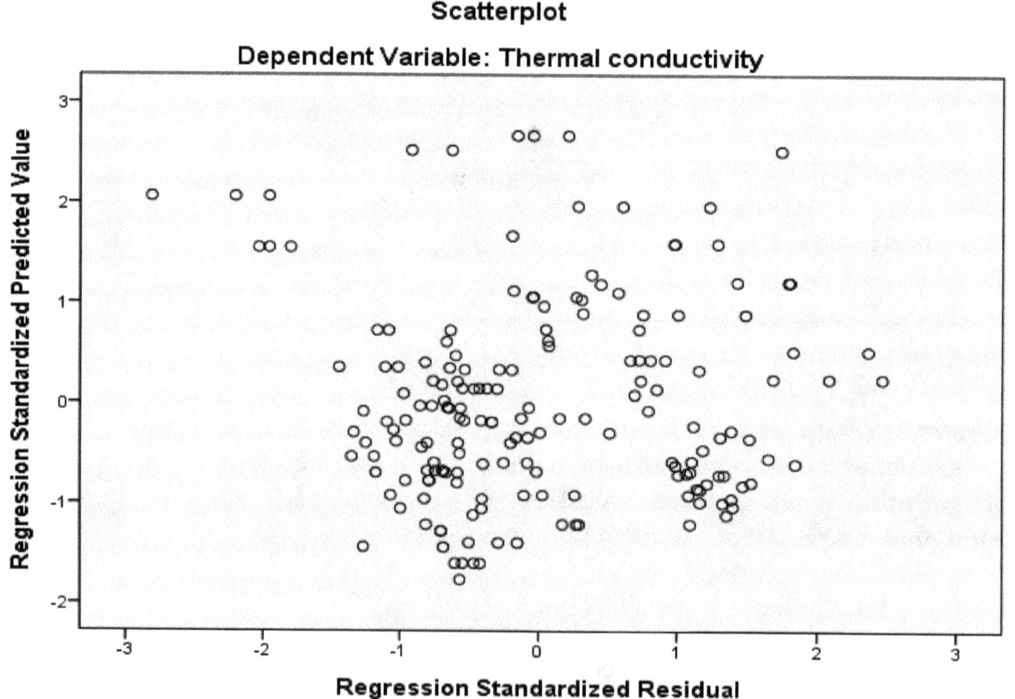

Figure 6. Homoscedasticity plot of residuals.

Table 6. Results of multiple regression analysis model for KJ-I.

	B	Standard Error	*t*	*p*-Value	*VIF*
Constant	−0.510	0.088	−5.761	<0.01	
X1 (dry density)	0.641	0.057	11.334	<0.01	1.001
X2 (degree of saturation)	0.624	0.034	18.611	<0.01	1.001
R^2	0.739				
adjR^2	0.736				

B: non-standardized coefficient, *t*: *B*/standard error, *VIF*: variance inflation factor.

Table 7. Results of ANOVA analysis.

	DF	SS	MS	F	*p*-Value
Regression	2	6.892	3.446	245.033	<0.01
Residuals	173	2.433	0.014		
Total	175	9.325			

4.3. Model Comparison with Na-Type Bentonite

Many studies have also investigated the thermal conductivity of Na-type bentonites as mentioned in the introduction. Tang et al. [11] also measured the thermal conductivity of MX-80 and suggested various predictive models. Among them, the volume fraction of air and thermal conductivity showed good correlations. The empirical formula can be derived as Equation (6):

$$\lambda = \alpha(V_a/V) + K_{sat} \tag{6}$$

Here, α is the slope of the K-V_a/V plot. K_{sat} is the thermal conductivity at the saturated condition. Equation (6) was also applied to predict thermal conductivity using the 176 datasets of the KJ-I and KJ-II bentonites. Once the dry density and the degree of saturation are obtained, the volume fraction of air can be easily obtained according to the basic geotechnical relation [26].

Table 8 summarizes the parameters from Equation (6) for the KJ, MX80 and Kunigel bentonites, and α and K_{sat} were = −2.05/1.22 for KJ-I and = −1.20/1.18 for KJ-II. Thermal conductivity is inversely proportional to the air fraction because the thermal conductivity of air is much smaller than that of water and soil particles [5,22]. On average, the thermal conductivity of the Kunigel bentonite showed the highest thermal conductivity. Since Kunigel contains about ten times more quartz mineral than MX80, and 6–8 times more than KJ, it is thought that this is the reason Kunigel had a higher thermal conductivity than KJ and MX80. Therefore, it can be inferred that the thermal conductivity of the compacted bentonite buffer materials does not depend on the type of exchangeable cation, such as Ca-type or Na-type.

Table 8. Parameters from Equation (6).

	KJ (Present Work)		MX80 [11]		Kunigel [11]	
	α	K_{sat}	α	K_{sat}	α	K_{sat}
Parameters	−1.76	1.19	−1.79	1.10	−2.36	1.39

5. Conclusions

This paper analyzed thermal conductivity results for Ca-type Korean bentonite buffer materials produced in the Gyeongju region, and suggested various predictive models with three influential independent variables: dry density, degree of saturation, and volume fraction of air. The main conclusions can be summarized as follows:

First, 142 thermal conductivity datasets for the KJ-I bentonite were collected according to dry density and degree of saturation, 34 datasets for KJ-II bentonite were measured using the QTM-500 equipment which is based on the transient hot-wire method. The KJ-I and KJ-II bentonites are composed of more than 60% montmorillonite, and 5% quartz. The thermal conductivity of KJ-I and KJ-II was proportional to dry density and degree of saturation. On average, the thermal conductivity of KJ-II was higher than that of KJ-I because KJ-II has a higher content of minerals with higher thermal conductivity, such as quartz, cristobalite, and calcite. Additionally, the thermal conductivity of KJ-II was measured by the drying path while KJ-I was measured by the wetting path. Compared to MX80, KJ-I and KJ-II had slightly higher thermal conductivities, and it is thought that this was mainly caused by the different mineral compositions of the KJ and MX80 bentonites.

The thermal conductivity estimation models for KJ-I and KJ-II were derived from this research. To satisfy the required functional criteria, the dry density of the buffer materials in Korean disposal systems was at least 1.6 Mg/m^3. This paper used three main equations which are known to be adequate to predict the thermal conductivity of the compacted bentonite buffer materials when the dry density is 1.6 Mg/m^3. In addition, this paper conducted a multiple regression analysis, and suggested a regression model for KJ-I and KJ-II which considered various dry densities and degrees of saturation as independent variables. The R^2 value was 0.739, and satisfied the statistical significance of the regression analyses. It is thought that the regression model proposed in this research can be used as an effective method to estimate the thermal conductivity of KJ-I and KJ-II. In order to compare with Na-type bentonites, such as MX80 and Kunigel, the empirical equation was applied with the volume fraction of air as the independent variable, since the volume fraction of air can be easily calculated using dry density and degree of saturation. Thermal conductivity is inversely proportional to the air fraction. The thermal conductivity of the Korean Ca-type bentonite was 10~15% higher than MX80, but lower than Kunigel, which had different thermal conductivity values because of its mineral composition. It is thought that Kunigel contains about ten times more quartz than MX80, and 6–8 times more than KJ.

The purpose of this paper was to investigate the thermal conductivity of Korean Ca-type bentonite, which can be considered one of the candidate buffer materials for safe HLW disposal in Korea. The predictive models of the Korean Ca-type bentonite suggested by this research can be applied in the design of such disposal systems, since they consider changes in saturation from the inflow of groundwater, and dry density from decay heat. The Na-type bentonite is known to have more swelling, but buffer materials are also required to have high thermal conductivity in order to release as much decay heat as possible from the spent fuel. It is thought that bentonite, which contains minerals with high thermal conductivity, will be adequate as a buffer material in terms of thermal properties. Furthermore, in the future, it will be necessary to determine optimum buffer materials which contain minerals with high thermal conductivity and swelling capacity.

Author Contributions: S.Y. and W.C. conducted experiment; C.L. and G.-Y.K. analyzed the data; and the authors made equal contribution and efforts on writing the manuscript.

References

1. Cho, W.J.; Kwon, S. Effects of Variable Saturation on the Thermal Analysis of the Engineered Barrier System for a Nuclear Waste Repository. *Nucl. Technol.* **2012**, *2*, 245–256. [CrossRef]
2. Gens, A.; Sánchez, M.; Do, L.; Guimarães, N.; Aloson, E.E.; Lloret, A.; Olivella, S.; Villar, M.V.; Huertas, F. A full-scale in situ heating test for high-level nuclear waste disposal: Observation, analysis and interpretation. *Geotechnique* **2009**, *59*, 377–399. [CrossRef]
3. Japan Nuclear Cycle Development Institute (JNC). *H12 Project to Establish Technical Basis for HLW Disposal in Japan (Supporting Report 2)*; JNC TN1400-99-020; Japan Nuclear Cycle Development Institute: Ibaraki, Japan, 1999.

4. Ye, W.M.; Chen, Y.G.; Chen, B.; Wang, Q.; Wang, J. Advances on the knowledge of the buffer/backfill properties of heavily-compacted GMZ bentonite. *Eng. Geol.* **2010**, *116*, 12–20. [CrossRef]
5. Cho, W.J.; Kwon, S. An empirical model for the thermal conductivity of compacted bentonite and a bentonite-sand mixture. *Heat Mass Transf.* **2011**, *47*, 1385–1393. [CrossRef]
6. Hoffmann, C.; Alonso, E.E.; Romero, E. Hydro-mechanical behavior of bentonite pellet mixtures. *Phys. Chem. Earth* **2007**, *32*, 832–849. [CrossRef]
7. Lee, J.O.; Choi, H.J.; Kim, G.Y. Numerical simulation studies on predicting the peak temperature in the buffer of an HLW repository/Numerical simulation studies on predicting the peak temperature in the buffer of an HLW repository. *Int. J. Heat Mass Trasnf.* **2017**, *115*, 192–204. [CrossRef]
8. Zheng, L.; Rutqvist, J.; Birkholzer, J.T.; Liu, H.H. On the impact of temperature up to 200 °C in clay repositories with bentonite engineer barrier systems: A study with coupled thermal, hydrological, chemical, and mechanical modeling. *Eng. Geol.* **2015**, *97*, 278–295. [CrossRef]
9. Börgesson, L.; Chijimatsu, M.; Fujita, T.; Nguyen, T.S.; Rutqvist, J.; Jing, L. Thermo-hydro-mechanical characterization of a bentonite-based buffer material by laboratory tests and numerical back analyses. *Int. J. Rock Mech. Min. Sci.* **2001**, *38*, 95–104. [CrossRef]
10. Tang, A.M.; Cui, Y.J. Effects of mineralogy on thermo-hydro-mechanical parameters of MX 80 bentonite. *Int. J. Rock Mech. Geotech. Eng.* **2010**, *2*, 91–96.
11. Tang, A.M.; Cui, Y.J.; Le, T.T. A study on the thermal conductivity of compacted bentonites. *Appl. Clay Sci.* **2008**, *41*, 181–189. [CrossRef]
12. Villar, M.V. *MX-80 Bentonite. Thermo-Hydro-Mechanical Characterization Performed at CIEMAT in the Context of the Prototype Project*; Centro de Investigaciones Energeticas, Medioambientalesy Tecnologicas: Madrid, Spain, 2005.
13. Hrubesova, E.; Lunackova, B.; Brodzki, O. Comparison of liquid limit of soils resulted from Casagrande test and modificated cone penetrometor methodology. *Procedia Eng.* **2016**, *142*, 364–370. [CrossRef]
14. Andrade, F.A.; Al-Qureshi, H.A.; Hotza, D. Measuring the plasticity of clays: A review. *Appl. Clay Sci.* **2011**, *51*, 1–7. [CrossRef]
15. *ASTM D2487/17—Standard Practice for Classification of Soils for Engineering Purpose (Unified Soil Classification System)*; ASTM International: West Conshohocken, PA, USA, 2017.
16. *ASTM C1113/C1113M-09—Standard Test Method for Thermal Conductivity of Refractories by Hot Wire (Platinum Resistance Thermometer Technique)*; ASTM International: West Conshohocken, PA, USA, 2013.
17. Ballarini, E.; Graupner, B.; Bauer, S. Thermal-hydraulic-mechanical behavior of bentonite and sand-bentonite materials as seal for a nuclear waste repository: Numerical simulation of column experiments. *Appl. Clay Sci.* **2017**, *135*, 289–299. [CrossRef]
18. Lee, J.P.; Choi, J.W.; Choi, H.J.; Lee, M.S. Increasing of Thermal Conductivity from Mixing of Additive on a Domestic Compacted Bentonite Buffer. *J. Nucl. Fuel Cycle Waste Technol.* **2013**, *11*, 11–21. [CrossRef]
19. Farouki, O.T. *Thermal Properties of Soils*; Series on Rock and Soil Mechanics; Trans Tech Publications: Zürich, Switzerland, 1986.
20. Cho, W.J.; Kim, G.Y. Reconsideration of thermal criteria for Korean spent fuel repository. *Ann. Nucl. Energy* **2016**, *88*, 73–82. [CrossRef]
21. Wilson, J.; Savage, D.; Bond, A.; Watson, S.; Pusch, R.; Bennet, D. *Bentonite: A Review of Key Properties, Process and Issues for Consideration in the UK Context*; QRS-1378zG-1.1; Quintessa Limited: Oxfordshire, UK, 2011.
22. Lee, J.O.; Choi, H.; Lee, J.Y. Thermal conductivity of compacted bentonite as a buffer material for a high-level radioactive waste repository. *Ann. Nucl. Energy* **2016**, *94*, 848–855. [CrossRef]
23. Hair, J.F., Jr.; Black, W.C.; Babin, B.J.; Anderson, R.E. *Multivariate Data Analysis*, 7th ed.; Prentice-Hall: Upper Saddle River, NJ, USA, 2009.
24. Yoon, S.; Lee, S.R.; Kim, Y.T.; Go, G.H. Estimation of saturated hydraulic conductivity of Korean weathered granite soils using a regression analysis. *Geomech. Eng.* **2015**, *9*, 101–113. [CrossRef]
25. Park, J.Y. A statistical Entrainment Growth Rate Estimation Model for Debris-Flow Runout Prediction. Master's Thesis, Korea Advanced Institute of Science and Technology, Daejeon, Korea, 2015; 84p.
26. Das, B.M. *Principle of Geotechnical Engineering*, 6th ed.; Nelson: Nelson, New Zealand, 2006.

6

Thermodynamic Modeling and Performance Analysis of a Combined Power Generation System based on HT-PEMFC and ORC

Hyun Sung Kang [1,2], Myong-Hwan Kim [3] and Yoon Hyuk Shin [1,*]

1 Eco-Friendly Vehicle R & D Division, Korea Automotive Technology Institute, 303 Pungse-Ro, Pungse-Myeon, Cheonan-Si 330-912, Korea; hskang@katech.re.kr
2 Department of Mechanical Engineering, Korea University, 409 Innovation Hall Building, Anam-Dong, Sungbuk-Gu, Seoul 02841, Korea
3 Hydrogen Fuel Cell Mobility R&D Center, Korea Automotive Technology Institute, 303 Pungse-Ro, Pungse-Myeon, Cheonan-Si 330-912, Korea; kimmh@katech.re.kr
* Correspondence: yhshin@katech.re.kr

Abstract: Recently, the need for energy-saving and eco-friendly energy systems is increasing as problems such as rapid climate change and air pollution are getting more serious. While research on a power generation system using hydrogen energy-based fuel cells, which rarely generates harmful substances unlike fossil fuels, is being done, a power generation system that combines fuel cells and Organic Rankine Cycle (ORC) is being recognized. In the case of High Temperature Proton Exchange Membrane Fuel Cell (HT-PEMFC) with an operating temperature of approximately 150 to 200 °C, the importance of a thermal management system increases. It also produces the waste heat energy at a relatively high temperature, so it can be used as a heat source for ORC system. In order to achieve this outcome, waste heat must be used on a limited scale within a certain range of the temperature of the stack coolant. Therefore, it is necessary to utilize the waste heat of ORC system reflecting the stack thermal management and to establish and predict an appropriate operating range. By constructing an analytical model of a combined power generation system of HT-PEMFC and ORC systems, this study compares the stack load and power generation performance and efficiency of the system by operating temperature. In the integrated lumped thermal capacity model, the effects of stack operating temperature and current density, which are important factors affecting the performance change of HT-PEMFC and ORC combined cycle power generation, were compared according to operating conditions. In the comparison of the change in power and waste heat generation of the HT-PEMFC stack, it was shown that the rate of change in power and waste heat generation by the stack operating temperature was clearly changed according to the current density. In the case of the ORC system, changes in the thermal efficiency of the ORC system according to the operating temperature of the stack and the environmental temperature (cooling temperature) of the object to which this system is applied were characteristic. This study is expected to contribute to the establishment of an optimal operation strategy and efficient system configuration according to the subjects of the HT-PEMFC and ORC combined power generation system in the future.

Keywords: high temperature proton exchange membrane fuel cell; thermal management; organic rankine cycle; plate heat exchanger; waste heat recovery; cooling system; thermodynamic modeling

---

1. Introduction

Today, the need to expand power generation systems utilizing eco-friendly and waste heat energy to tackle climate changes is increasing, and active research on hydrogen fuel cell generation (electricity generation) and a cogeneration system capable of utilizing waste heat is being done. Unlike engine or boiler-based power generation systems that generate power through a combustion process that uses fossil fuels to produce thermal energy and emission, hydrogen fuel cell systems is an eco-friendly power generator of electricity, thermal energy, and water through the chemical bonding process of hydrogen and oxygen.

The Solid Oxide Fuel cell (SOFC) based micro-cogenerative power system is being actively researched, and modeling research for predicting appropriate operating conditions is being considered for important research project purposes. Arpino et al. investigated the factors that influence the measurement uncertainty for combined heat and power design using SOFC [1]. In addition, they studied the correlation between the 0D model of those SOFC-based systems and the collected data. Additionally, an effective thermal management strategy through fuel utilization adjustments was presented for optimizing cogenerative power system operation [2]. Duhn et al. conducted an analytical study of the cooling plate design to improve operational efficiency by ensuring the pressure drop uniformity of the gas distributor of the SOFC system [3]. As described above, in a fuel cell-based power generation system having a high operating temperature, optimum control of the working fluid is important in addition to proper operating temperature and pressure drop formation in order to improve the efficiency and performance of the system.

In the case of High Temperature Proton Exchange Membrane Fuel Cell (HT-PEMFC), there is an advantage in that it can utilize waste heat at a relatively high temperature (150 °C or higher) with highly efficient power generation. In order to secure the power efficiency and reliability of such HT-PEMFC, a thermal management system is essential to maintain a high operating temperature [4,5]. As the operating temperature of the stack must be kept constant, the stack coolant must be used within a controlled temperature. Consequently, a thorough examination of the appropriate operating range of waste heat utilization (heat exchange) reflecting the respective stack thermal management and system control thereto, should be performed for the optimal design of cogeneration using HT-PEMFC and waste heat recovery [6,7].

Among the fuel cells, the PEMFC exhibits a relatively high power density and power efficiency, and it can minimize noise and residual emissions. It is divided into Low Temperature Proton Exchange Membrane Fuel Cell (LT-PEMFC) with an operating temperature of 60 to 80 °C and HT-PEMFC with an operating temperature of 100 °C or more. The power efficiency of HT-PEMFC appears to be approximately 45 to 60% [8]. Currently, research is being carried out on the cogeneration system suitable for each operation characteristic of each PEMFC type [9]. The advantage of HT-PEMFC is the simplification of the water management device configuration due to the high operating temperature as well as the generation of highly useful waste heat. Specifically, if the liquid cooling system is applied to HT-PEMFC thermal management with an operating temperature of 100 °C or higher, waste heat exchange with higher utilization is possible, which is advantageous for cogeneration and trigeneration systems [10]. Najafi et al. compared the performance and efficiency of the HT-PEMFC trigeneration system according to operation strategies during a certain operating period while the research team carried out a study on a trigeneration system to which both LT-PEMFC and HT-PEMFC were applied. Furthermore, research on warm-up strategies to quickly increase the stack operating temperature when the HT-PEMFC starts up was conducted [11]. Thus, based on the previous studies, it seems that the HT-PEMFC-based cooling and heating system can be selectively applied according to the operation strategy and subject.

It is important to secure the performance and efficiency of the waste heat recovery system in order to expand the subjects for application and functionality of the combined power generation system using such HT-PEMFC. This is one of the most important factors in selecting a target building and power system to secure electric energy with high utilization at a certain level depending on the operating environment [12,13]. Therefore, today, active research on Organic Rankine Cycle (ORC) system using stack waste heat energy in addition to fuel cell systems is being done [14]. Dickes' research team experimentally examined the temperature distribution of the working fluid for power generation in the evaporator heat exchanger in the ORC power system, and Jang's research team conducted a study on the performance of the compact ORC system at the 1 kW-level using a heat source in the range of 100 to 140 °C [15,16]. In addition, Jeong's research team conducted a study on the heat exchange performance and characteristics of the plate heat exchanger for each working fluid operating condition applied to the ORC system [17]. S.C. Yang et al. conducted a pilot study on an ORC power generation system at the 3 kW-level, capable of utilizing waste heat at 100 °C [18]. In the case of HT-PEMFC, the inlet temperature and mass flow rate of the coolant must be kept relatively constant in order to secure appropriate power efficiency at a relatively high operating temperature [19]. This is a limiting factor that must be reflected in the operational design of waste heat utilization systems such as ORC power generation and is the reason for the need to optimize the integrated system linked to the stack thermal management system. To this end, it may be useful to introduce a HT-PEMFC and ORC power generation integrated system modeling, as well as a confirmation and verification process of power generation performance and efficiency range according to the application subject and operating conditions.

In this study, through an analytical method based on the existing HT-PEMFC and ORC power generation system model, the power generation performance and efficiency range according to the operating conditions of the HT-PEMFC and ORC combined power generation system considering stack thermal management were confirmed, and the rate of change of power generation performance (effect on power generation performance change) for each control factor for the combined power generation system was presented. For this, a system analysis was conducted to predict system performance and efficiency according to changes in operating conditions such as stack operating temperature, current density, and ORC cooling temperature for a combined power generation system composed of a lumped thermal capacity model.

2. System Description Based on HT-PEMFC and ORC

Figure 1 shows the model composition for performance prediction and comparative analysis of the combined system consisting of the HT-PEMFC subsystem for cooling of the HT-PEMFC and the ORC subsystem for waste heat recovery power generation. The HT-PEMFC subsystem and the ORC subsystem, each with a fluid flow diagram, share an evaporator. The coolant of the HT-PEMFC subsystem was selected as Tri-ethylene glycol (TEG) since its phase does not change at the operating temperature of HT-PEMFC, which is 423~463 K. It follows the black solid line in Figure 1. The HT-PEMFC subsystem consists of an auxiliary heater/cooler, thermal storage, 3-way valve, cooling pump, and evaporator heat exchanger. The auxiliary heater/cooler is configured to keep the inlet temperature of PEMFC constant regardless of the influence of the ORC subsystem when the thermal power of the HT-PEMFC and the heat supplied to the ORC is not the same during the initial system startup. The 3-way valve is configured to meet the same flow conditions as the thermal power of HT-PEMFC and the heat exchange amount of the ORC evaporator. The cooling pump was controlled so that it would meet the flow condition in which the temperature difference between the inlet and outlet of HT-PEMFC satisfies 5 K.

To simplify the analysis of this system, the study followed subsequent assumptions:

1. All equipment of the system follows the lumped model and ignores heat loss.
2. The pressure loss in the pipe through which the stack coolant and working fluid travel is ignored.
3. Temperature and cell voltage are evenly distributed over the entire electrode of HT-PEMFC, and the reaction gas mixture is an ideal gas fluid.
4. The cathode charge transfer coefficient and the anode charge transfer coefficient are the same.
5. The isentropic efficiency of the expander is 60%, and the overall efficiency of the refrigerant pump is 50%.
6. In the condenser of the ORC system, the working fluid is sufficiently subcooled, and the existing superheat of the evaporator is 5 K.

7. The pressure loss of the evaporator reflected only the pressure loss on the heat source side, and only the loss due to friction was considered.

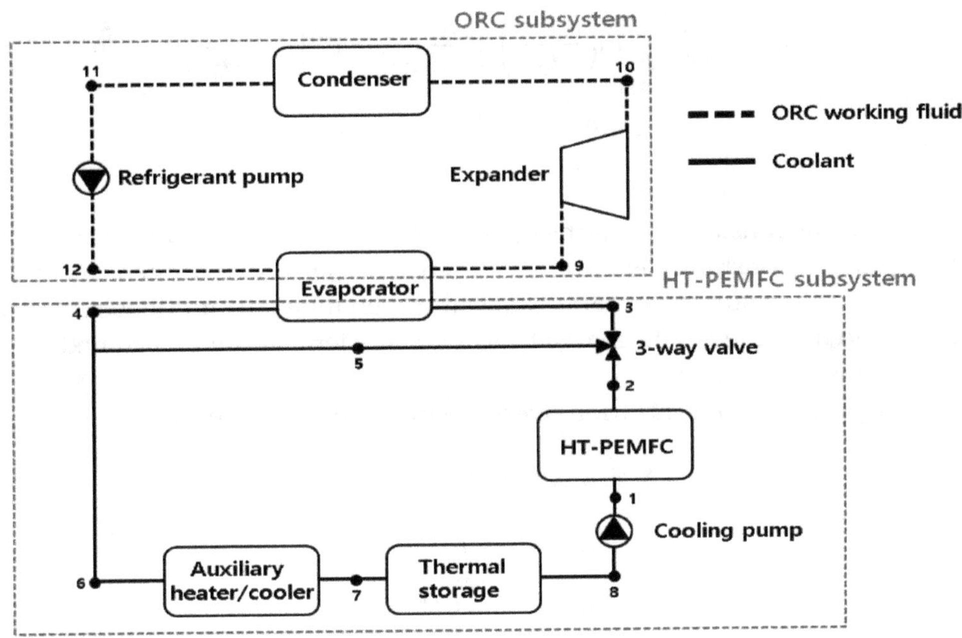

Figure 1. Schematic diagram of combined system.

3. Combined HT-PEMFC and ORC System Modeling

3.1. Thermodynamic Model of HT-PEMFC

The HT-PEMFC model applied in this analysis can be used by setting the operating temperature range of 120 to 200 °C, the cathode stoichiometric ratio range of 2 to 5, and the CO concentration range of 0.1 to 10%. Using a previously studied model as a reference [20,21], it is a one-dimensional isothermal model, HT-PEMFC based on PBI, and the electrochemical reaction in the fuel cell is as follows [22]:

$$Anode: H_2 \rightarrow 2H^+ + 2e^- \tag{1}$$

$$Cathode\ 2H^+ + \frac{1}{2}O_2 + 2e^- \rightarrow H_2O \tag{2}$$

$$Overall\ reaction\ H_2(g) + \frac{1}{2}O_2(g) \rightarrow H_2(g) \tag{3}$$

The overall cell voltage of the stack is calculated through the overpotential loss acting on the cathode and the overpotential loss acting on the anode at the ideal standard potential as shown in the Equation (4). The overpotential is divided into the activation overpotential (η_{act}), ohmic overpotential (η_{ohmic}), and concentration overpotential (η_{conc}) [23]. The activation overpotential is affected by the Tafel equation and the charge transfer coefficient while the ohmic overpotential is affected by the thickness of the membrane and the catalyst. The last term, concentration overpotential, represents the effect of cathode stoichiometry.

The ideal standard potential (V_{ideal}) is calculated by the variation of Gibbs free energy (Δgf) through electrochemical reaction and Faraday Constant (F) as shown in Equations (5) and (6), but in this model, the open circuit voltage of a reference [20] is used.

$$V_{cell} = V_{ocv} - \eta_{act} - \eta_{ohmic} - \eta_{conc} \tag{4}$$

$$\Delta gf = gf_{H_2O} - gf_{H_2} - gf_{O_2} \tag{5}$$

$$V_{ideal} = -\frac{\Delta gf}{2F} \tag{6}$$

Activation overpotential acting on cathode and anode is obtained from the following equations:

$$\eta_{act} = \frac{RT_{cell}}{4\alpha_c F}\ln\left(\frac{I+I_0}{I_0}\right) + \frac{RT_{cell}}{\alpha_a F} sinh^{-1}\left(\frac{I}{2k_{eh}\theta_{h2}}\right) \tag{7}$$

$$\alpha_c = a_0 T_{cell} + b_0 \tag{8}$$

$$I_0 = a_1 e^{-b_1 T_{cell}} \tag{9}$$

where R is the universal gas constant, T_{cell} is the operating temperature, α_c is the cathode charge transfer coefficient, F is the faraday constant, I is the current density, k_{eh} is the hydrogen electro-oxidation rate constant, θ_{h2} is the hydrogen surface coverage, I_0 is the exchange current density, λ_{air} is the cathode stoichiometry ratio, and α_a is the anode charge transfer coefficient, and it is assumed to be the same as the cathode charge transfer coefficient.

Ohmic overpotential and concentration overpotential acting on the cathode is given by:

$$\eta_{ohmic} = R_{ohmic} I \tag{10}$$

$$R_{ohmic} = a_2 T_{cell} + b_2 \tag{11}$$

$$\eta_{conc} = \frac{R_{conc}}{\lambda_{air} - 1} I \tag{12}$$

$$R_{conc} = a_3 T_{cell} + b_3 \tag{13}$$

Linear regression was used for the cathode charge transfer coefficient, ohmic resistance (R_{ohmic}), and concentration resistance (R_{conc}), and the exchange current density was expressed as an exponential function type. The values of the regressions used are shown in Table 1.

Table 1. Numerical value for regressions used in the High Temperature Proton Exchange Membrane Fuel Cell (HT-PEMFC) model.

Parameters	Values	Unit
Charge transfer constant, a_0	2.761×10^{-3}	[K^{-1}]
Charge transfer constant, b_0	−0.9453	-
Limiting current constant, a_1	3.3×10^{3}	[A]
Limiting current constant, b_1	−0.04368	-
Ohmic loss constant, a_2	-1.667×10^{-4}	[Ω/K]
Ohmic loss constant, b_2	0.2289	[Ω]
Diffusion limitation constant, a_3	-8.203×10^{-4}	[Ω/K]
Diffusion limitation constant, b_3	0.4306	[Ω]

It was assumed that all cell unit performances of the HT-PEMFC were the same, and the electric power (W_{FC}) and thermal power (Q_{FC}) of HT-PEMFC were calculated in proportion to the number of cells (N_{cell}) and single cell active area (A_{cell}) as in Equations (14) and (15). Moreover, the power efficiency (η_{FC}) of HT-PEMFC can be obtained as in Equation (16) based on the lower heating value (LHV) of hydrogen. Table 2 shows the parameters used in the HT-PEMFC model

$$W_{FC} = N_{cell} V_{cell} I A_{cell} \tag{14}$$

$$Q_{FC} = N_{cell}\left(\frac{LHV}{2F} - V_{cell}\right) I A_{cell} \tag{15}$$

$$\eta_{FC} = \frac{W_{FC}}{N_{cell}\frac{LHV}{2F} I A_{cell}} \tag{16}$$

Table 2. Operating parameters and empirical parameters used in the HT-PEMFC model.

Parameters	Values	Unit
Open circuit Voltage, V_{ocv}	0.95	[V]
Number of cells, N_{cell}	880	-
Single cell active area, A_{cell}	300	[cm^2]
Operating temperature, T_{cell}	433	[K]
Current density, I	0.4	[A/cm^2]
Universal gas constant, R	8.314	[J/mol·K]
Faraday constant, F	96485.3	[C/mol]
Cathode stoichiometry ratio, λ_{air} [24]	3	-
Hydrogen electro-oxidation rate constant, k_{eh} [24]	1.63818	[A/cm^2]
Hydrogen surface coverage, θ_{h2} [24]	0.14212	-
Low heating Value of hydrogen, LHV	239.92	[kJ/mol]

The thermal power generated by the HT-PEMFC was assumed to be heat-exchanged with the coolant by the Dittus-Boelter Equation (17), and the heat generated by auxiliary devices such as the cooling pump was ignored.

$$Nu = 0.023Re^{0.8}Pr^{0.4} \tag{17}$$

Furthermore, the pressure drop on the coolant side of the HT-PEMFC was reflected by the curve fitting the pressure drop according to the flow rate based on the experimental value. In general, the pressure drop on the coolant side of the stack is dependent on the flow path design of the cooling plate, and in this study, the pressure drop test value of the most widely commercialized vehicle stack with a level similar to the reaction area was applied to the model.

$$\Delta P_{FC} = 4074 + 1.86 \times 10^6 Q_{FC} + 3.184 \times 10^9 {Q_{FC}}^2 \tag{18}$$

As a cooling pump model used to transport Triethylene glycol (TEG), which is a stack coolant, a commercial pump for cooling of a maximum 100 kW stack that has a performance curve as shown in Figure 2 was applied [25].

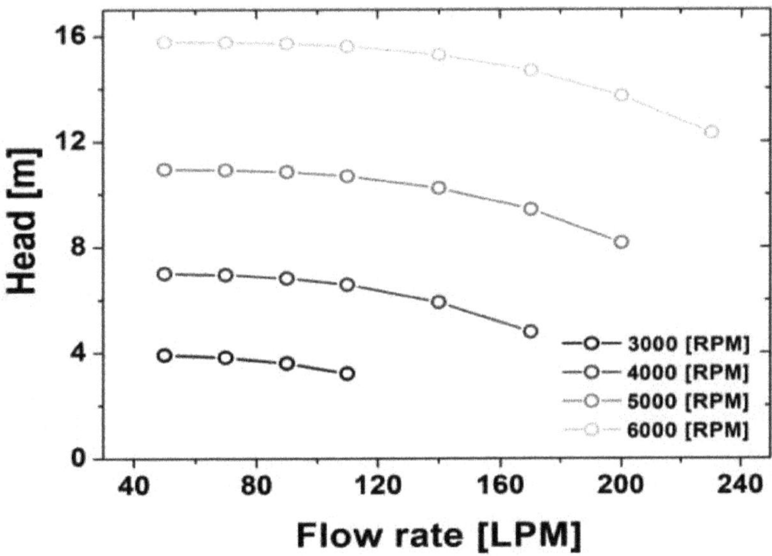

Figure 2. The performance curve of the cooling pump.

3.2. Thermodynamic Model of ORC

Figure 3 shows the T-s diagram of the ideal ORC cycle and conceptually shows each state and system. In Figure 3, the movement from point 12 to point 9 refers to the section in which the liquid working fluid changes to the gaseous state through the evaporator and is the section to recover waste heat from HT-PEMFC in the combined system. The vaporized working fluid generates power through the expander, which is the travel section from point 9 to 10, reducing the pressure. The working fluid with reduced pressure and temperature is liquefied in the travel section from point 10 to 11 through the condenser and maintains the pressure difference while transporting the liquefied working fluid through the pump.

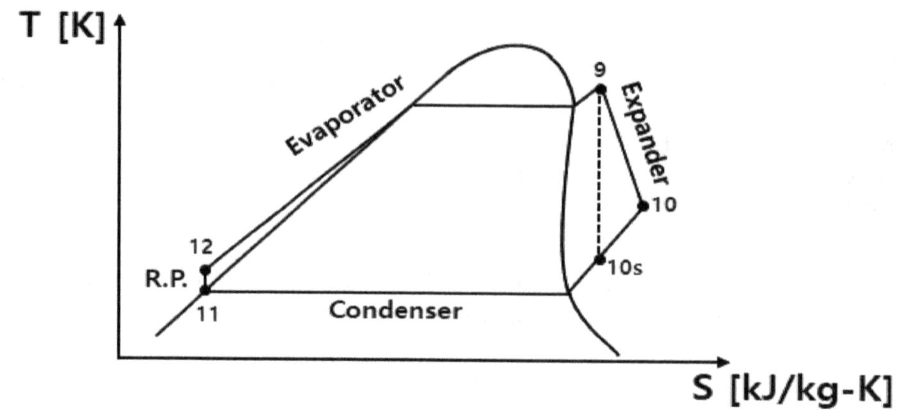

Figure 3. Thermodynamic T-s diagram for ideal Organic Rankine Cycle (ORC) cycle (R.P. is the meaning of refrigerant pump).

R245fa was selected as the working fluid used in this study considering the operating temperature range of the HT-PEMFC. The pressure on the evaporator side of the ORC power generation of the combined power generation system was set to 12 bar, and by setting the temperature range at the outlet side of the condenser to 293~308 K, the saturation pressure appropriate for the respective temperature was used.

All waste heat generated from the HT-PEMFC is heat-exchanged through the evaporator and is calculated as shown in Equation (19). $\dot{Q}_{eva}$ is the amount of heat exchange of the evaporator, and $\dot{m}_{ORC}$ is the mass flow rate of the working fluid of the ORC system. h means the enthalpy according to the temperature and pressure for each location indicated by each number on the T-s diagram.

$$\dot{Q}_{eva} = \dot{m}_{ORC}(h_9 - h_{12}) \tag{19}$$

The amount of power generated through the expander is W_{exp} and is calculated as in the Equation (20).

$$W_{exp} = \dot{m}_{ORC}(h_9 - h_{10}) \tag{20}$$

The amount of heat dissipated through the condenser is $\dot{Q}_{con}$ and is calculated as in the Equation (21).

$$\dot{Q}_{con} = \dot{m}_{ORC}(h_{10} - h_{11}) \tag{21}$$

The power consumption of the refrigerant pump is W_{rp} and is calculated as in the Equation (22), taking into account the overall efficiency η_{rp}.

$$W_{rp} = \frac{\dot{m}_{ORC}(h_{12} - h_{11})}{\eta_{rp}} \tag{22}$$

The net power generated through the ORC system is calculated by the power generated by the expander and the power consumed by the refrigerant pump as shown in the Equation (23).

$$W_{ORC} = W_{exp} - W_{rp} \tag{23}$$

The thermal efficiency of the ORC system is η_{ORC} and is calculated by the net power of the ORC and the endothermic reaction through the evaporator as in the Equation (24).

$$\eta_{ORC} = \frac{W_{ORC}}{\dot{Q}_{eva}} \quad (24)$$

3.3. Analytical Model of the Evaporator Heat Exchanger

A heat exchanger model for the evaporator was constructed to calculate the mass flow rate required by the heat source according to the mass flow rate of the working fluid of the ORC power generation system and the inlet temperature of the heat source (stack coolant) side. The plate heat exchanger used in the experiment in Jeong's study was used as a reference for the shape information of the respective evaporator and is shown in Table 3. [17]. The basic geometric characteristics of the chevron plate heat exchanger are shown in Figure 4.

Table 3. Operating parameters of the chevron plate heat exchanger.

Parameters	Values	Unit
Effective width of plate, L_h	0.111	[m]
Vertical distance between ports, L_v	0.466	[m]
Plate thickness, t	0.0004	[m]
Chevron configuration pitch, λ_p	0.007	[m]
Plate channel gap, g	0.002	[m]
Flow channel hydraulic diameter, D_h	0.003389	[m]
Effective heat transfer area, A_{hx}	0.06105	[m^2]
Plate chevron angle, β	35	[°]
Plate thermal conductivity, k	15	[w/m-K]
Surface enlargement factor, φ	1.18	-
Working fluid channel number, N_{wf}	21	-
Heat source channel number, N_{hs}	22	-

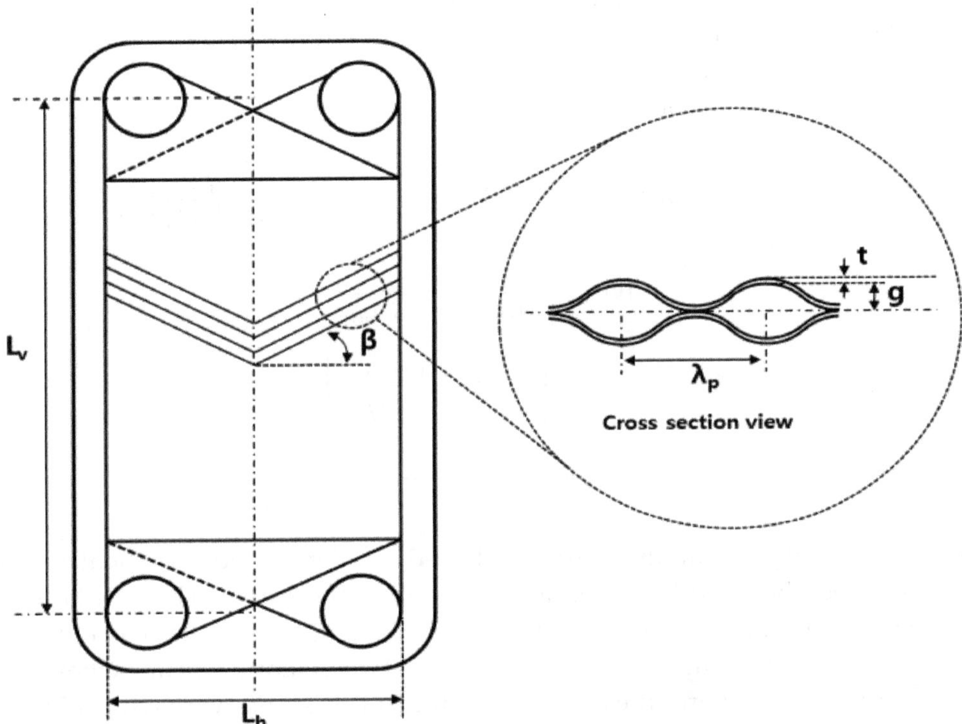

Figure 4. Basic geometric characteristics of chevron plate heat exchanger.

The overall heat transfer coefficient (U) of the Evaporator Heat Exchanger is calculated by the convective heat transfer coefficient of the working fluid (h_{wf}), the convective heat transfer coefficient of the heat source (h_{hs}), and the conduction heat transfer coefficient of the heat exchanger (k_p) as shown in the Equation (25).

$$\frac{1}{U} = \frac{1}{h_{wf}} + \frac{t}{k_p} + \frac{1}{h_{hs}} \tag{25}$$

The heat source is in a single-phase state in all operating areas, and the convective heat transfer coefficient in the single-phase state followed the Muley and Manglik correlation. Based on the Reynolds number, it followed the Equation (26) at 400 or below and followed the Equation (27) at 800 or more [26]. Moreover, the Nusselt number was interpolated using the transitional algorithm in the transition zone. Re_L is the Reynolds number in the liquid state, whereas Pr_L is the Prantle number in the liquid state. In addition, k_L is the heat transfer coefficient in the liquid state, while D_h is the hydraulic diameter of the plate heat exchanger.

$$h_{hs} = 0.44\left(\frac{\beta}{30}\right)^{0.38} Re_L{}^{0.5} Pr_L{}^{0.33}\left(\frac{k_L}{D_h}\right) \tag{26}$$

$$D_0 = 0.2688 - 0.006967\beta + 7.244 \times 10^{-5}\beta^2$$

$$D_1 = 20.78 - 50.94\varphi + 41.1\varphi^2 - 10.51\varphi^3$$

$$D_2 = 0.728 + 0.0543 sin\left(\tfrac{\pi\beta}{45} + 3.7\right)$$

$$h_{hs} = D_0 D_1 Re_L{}^{D_2} Pr_L{}^{0.33}\left(\frac{k_L}{D_h}\right) \tag{27}$$

The convective heat transfer coefficient in the two phase region of the evaporator's working fluid follows the Yan and Lin correlation and is as shown in Equation (28) [27]. Re_{eq} is the equivalent Reynolds number, Bo is the boiling number, G_{eq} is the equivalent mass flux. G is the channel mass flux. q is the heat flux. x is the vapor quality. μ_L is the dynamic viscosity of the liquid. ρ_L is the density of the liquid, and ρ_v is the density of the vapor.

$$h_{wf} = 1.926 Re_{eq}{}^{0.5} Pr_L{}^{0.33} Bo^{0.3}\left[1 - x + x\left(\frac{\rho_L}{\rho_v}\right)^{0.5}\right]\left(\frac{k_L}{D_h}\right) \tag{28}$$

$$Bo = \frac{q}{GA_{hx}} Bo = \frac{q}{GA_{hx}} \tag{29}$$

$$G_{eq} = G\left[1 - x + x\left(\frac{\rho_L}{\rho_v}\right)^{0.5}\right] \tag{30}$$

$$Re_{eq} = \frac{G_{eq} D_h}{\mu_L} \tag{31}$$

To compare and verify the analysis results based on the evaporator heat exchanger model constructed as described above and the previous experimental results, The validation analysis was conducted using the same conditions as in Jeong's study using R245fa as the working fluid and water as the heat source. As shown in Figure 5, the difference between Jeong's heat exchanger performance data that was previously studied and the analysis results in this study were within 3% approximately, and the analysis model constructed accordingly was confirmed to have a certain level of reliability.

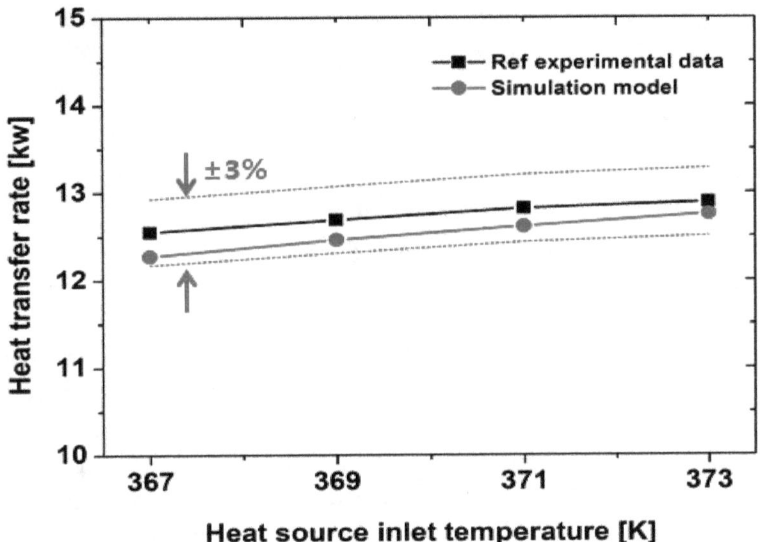

Figure 5. Validation of heat exchanger performance.

The pressure loss on the heat source side of the evaporator heat exchanger is ΔP_{eva} and is defined as the Equation (32) considering only the loss due to friction.

$$\Delta P_{eva} = f_{hs} \frac{L_v N_{hs} G^2}{2 D_h \rho_L} \tag{32}$$

f_{hs} follows the Darcy friction factor and is expressed as in the Equation (33) [17].

$$f_{hs} = 72.5 {Re_L}^{-0.045} \tag{33}$$

3.4. Performance of the Combined System (HT-PEMFC and ORC)

The total electric power produced by the combined system consisting of the HT-PEMFC and ORC is expressed as the sum of the power generated by the HT-PEMFC (W_{FC}), the power consumed by the cooling pump for the HT-PEMFC (W_{cp}), the power generated through ORC (W_{exp}), and the power consumed by the refrigerant pump (W_{rp}) as shown in the Equation (34).

$$W_{total} = W_{FC} - W_{cp} + W_{exp} - W_{rp} \tag{34}$$

In addition, the power efficiency of the combined system is expressed as the ratio of the total power of the combined system according to the *LHV* of HT-PEMFC, as shown in the Equation (35).

$$\eta_{system} = \frac{W_{total}}{N_{cell} \frac{LHV}{2F} I A_{cell}} \tag{35}$$

The whole system is analyzed using the commercial program Flomaster based on the law of conservation of energy (36), the law of conservation of mass (37), and the law of conservation of species (38).

$$\sum \left(\dot{m}h\right)_{in} + \sum \dot{Q}_{in} = \sum \left(\dot{m}h\right)_{out} + \sum \dot{Q}_{out} \tag{36}$$

$$\sum \left(\dot{m}\right)_{in} = \sum \left(\dot{m}\right)_{out} \tag{37}$$

$$\sum \left(\dot{m}x\right)_{in} = \sum \left(\dot{m}x\right)_{out} \tag{38}$$

4. Results and Discussion

4.1. Effect of Stack Temperature

In order to check the performance change according to the operating temperature and current density of the stack, the performance curves for each operating temperature (433 K, 443 K, 453 K, 463 K) and current density (0~0.5 A/cm^2) were verified. As shown in Figure 6a, as the temperature increased, the stack's single cell voltage and efficiency increased as well because of the decrease in the cell activation overpotential. In addition, it tended to decrease when the current density increased. Furthermore, the stack electric power and thermal power showed a tendency to increase as the current density increased, but the percentage of increase in the electric power decreased although the percentage of increase in the thermal power increased. As the temperature of the stack increased, the stack electric power increased thanks to the increase in power efficiency, whereas the stack thermal power decreased. When the current density was 0.1 A/cm^2 and 0.4 A/cm^2 at a stack temperature of 433 K, the single cell voltage was 0.66 V and 0.47 V, the stack power efficiency was 53.7% and 38.1%, the stack electric power was 17.6 kW and 50 kW, and the stack thermal power was 15.2 kW and 81.3 kW.

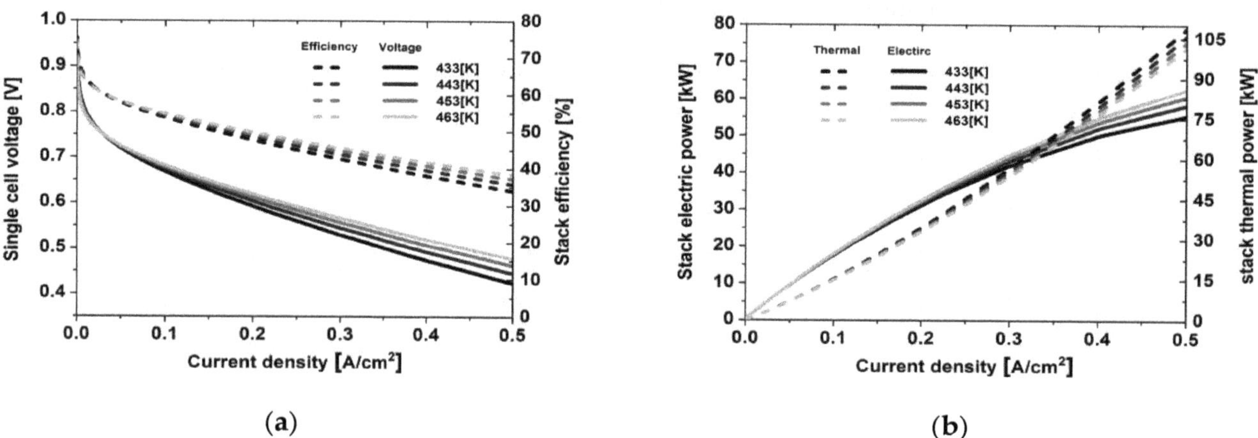

Figure 6. HT-PEMFC performance curve according to stack temperature and current density. (**a**) Single cell voltage and stack power efficiency; (**b**) Stack electric power and thermal power.

4.2. Effect of Working Fluid Mass Flow Rate in the ORC System

The performance change was analyzed by applying the evaporator model configured to calculate the performance according to the mass flow rate of R245fa, the working fluid of the ORC system, and the heat exchange amount of the evaporator. The evaporator pressure was selected as 12 bar considering the temperature level of the waste heat of the stack, and the condenser pressure was selected as 2.2 bar considering the extreme summer outdoor temperature. Moreover, the flow rate of the heat source (stack coolant, TEG) in which the superheat of the evaporator satisfies 5 K was calculated according to the corresponding inlet temperatures of 428 K, 448 K, and 468 K.

As shown in Figure 7a, as the mass flow rate of R245fa increased, the ORC net power increased linearly by the power consumption of the expander and the power consumption of the refrigerant pump. Although there was a change in performance according to the mass flow rate of the working fluid, the efficiency of the ORC system was relatively constant at about 7.69%, because all the conditions satisfied 5 K of superheat. In addition, as shown in Figure 7c, since the mass flow rate of the heat source (TEG) side where the superheat of the evaporator satisfies 5 K required a higher heat transfer coefficient as the inlet temperature of the heat source decreased, the mass flow rate increased. For the R245fa mass flow rate of 0.3 kg/s, the TEG-required mass flow rate was a maximum of 0.83 kg/s.

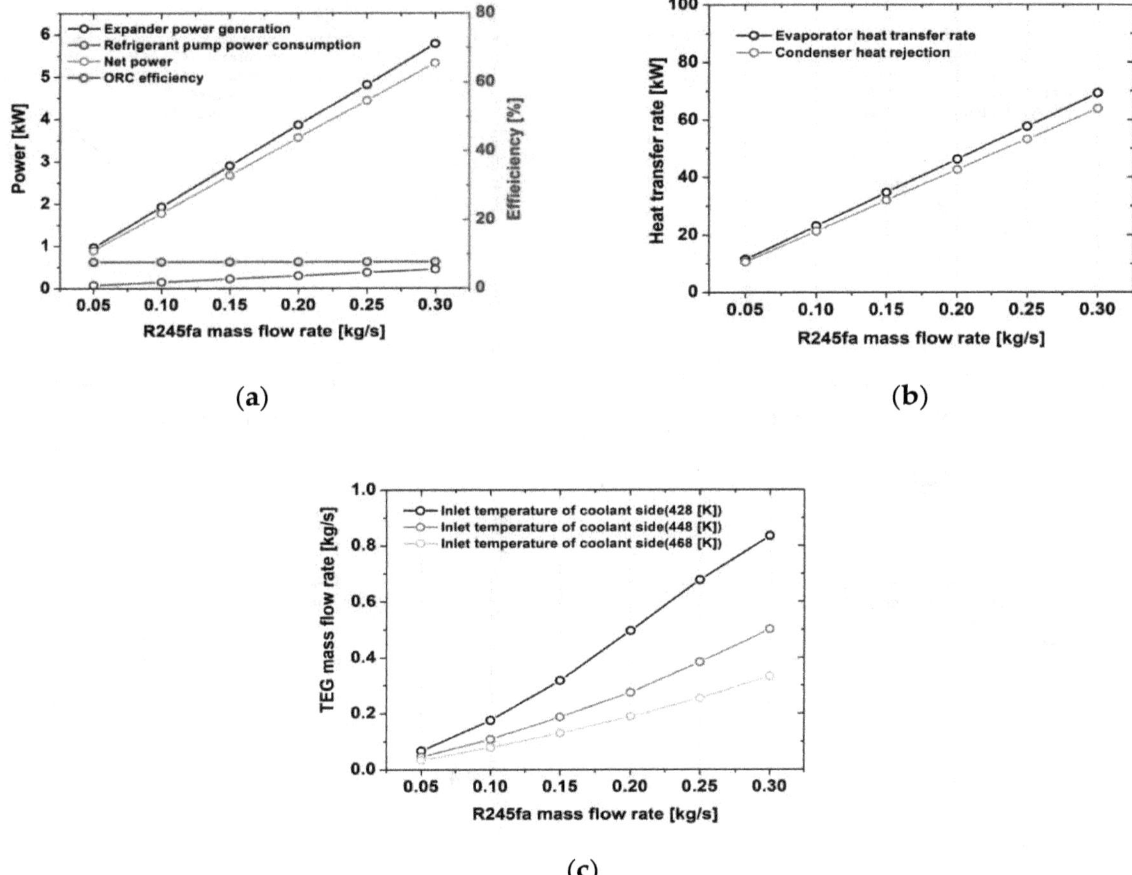

Figure 7. ORC system performance curve according to R245fa mass flow rate. (**a**) Electric power; (**b**) Evaporator and condenser heat transfer rate; (**c**) Mass flow rate of Tri-ethylene glycol (TEG) that satisfies superheat 5 K in the evaporator.

4.3. Effect of Stack Inlet Temperature in the Combined System

In order to analyze the combined system that merged the HT-PEMFC subsystem and the ORC subsystem, the transport pump controlled the mass flow rate so that the temperature difference at the inlet and outlet of the stack was 5 K. The mass flow rate was controlled through a 3-way valve so that all thermal power generated from the stack was exchanged with the evaporator of the ORC subsystem. The system performance was compared and analyzed after the inlet temperature conditions of the stack were selected as 433 K, 443 K, 453 K, and 463 K, and the current densities of the stack were 0.15 A/cm^2, 0.2 A/cm^2, 0.25 A/cm^2, 0.3 A/cm^2, 0.35 A/cm^2 and 0.4 A/cm^2.

As shown in Figure 8a, the mass flow rate of the cooling pump that satisfies the temperature difference between the inlet and outlet of the stack as 5 K is proportional to the current density. As the thermal power of the stack increased as shown in Figure 9d, the required convective heat transfer coefficient also increased, resulting in an increase in the mass flow rate that satisfied the operating conditions. As the inlet temperature of the stack increased, the physical properties of TEG changed, which influenced the formation of the mass flow rate of the cooling pump. The mass flow rate at the evaporator heat source (TEG) side of the ORC subsystem increased as the current density increased, but it decreased as the inlet temperature of the stack increased. The results of the mass flow rate of the cooling pump and the mass flow rate of the evaporator's heat source (TEG) according to the operating conditions, as well as the pressure drop of the stack and the pressure drop of the evaporator are shown in Figure 8c,d, respectively.

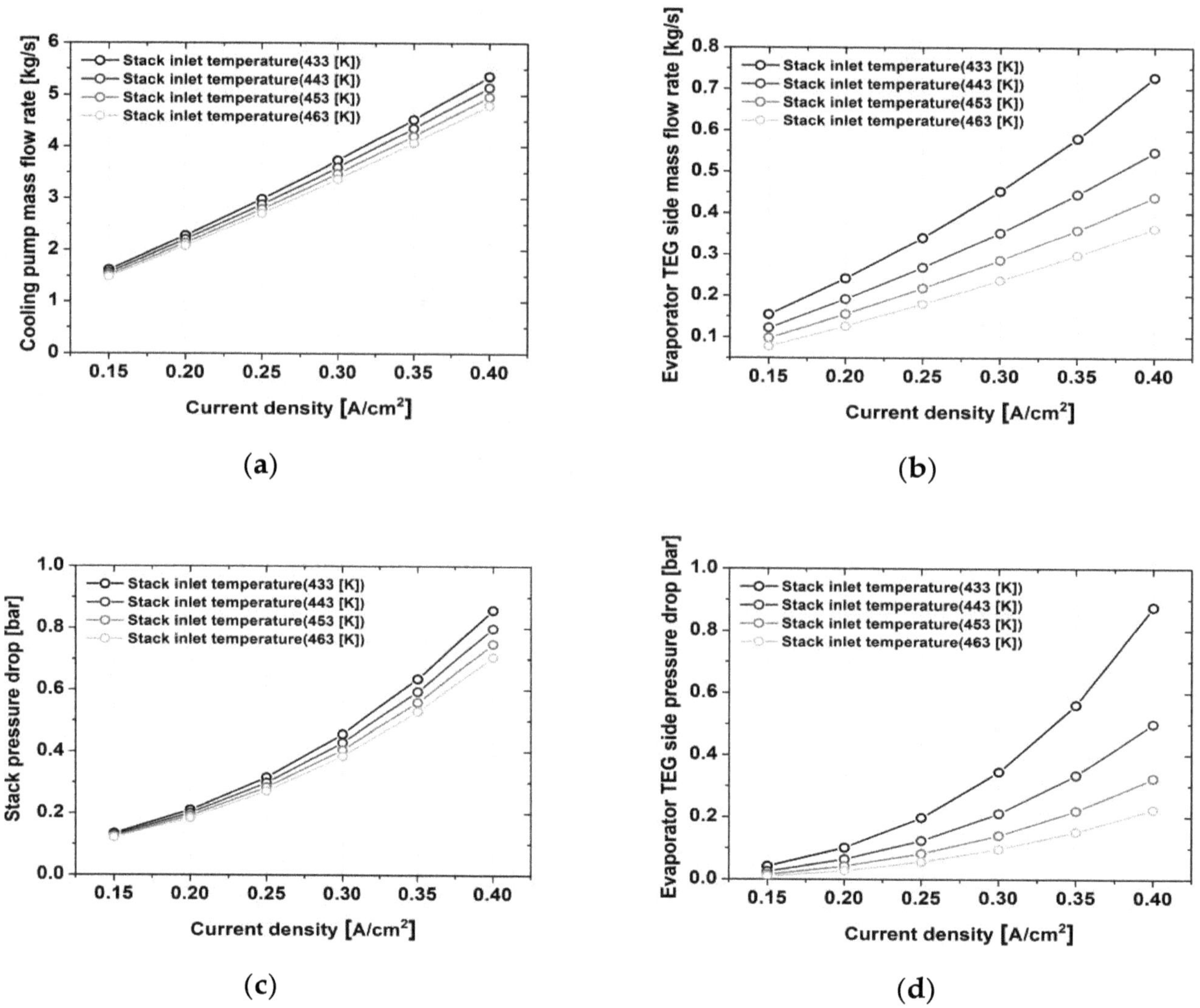

(a) (b)

(c) (d)

Figure 8. Validation trends of system loss and mass flow rate with stack inlet temperature. (**a**) Cooling pump mass flow rate; (**b**) Evaporator TEG side mass flow rate; (**c**) Stack pressure drop; (**d**) Evaporator TEG side pressure drop.

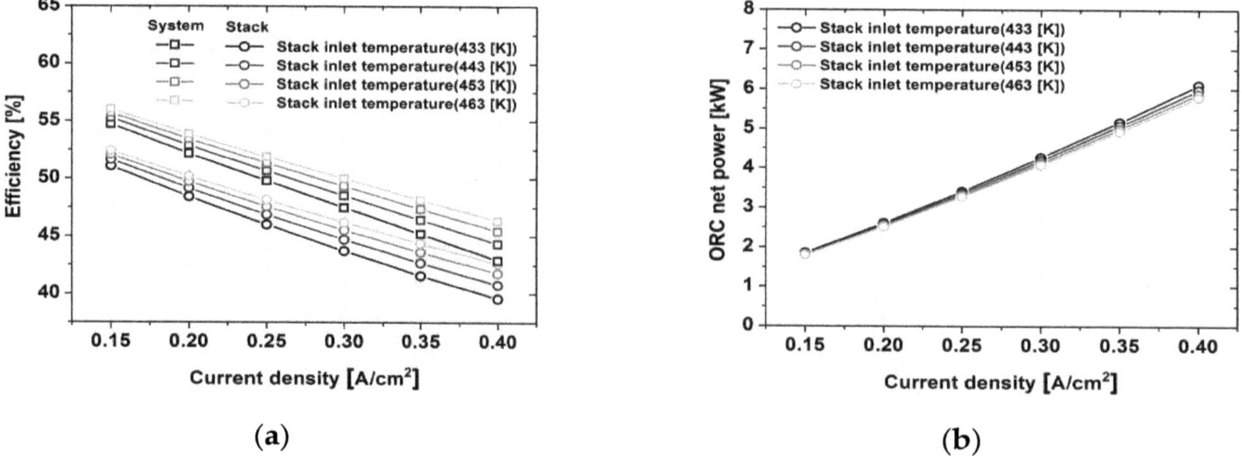

Figure 9. *Cont.*

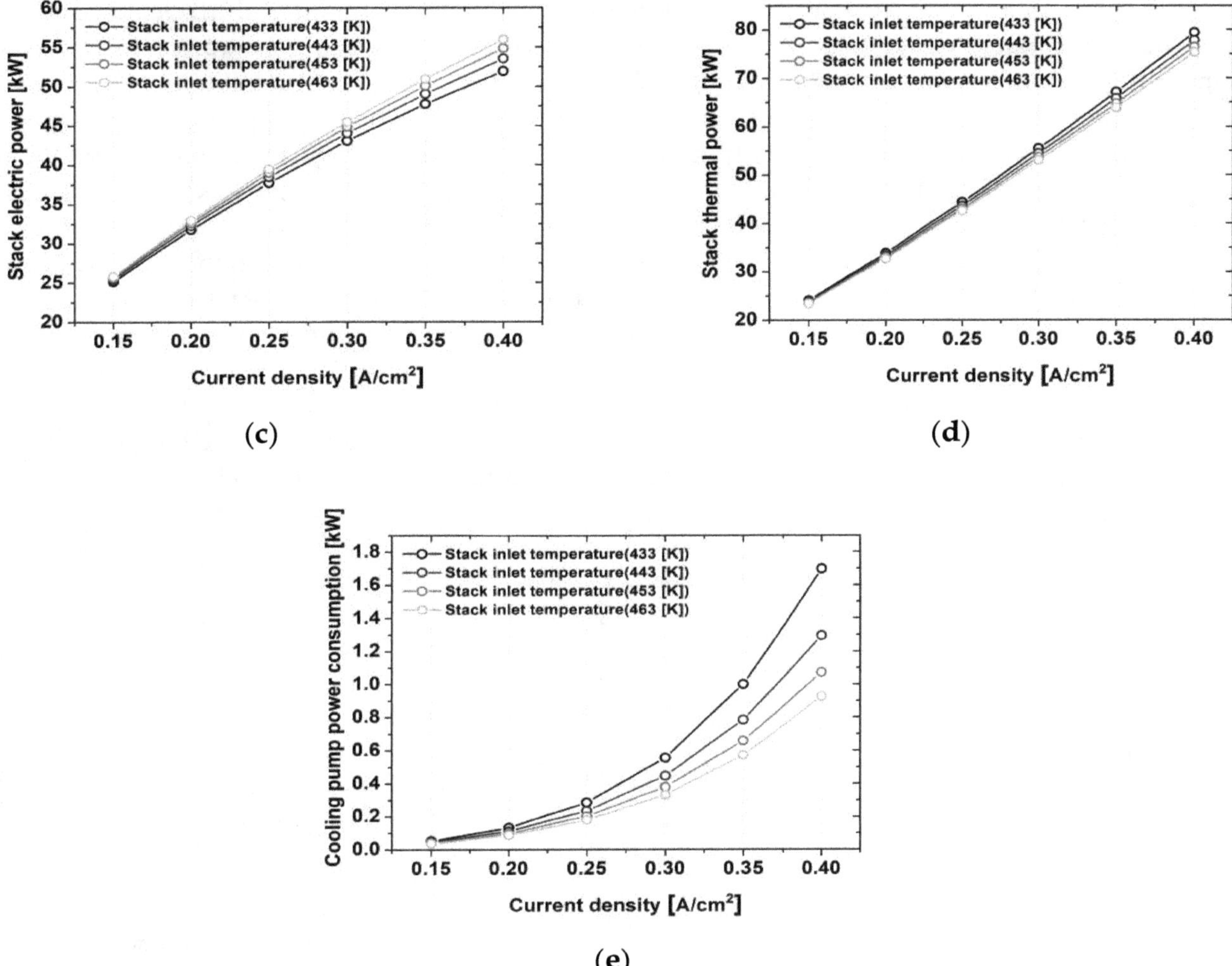

Figure 9. Validation trends of system performance with stack inlet temperature. (**a**) Combined system and stack power efficiency; (**b**) ORC net power; (**c**) Stack electric power; (**d**) Stack thermal power; (**e**) Cooling pump power consumption.

The efficiency of the combined system and the stack power efficiency are shown in Figure 9a, and the highest efficiency was shown as 56.03% and 52.45% at a current density of 0.15 A/cm^2 and a stack inlet temperature of 463 K. Additionally, the percentage increase in the combined system power efficiency compared to the stack power efficiency increased by up to 3.81% at a current density of 0.25 A/cm^2 and a stack inlet temperature of 433 K. In the case of the ORC net power, as the current density increased, the thermal power of the stack and the heat exchange amount of the evaporator increased, resulting in an increase in power generation. However, when the inlet temperature of the stack increased, the power generation decreased, and up to 0.3 kW decreased at a current density of 0.4 A/cm^2. In terms of the stack electric power, it reached a maximum of 55.96 kW at a current density of 0.4 A/cm^2 and 79.36 kW and a stack inlet temperature of 463 K as shown in Figure 9c, whereas in terms of the stack thermal power, it reached a maximum of 79.36 kW at a current density of 0.4 A/cm^2 and a stack inlet temperature of 433 K as shown in Figure 9d. As the current density increased, the power consumption of the cooling pump increased the required mass flow rate on the stack and the TEG side of the evaporator as shown in Figure 8, resulting in the increase in the corresponding pressure drop. As shown in Figure 9e, the power consumption of the cooling pump required a maximum of 1.69 kW at a current density of 0.4 A/cm^2 and a stack inlet temperature of 433 K.

Figure 10a shows the rate of change in the stack power, waste heat generation, and ORC power generation performance according to the difference of the stack coolant inlet temperature for each stack current density. As the current density is relatively higher, the rate of change in the stack power and waste heat generation amount according to difference of the stack inlet temperature clearly increases. Additionally, the rate of change in power generation of stack considering power consumption of cooling pump is increased by up to 20% at current density of 0.4 A·cm^{-2} compared to rate of change in power considering only the stack model. On the other hand, the rate of change in the stack power, heat generation, and ORC power generation performance according to the current density for each inlet temperature showed a relatively low difference as shown in Figure 10b. Based on these system analysis results, the stack inlet temperature of the HT-PEMFC power generation system is judged as an important operating condition that affects the power generation performance change characteristics. In addition, while the effect of the difference of the current density for each the stack inlet temperature is relatively constant, the effect of the difference of the stack inlet temperature is expected to increase as the current density increases. Additionally, when cooling actuators such as a water pump and loss factors are added, the effect of stack operating temperature is expected to increase. In the case of the ORC system, the rate of change in power generation performance according to the temperature and current density was relatively low in the operating temperature range of this stack model.

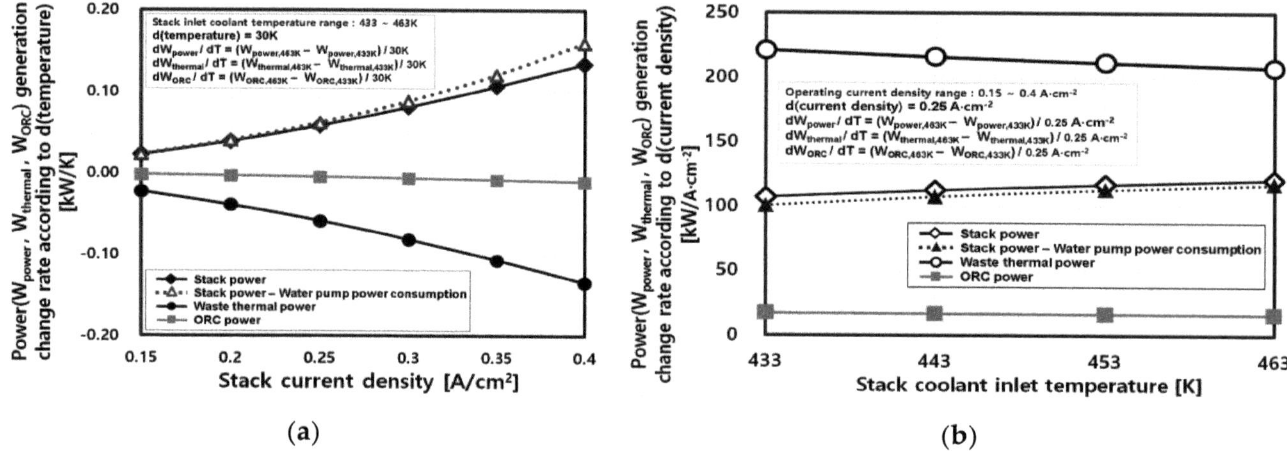

Figure 10. (**a**) The rate of change of stack power, waste heat generation, and ORC power generation according to operating temperature for each stack current density; (**b**) The rate of change in stack power, heat generation, and ORC power generation performance according to the current density for each stack coolant inlet temperature.

As shown in Figure 11, the system efficiency was compared excluding the stack power efficiency as a result of the coolant inlet temperature of the stack and condensing temperature of the working fluid formed at the condenser outlet for the ORC system. This is shown by excluding only the stack power generation from the overall efficiency of the combined power generation system. Through this, the efficiency changes of the ORC power generation system by pumps and heat exchangers excluding the stack were compared, and within the current density range, overall system efficiency except the stack tended to increase as the working fluid condensing temperature decreased. When the working fluid condensing temperature was 20 °C, the maximum efficiency was about 4.75%, which was a 25% increase compared to the case where the working fluid condensing temperature was 35 °C. Additionally, as the operating temperature of the stack increased, the deviation of the system efficiency except the stack tended to decrease relatively according to the change in current density. When the current density was 0.4 A/cm^2, the change in the system efficiency except the stack appeared to be the biggest according to the change of the stack operation temperature. This is believed to indicate that the influence of the operating temperature gradually increases under the power generation condition with the stack high load.

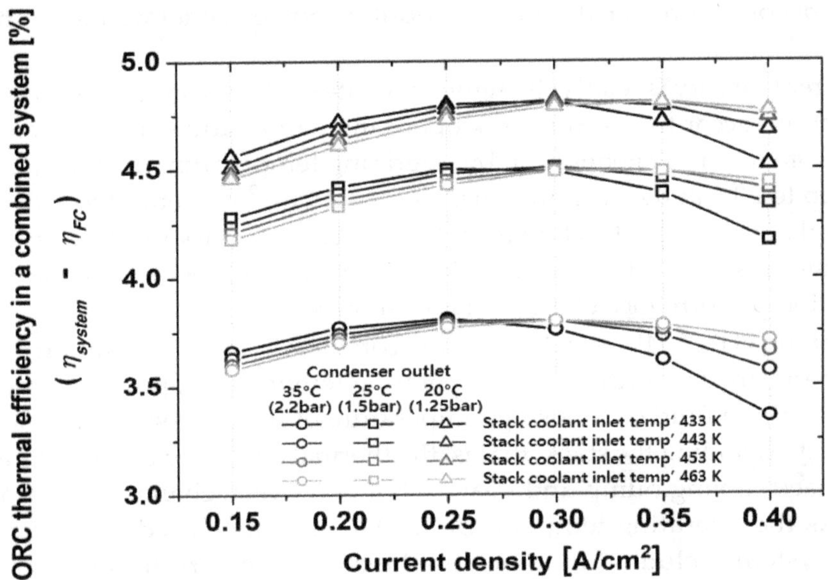

Figure 11. Difference between system and stack power efficiency with stack inlet temperature and condensing temperature.

5. Conclusions

In the case of HT-PEMFC, thermal management is formed as important as a relatively high operating temperature, and in the case of a heat exchange system that utilizes waste heat, since the operating range and strategies taking account of the thermal management of the stack must be selected, it is necessary to predict the power generation performance and efficiency according to the operating conditions. In this study, a model for a combined power generation system composed of a HT-PEMFC stack and an ORC power generation system was established, and the power generation performance and system efficiency were analytically compared according to the stack and ORC operating conditions. Each system was made of a model using the existing research contents and in the case of the plate heat exchanger for the Evaporator of the ORC system, which is the most important element for stack thermal management and waste heat recovery, reliability was secured by comparing the experimental results and the analysis results of the model. Through the analysis using the final combined power generation system model, the system power generation performance and efficiency were compared and predicted according to the operating temperature of the stack, the power generation load, and the ORC system working fluid condensing temperature, and the results are summarized as follows.

(1) For the analytical comparative study, modeling of each of the HT-PEMFC stack and ORC combined power generation system was conducted. In particular, in order to secure the reliability of the plate heat exchanger for the ORC power generation system, previous experimental results under the same operating conditions and the model-based analysis results established in this study were compared. The reliability of the combined power generation system model was secured through this process.

(2) Using the established combined power generation system model, the power generation performance and system efficiency of each stack and ORC system according to the power generation load and operating temperature of the HT-PEMFC stack were compared analytically. It is believed that the model has a higher degree utilization in a HT-PEMFC stack in which the higher the operating temperature within the allowable range, the higher the power generation and efficiency is. It also has higher degree utilization at a stack operating temperature where the waste heat that is proportional to the power generation load is relatively high. Furthermore, it is determined that the stack capacity and rated power generation section (current density range during power generation) need to be selected considering the target subjects as the amount

of waste heat becomes greater than the amount of power generated at points above a certain current density.

(3) And as the current density is relatively higher, the rate of change in the stack power and waste heat generation amount according to the stack operating temperature clearly increases. Additionally, the rate of change in power generation by operating temperature of stack with cooling pump is increased by up to 20% at the current density of 0.4 $A{\cdot}cm^{-2}$ compared to rate of change in power considering only stack model. Therefore, the operating temperature of the HT-PEMFC stack generation system is able to be considered as an important operating condition that affects the power generation performance change characteristics.

(4) In the model of the HT-PEMFC stack and ORC combined power generation system, comparative analysis was performed according to the operating temperature, power generation load (current density), and working fluid condensing temperature of the ORC system in order to compare the system efficiency excluding the stack, that is, the thermal efficiency of the ORC and subsystem that includes the stack cooling pump and heat exchanger, which change according to the operating conditions. As the operating temperature of the stack increased, the efficiency deviation of ORC and subsystem excluding the stack by the change in current density tended to decrease. Considering the energy load consumed by the thermal management part, it was shown that, under a certain current density, the lower the stack operation temperature was, and the more the efficiency of the ORC and subsystem except the stack improved. Moreover, as the working fluid condensing temperature decreased, the efficiency of the combined power generation system except for the stack tended to increase as well.

The HT-PEMFC stack and ORC combined power generation system require an appropriate operation strategy according to the target subjects and operating environment. To this end, this study constructed a combined power generation system model that considered the thermal management of the stack and the heat exchange process of waste heat and verified the operation range according to operating. It is believed that the results of this study will contribute to the selection of stacks and establishment of the strategies according to the target subjects and operation environments of the HT-PEMFC stack and ORC combined power generation system. In the future, an improvement on the model will be made regarding the target subjects of specific combined power generation, and analytical and experimental comparative studies will be conducted.

Author Contributions: H.S.K. designed the research, H.S.K., M.-H.K. and Y.H.S. discussed the results and contributed to writing the paper. All authors have read and agreed to the published version of the manuscript.

References

1. Arpino, F.; Massarotti, N.; Mauro, A.; Vanoli, L. Metrological analysis of the measurement system for a micro-cogenerative SOFC module. *Int. J. Hydrogen Energy* **2011**, *36*, 10228–10234. [CrossRef]
2. Arpino, F.; Dell'Isola, M.; Maugeri, D.; Massarotti, N.; Mauro, A. A new model for the analysis of operating conditions of micro-cogenerative SOFC unit. *Int. J. Hydrogen Energy* **2013**, *38*, 336–344. [CrossRef]
3. Duhn, J.D.; Jensen, A.D.; Wedel, S.; Wix, C. Optimization of a new flow design for solid oxide cells using computational fluid dynamics modelling. *J. Power Sources* **2016**, *336*, 261–271. [CrossRef]
4. Supra, J.; Janßen, H.; Lehnert, W.; Stolten, D. Temperature distribution in a liquid-cooled HT-PEFC stack. *Int. J. Hydrogen Energy* **2013**, *38*, 1943–1951. [CrossRef]
5. Lüke, L.; Janßen, H.; Kvesić, M.; Lehnert, W.; Stolten, D. Performance analysis of HT-PEFC stacks. *Int. J. Hydrogen Energy* **2012**, *37*, 9171–9181. [CrossRef]
6. Kandidayeni, M.; Macias, A.; Boulon, L.; Trovão, J.P.F. Online Modeling of a Fuel Cell System for an Energy Management Strategy Design. *Energies* **2020**, *13*, 3713. [CrossRef]
7. Dudek, M.; Raźniak, A.; Rosół, M.; Siwek, T.; Dudek, P. Design, Development, and Performance of a 10 kW Polymer Exchange Membrane Fuel Cell Stack as Part of a Hybrid Power Source Designed to Supply a Motor Glider. *Energies* **2020**, *13*, 4393. [CrossRef]

8. Jannelli, E.; Minutillo, M.; Perna, A. Analyzing microcogeneration systems based on LT-PEMFC and HT-PEMFC by energy balances. *Appl. Energy* **2013**, *108*, 82–91. [CrossRef]
9. Chen, X.; Li, W.; Gong, G.; Wan, Z.; Tu, Z. Parametric analysis and optimization of PEMFC system for maximum power and efficiency using MOEA/D. *Appl. Therm. Eng.* **2017**, *121*, 400–409. [CrossRef]
10. Bargal, M.H.; Abdelkareem, M.A.; Tao, Q.; Li, J.; Shi, J.; Wang, Y. Liquid cooling techniques in proton exchange membrane fuel cell stacks: A detailed survey. *Alex. Eng. J.* **2020**, *59*, 635–655. [CrossRef]
11. Najafi, B.; Mamaghani, A.H.; Rinaldi, F.; Casalegno, A. Long-term performance analysis of an HT-PEM fuel cell based micro-CHP system: Operational strategies. *Appl. Energy* **2015**, *147*, 582–592. [CrossRef]
12. Cardona, E.; Piacentino, A. A methodology for sizing a trigeneration plant in Mediterranean areas. *Appl. Therm. Eng.* **2003**, *23*, 1665–1680. [CrossRef]
13. Ziher, D.; Poredos, A. Economics of a trigeneration system in a hospital. *Appl. Therm. Eng.* **2006**, *26*, 680–687. [CrossRef]
14. He, T.; Shi, R.; Peng, J.; Zhuge, W.; Zhang, Y. Waste heat recovery of a PEMFC system by using organic rankine cycle. *Energies* **2016**, *9*, 267. [CrossRef]
15. Dickes, R.; Dumont, O.; Lemort, V. Experimental assessment of the fluid charge distribution in an organic Rankine cycle (ORC) power system. *Appl. Therm. Eng.* **2020**, *179*, 115689. [CrossRef]
16. Jang, Y.; Lee, J. Comprehensive assessment of the impact of operating parameters on sub 1-kW compact ORC performance. *Energy Convers. Manag.* **2019**, *182*, 369–382. [CrossRef]
17. Jeong, H.; Oh, J.; Lee, H. Experimental investigation of performance of plate heat exchanger as organic Rankine cycle evaporator. *Int. J. Heat Mass Transf.* **2020**, *159*, 120158. [CrossRef]
18. Yang, S.C.; Hung, T.C.; Feng, Y.Q.; Wu, C.J.; Wong, K.W.; Huang, K.C. Experimental investigation on a 3 kW organic Rankine cycle for low-grade waste heat under different operation parameters. *Appl. Therm. Eng.* **2017**, *113*, 756–764. [CrossRef]
19. Rosli, R.E.; Sulong, A.B.; Daud, W.R.W.; Zulkifley, M.A.; Husaini, T.; Rosli, M.I.; Majlan, E.H.; Haque, M.A. A review of high-temperature proton exchange membrane fuel cell (HT-PEMFC) system. *Int. J. Hydrogen Energy* **2017**, *42*, 9293–9314. [CrossRef]
20. Korsgaard, A.R.; Refshauge, R.; Nielsen, M.P.; Bang, M.; Kær, S.K. Experimental characterization and modeling of commercial polybenzimidazole-based MEA performance. *J. Power Sources* **2006**, *162*, 239–245. [CrossRef]
21. Korsgaard, A.R.; Nielsen, M.P.; Kær, S.K. Part one: A novel model of HTPEM-based micro-combined heat and power fuel cell system. *Int. J. Hydrogen Energy* **2008**, *33*, 1909–1920. [CrossRef]
22. Kang, H.S.; Shin, Y.H. Analytical Study of Tri-Generation System Integrated with Thermal Management Using HT-PEMFC Stack. *Energies* **2019**, *12*, 3145. [CrossRef]
23. Pukrushpan, J.T. Modeling and control of fuel cell systems and fuel processors. Ph.D. Thesis, University of Michigan, Ann Arbor, MI, USA, 2003.
24. Chang, H.; Wan, Z.; Zheng, Y.; Chen, X.; Shu, S.; Tu, Z.; Chan, S.H.; Chen, R.; Wang, X. Energy- and exergy-based working fluid selection and performance analysis of a high-temperature PEMFC-based micro combined cooling heating and power system. *Appl. Energy* **2017**, *204*, 446–458. [CrossRef]
25. Lee, H.S.; Lee, M.Y.; Cho, C.W. Analytic study on thermal management operating conditions of balance of 100 kW fuel cell power plant for a fuel cell electric vehicle. *J. Korea Acad. Ind. Coop. Soc.* **2019**, *16*, 1–6.
26. Muley, A.; Manglik, R.M. Experimental study of turbulent flow heat transfer and pressure drop in a plate heat exchanger with chevron plates. *J. Heat Transf.* **1999**, *121*, 110–117. [CrossRef]
27. Yan, Y.Y.; Lin, T.F. Evaporation Heat Transfer and Pressure Drop of Refrigerant R-134a in a Plate Heat Exchanger. *J. Heat Transf.* **1999**, *121*, 118–127. [CrossRef]

7

Comprehensive Electric Arc Furnace Electric Energy Consumption Modeling

Miha Kovačič [1,2], Klemen Stopar [1], Robert Vertnik [1,2] and Božidar Šarler [2,3,*]

1 Štore Steel Ltd., Železarska cesta 3, SI-3220 Štore, Slovenia; miha.kovacic@store-steel.si (M.K.); klemen.stopar@store-steel.si (K.S.); robert.vertnik@store-steel.si (R.V.)
2 Faculty of Mechanical Engineering, University in Ljubljana, Aškerčeva 6, SI-1000 Ljubljana, Slovenia
3 Institute of Metals and Technology, Lepi pot 11, SI-1000 Ljubljana, Slovenia
* Correspondence: bozidar.sarler@fs.uni-lj.si

Abstract: The electric arc furnace operation at the Štore Steel company, one of the largest flat spring steel producers in Europe, consists of charging, melting, refining the chemical composition, adjusting the temperature, and tapping. Knowledge of the consumed energy within the individual electric arc operation steps is essential. The electric energy consumption during melting and refining was analyzed including the maintenance and technological delays. In modeling the electric energy consumption, 25 parameters were considered during melting (e.g., coke, dolomite, quantity), refining and tapping (e.g., injected oxygen, carbon, and limestone quantity) that were selected from 3248 consecutively produced batches in 2018. Two approaches were employed for the data analysis: linear regression and genetic programming model. The linear regression model was used in the first randomly generated generations of each of the 100 independent developed civilizations. More accurate models were subsequently obtained during the simulated evolution. The average relative deviation of the linear regression and the genetic programming model predictions from the experimental data were 3.60% and 3.31%, respectively. Both models were subsequently validated by using data from 278 batches produced in 2019, where the maintenance and the technological delays were below 20 minutes per batch. It was possible, based on the linear regression and the genetically developed model, to calculate that the average electric energy consumption could be reduced by up to 1.04% and 1.16%, respectively, in the case of maintenance and other technological delays.

Keywords: steelmaking; electric arc furnace; consumption; electric energy; melting; refining; tapping; modeling; linear regression; genetic programming

---

1. Introduction

The electric arc furnace (EAF) is a central element and the highest energy consumer in the recycled steel processing industry. The EAF contains electric energy, with a moderate addition of chemical energy, that is used for generating the required heat for the melting of recyclable scrap. The heat energy is primarily generated by the burning arc between the electrodes and the scrap, or its melt. The EAF consists of a shell (walls with water cooled panels and lower vessel), a heart (refractory material that covers lower vessel), and a roof with the electrodes. A scheme of the EAF is presented in Figure 1 [1–3].

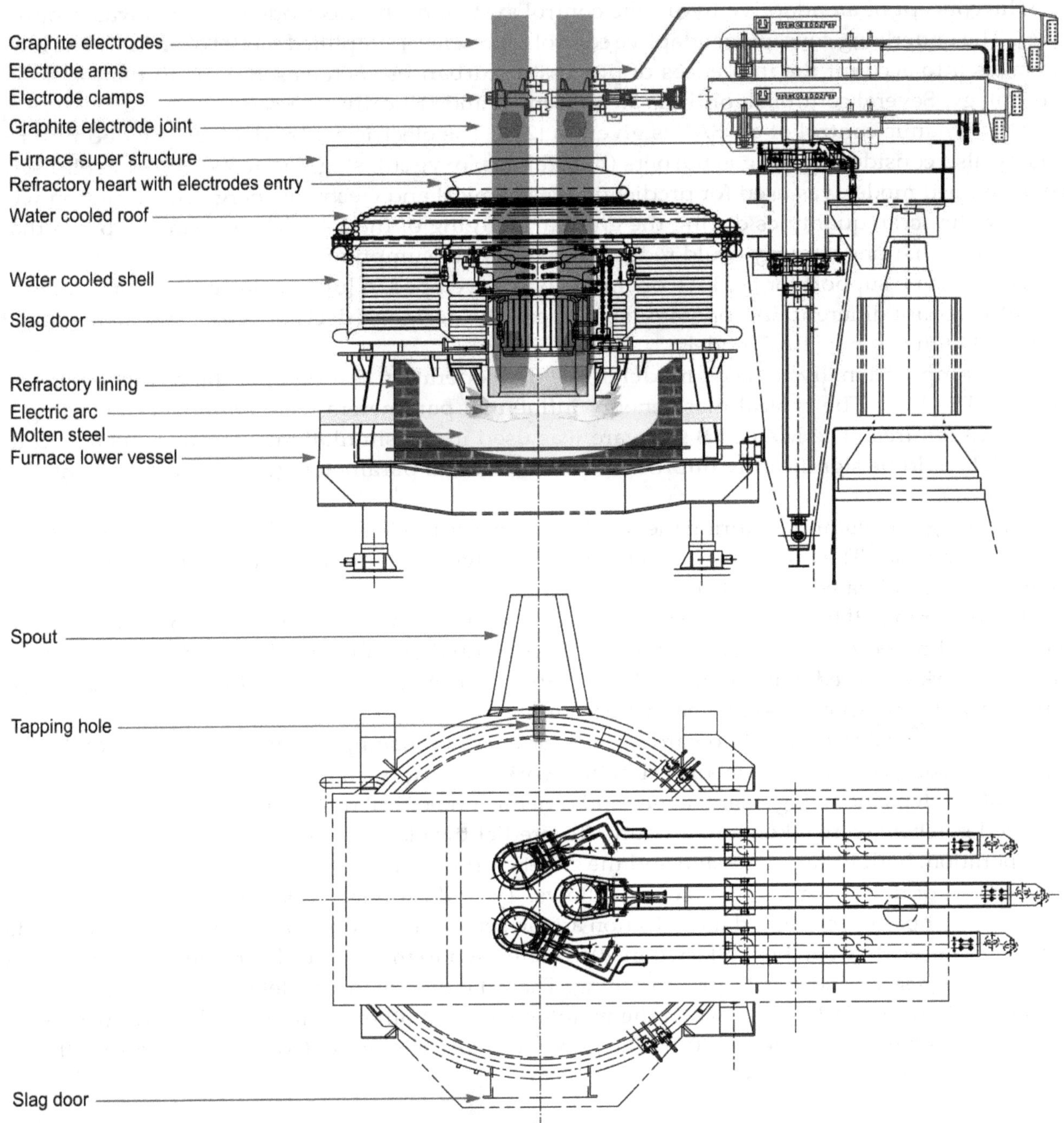

Figure 1. Scheme of the electric arc furnace.

The main EAF operation steps are as follows [1–3]: charging and melting, refining (oxidizing of the melt), chemical composition and temperature adjusting, and tapping (discharging of the furnace).

With respect to energy consumption, the contemporary research has mainly focused on the total (electric and chemical) consumed energy [2,4–6] and individual (electric or chemical) consumed energy [7,8] including other aspects of EAF operation such as transformer optimization [1,9–11], molten steel residue [12,13], scrap type [14,15], scrap management [14,16,17], electrode regulation [18–20], oxygen injectors [13,21–23], and slag cover [24].

The influences of maintenance on the power, steel, and cement industries were analyzed in [25]. The authors found that maintenance and rehabilitation were the key factors only when producing steel using the blast furnace. However, the influence of maintenance on producing steel from scrap through an EAF was not deduced due to insufficient data.

The concept of an adaptive hydraulic control system of the electrode positions was proposed in [26]. The underlying concept for adaptive control represents a simplified model of an EAF. The model also takes into account the influences of process disturbances such as scrap manipulation and its morphology. Several control algorithms are presented and critically assessed.

The dynamic control of an EAF is given in [27]. The electric arc model was divided into four parts by also considering the gas burners (natural gas, oxygen), slag, molten steel, and solid scrap. The developed model was used for predicting the chemical and electrical energy consumption while changing the scrap quantities during the gradual charging of the EAF. The research showed that a proper scrap charging strategy could reduce the energy consumption.

The decision support for the EAF operation was developed in [6] by using open source tools and took into account different EAF operator strategies. The designed decision support system could be integrated with complex EAF models.

The computationally reduced model of the EAF operation during only the refining stage was elaborated in [28]. The typical mass-energy influential parameters were employed including the equipment failures. The MATLAB software was used in the simulations. The authors stated that the model could be significantly improved with additional parameters (e.g., carbon concentration, temperature).

The energy consumption during the refining stage was modeled in [21] by using a comprehensive parameter analysis. The scrap melting evolution (i.e., quantities, timing) was also taken into account. The model was validated in practice on a 40 t EAF.

The paper in [29] focused on modeling the tapping temperature. The energy consumption could be optimized based on the consideration of the influential parameters. For modeling, an artificial neural network was used that combined the final fuzzy interference function. In addition, the operator strategies and experiences were taken into account.

A comprehensive approach toward the electric energy consumption of the EAF, used at the Štore Steel steelmaking company, is elaborated in this work. The entire set of influential parameters during all operation steps including maintenance and other technological delays in 2018 (3248 consecutively produced batches) were taken into account. To predict the electric energy consumption during the EAF operation, both linear regression and the genetic programming were used.

The rest of this paper is organized as follows. Typical processes related to the EAF used at the Štore Steel company including data collection are presented first. Afterward, the related process data from 3248 consecutive batches collected in 2018 were used to model the electric energy consumption with linear regression and genetic programming. The validation of the modeling results was conducted by using data from 278 batches (when the maintenance and other technological delays were below 20 minutes per batch), collected in 2019. The importance of the represented developments for the steel industry is given in the conclusions.

2. Materials and Methods

The Štore Steel company is one of the major flat spring steel producers in Europe. The company produces more than 1000 steel grades with different chemical compositions. The scrap is melted, ladle treated, and continuously cast in billets. The cooled-down billets are reheated and rolled in the continuous rolling plant. The rolled bars can be additionally straightened, examined, cut, sawn, chamfered, drilled, and peeled in the cold finishing plant. The Štore Steel company is known for its application of advanced artificial intelligence modeling tools [30] for better understanding and optimization of the processes.

The production process at the Štore Steel company starts with a 60 t EAF. The scrap is delivered in baskets by train from a scrapyard, located 300 m from the steel plant. The following types of scrap steel are used for melting: E1 (old thick steel scrap); E2 (old thin steel scrap); E3 (thick new production steel scrap); E8 (thin new production steel scrap); E40 (shredded steel scrap); scrapped non-alloyed steel; low-alloyed steel (moderate content of Cr); and pig iron.

The electric arc furnace is typically charged with three baskets. The first, second, and third baskets have the capacity of 22–30 t, 15–20 t, and 6–15 t, respectively. Each individual charging lasts approximately three minutes. The melting of the scrap after charging with the first, second, and third baskets lasts approximately 20 min, 15 min, and 10 min, respectively.

The following activities are conducted before charging with the first basket: examination, cleaning and reparation of the slag door and tapping spout with its refractory material; examination of the EAF refractory linings and reparation of the linings with the dolomite or magnesite; examination of the water-cooled panels; examination of the mast arm (which holds the electrodes); and the changing and settings of the electrodes.

For the slag formation, coke, lime, and dolomite are also used, which are deposited before melting the first basket. The slag insulation and protective ability expands the lifespan of the refractory material, preventing the EAF roof from exposure, and shielding the cooling panels from the intensive heat radiation.

Melting is conducted after swinging back the furnace roof. After lowering the electrodes, the burning arc between the graphite electrodes and the scrap or the molten steel is established. After the last basket has been melted, the EAF roof is swung off, and the remaining non-melted scrap is pushed into the melt bath.

In order to speed up the melting process, oxygen and natural gas from wall-mounted combined burners (natural gas) and injectors (oxygen, coke) are also used, in addition to the electric arc, to generate the complementary chemical heat. After melting the last basket during the refining process, the oxygen jets from the lances penetrate the slag and react with the liquid bath. In particular, the oxidation with the carbon, phosphorous, and sulfur is important. The oxidized products are trapped by the slag, which is removed through the slag doors by tilting the EAF backward. Afterward, the chemical composition analysis is conducted. After the chemical composition changes, the tapping (i.e., tilting the EAF forward) is conducted. The molten steel is charged into the ladle and consequently, the ladle treatment is conducted (e.g., slag formation, chemical composition adjustments, melt stirring). Typical delays during the refining process are connected with the chemical and temperature analysis, oxygen blowing, changing of the steel grade (especially Ca-treated steels for its improved machinability), and waiting for the lower electricity tariff.

In the present research, 26 process parameters including the electric energy consumption were considered. The data were taken from 3248 consecutively produced batches at the Štore Steel company during 2018. The dataset was composed of:

- Melting:
  - the considered process parameters were:
    - coke (kg): used for protective slag formation,
    - lime (kg): used for protective slag formation,
    - dolomite [kg]: used for protective slag formation,
    - E-type scrap (kg),
    - low-alloyed steel (moderate content of Cr) (kg),
    - packets of scrap (kg),
    - oxygen consumption (Nm^3) used for cutting the scrap and its combustion and forming the slag (important component of slag is FeO), and
    - natural gas consumption (Nm^3) used for heating the scrap.
  - The considered maintenance and other technological delays are:
    - lime addition (min): the additional time needed for lime addition,

    - scrap charging (min): the additional time needed for charging of the electric arc furnace with scrap,
    - reparation of the linings with the dolomite or magnesite (min): the additional time needed for reparation of the refractory linings of the heart of the electric arc furnace,
    - electrode settings (min): the additional time needed for electrode settings and replacing,
    - other technological delays (min): the additional delays due to, for example, the maintenance of a dust collector, water cooling system, or overhead cranes,

- Refining and tapping:
  - the considered process parameters are:
    - oxygen consumption (Nm^3),which is used for uniform melt temperature distribution for removing the unwanted chemical elements such as sulfur or phosphorus,
    - limestone (kg), which is used for slag creation,
    - carbon content obtained by the first chemical composition analysis (%),
    - nominal final carbon content (%) where the melt can be used for producing several different grades of steel in further processing steps; the possibilities are determined from the first chemical composition analysis, and
    - carbon powder (kg), which is used for carbonizing and additional slag formation,
  - the considered maintenance and other technological delays:
    - chemical analysis delay (min): there can be problems with the sampling or the chemical analysis has to be repeated,
    - temperature and oxygen analysis delay (min): there can be problems with the sampling or the automatic lance used for the analysis,
    - extended refining (min): due to the chemical analysis and the temperature adjustments, the refining process needs to be extended in order to achieve a proper chemical composition and a proper temperature before tapping,
    - delay due to Ca-treated steel production (min): to produce Ca-treated steel, proper oxygen content is needed before tapping; in addition, the spout wear and geometry are important,
    - delay due to waiting for a lower electricity tariff (min): during the higher electricity tariff period (from 6:00 to 8:00 a.m.), the production in the steel plant stops,
    - delay due to steel grade changing (min): based on the first chemical analysis, the steel grade can be changed according to the foreseen planned production,
    - delay during tapping (min): delays can occur due to spout maintenance or spout blocking, ladle treatment and casting coordination and management, and, last but not least,
- Electric energy consumption (MWh).

The average values and the standard deviation of the individual parameters are presented in Table 1.

Table 1. The average values and the standard deviation of the individual parameters from 3248 consecutively produced batches at the Štore Steel company in 2018.

Parameter	Abbreviation	Average	Standard Deviation
Coke (kg)	COKE	814.27	89.35
Lime (kg)	CAO	998.16	90.20
Dolomite (kg)	CAOMGO	703.74	123.23
E-type scrap (kg)	E_SCRAP	42.54	5.32
Low-alloyed steel (moderate content of cr) (kg)	SCRAP_BLUE	6.19	5.17
Packets of scrap (kg)	SCRAP_PACK	7.03	3.99
Oxygen consumption (Nm^3)	OXYGEN_MELTING	1220.50	117.67
Natural gas consumption (Nm^3)	GAS	442.01	61.36
Lime addition (min)	CACO3_T	0.13	0.82
Scrap charging (min)	SCRAP_MANIPULATION_T	0.93	1.75
Reparation of the linings with the dolomite or magnesite (min)	REPARATION_MAINT	1.23	7.03
Electrode settings (min)	ELECTRODE_MANIPULATION_T	1.99	6.58
Other technological delays (min)	OTHER_T	5.48	42.44
Oxygen consumption (Mm^3)	OXYGEN_REFINING	459.00	115.81
Limestone (kg)	CACO3	72.75	185.92
Carbon content obtained by the first chemical composition analysis (%)	C_1	0.23	0.14
Required, final carbon content (%)	C_REQUIRED	0.41	0.16
Carbon powder (kg)	C	175.11	103.09
Chemical analysis delay (min)	CHEMICAL_ANALYSIS_T	4.02	3.48
Temperature and oxygen analysis delay (min)	OXYGEN_TIME_ANALYSIS_T	1.00	3.42
Extended refining (min)	REFINING_T	1.28	2.75
Delay due to Ca-treated steel production (min)	CA_TREATMENT_T	1.84	9.04
Delay due to waiting for lower electricity tariff (min)	PEAK_TARIFFE_T	5.20	27.76
Delay due to steel grade changing (min)	GRADE_CHANGING_T	2.87	9.13
Delay during tapping (min)	TAPPING_T	0.97	3.95

3. EAF Electric Energy Consumption Modeling

Based on the collected data (Table 1), the prediction of the EAF electric energy consumption was conducted by using linear regression and genetic programming. For the fitness function, the average relative deviation between the predicted and the experimental data was selected. The fitness function is defined as:

$$\Delta = \frac{\sum_{i=1}^{n} \frac{|Q_i - Q'_i|}{Q_i}}{n}, \qquad (1)$$

where n is the size of the collected data and Q_i and Q'_i stand for the actual and the predicted electric energy consumption, respectively.

3.1. Linear Regression Modeling

The linear regression analysis results demonstrated that the model significantly predicted the electric energy consumption ($p < 0.05$, ANOVA) and that 63.60% of the total variances could be explained by independent variables variances (R-square). Out of the 25 independent parameters considered, only the following were not significantly influential ($p > 0.05$): lime, dolomite, scrap charging, chemical analysis delay, temperature and oxygen analysis delay, and delay during tapping.

The deduced linear regression model is:

$$\begin{aligned} COKE\cdot\ & 0.002 + E\_SCRAP\cdot 0.152 + SCRAP\_BLUE\cdot 0.198 + SCRAP\_PACK\cdot 0.195 \\ & + OXYGEN\_MELTING\cdot 0.003 + GAS\cdot 0.005 + CACO3\_T\cdot 0.075 \\ & + SCRAP\_MANIPULATION\_T\cdot 0.003 + REPARATION\_MAINT\cdot 0.015 \\ & + ELECTRODE\_MANIPULATION\_T\cdot 0.015 + OTHER\_T\cdot 0.004 \\ & + OXYGEN\_REFINING\cdot(-0.003) + CACO3\cdot 0.001 + C\_1\cdot 0.73 + C\_REQUIRED \\ & \cdot(-0.45) + C\cdot 0.007 + OXYGEN\_TIME\_ANALYSIS\_T\cdot 0.007 + REFINING\_T \\ & \cdot 0.041 + CA\_TREATMENT\_T\cdot 0.013 + PEAK\_TARIFFE\_T\cdot 0.011 \\ & + GRADE\_CHANGING\_T\cdot 0.012 + TAPPING\_T\cdot 0.005 + 8.2872. \end{aligned} \qquad (2)$$

The average and maximal relative deviation from the experimental data was 3.60% and 36.75%, respectively. The calculated influences of the individual parameters (individual variables) on the electric energy consumption are presented in Figure 2. It is possible to conclude that E-type scrap, low-alloyed steel (moderate content of Cr), packets of scrap, oxygen consumption during melting, natural gas consumption, limestone, other technological delays, and coke injection during refining were the most influential factors. Based on the linear regression model, it was possible to calculate that the average electric energy consumption could be reduced by up to 1.04% in the case of the maintenance and other technological delays that we wanted to avoid. On the other hand, the time savings represented 24.89% of the average tapping time. As above-mentioned, during the higher electricity tariff period from 6:00 to 8:00 a.m., the production in the steel plant stopped.

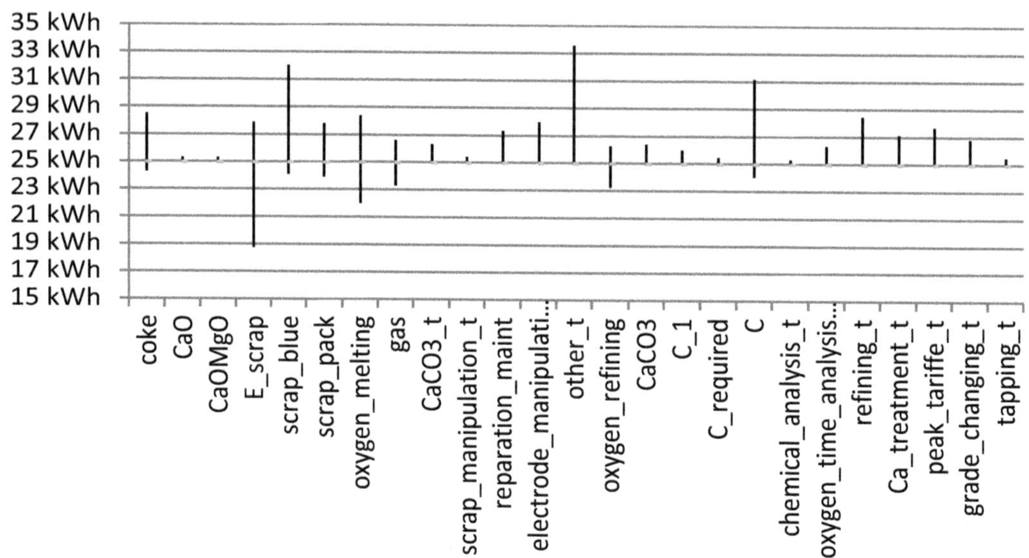

Figure 2. The calculated influences of the individual parameters on the electric energy consumption using the linear regression model.

3.2. Genetic Programing Modeling

Genetic programming is probably the most general evolutionary optimization method [31,32]. The organisms that undergo adaptation are in fact the mathematical expressions (models) for predicting the ratio between the material with the surface defects and the examined material. The models, i.e., the computer programs, consist of the selected function (i.e., basic arithmetical functions) and terminal genes (i.e., independent input parameters and random floating-point constants). Typical function genes are: addition (+), subtraction (−), multiplication (*), and division (/), and terminal genes (e.g., x, y, z). Random computer programs (Figure 3) for calculating various forms and lengths are generated by means of the selected genes at the beginning of the simulated evolution.

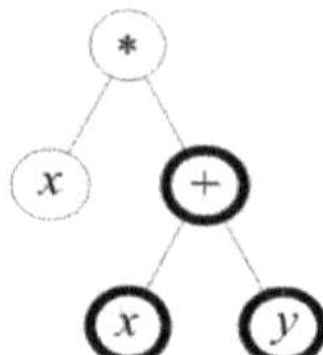

Figure 3. Random computer program as mathematical expression $x(x + y)$.

The varying of the computer programs is carried out by means of genetic operations (e.g., crossover, mutation) during several iterations, called generations. The crossover operation is presented in Figure 4. After the completion of the variation of the computer programs, a new generation is obtained. Each result, obtained from an individual program from a generation, is compared with the experimental data. The process of changing and evaluating the organisms is repeated until the termination criterion of the process is fulfilled.

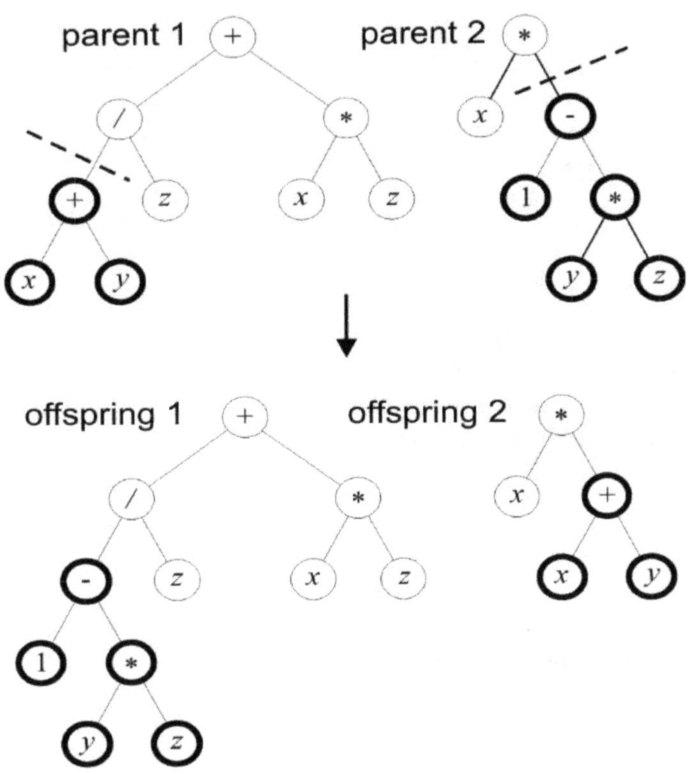

Figure 4. Crossover operation (out of two parental organisms, the offspring with randomly distributed genetic material are evolved).

An in-house genetic programming system, coded in the AutoLISP programming language, which is integrated into AutoCAD (i.e., commercial computer-aided design software), was used [33–35]. Its settings included the size of the population of organisms: 500; the maximum number of generations: 100; reproduction probability: 40%; crossover probability: 60%; maximum permissible depth in the creation of the population: 6; maximum permissible depth after the operation of crossover of two organisms: 10, and the smallest permissible depth of organisms in generating new organisms: 2.

The genetic operations of the reproduction and the crossover were used. To select the organisms, the tournament method with a tournament size 7 was used.

The in-house genetic programming system was run 100 times in order to develop 100 models for the prediction of electric energy consumption. Each run lasted approximately two and a half hours on an I7 Intel processor and 8 GB of RAM.

It must be emphasized that during the random generation of the computer programs (models for electric energy consumption), the already developed linear regression model (Equation (2)) was employed. The population size was 500. Out of these 500 organisms (computer programs), 50 were the same linear regression model, and the remaining 450 organisms were randomly generated at the beginning of the simulated evolution. Afterward, the population was changed with the genetic operations (e.g., crossover) without introducing any additionally developed linear regression models.

The best mathematical model obtained from 100 runs of genetic programming system was:

$$\begin{gathered}8.39083 + 0.001 \cdot CACO3 + 0.00133 \cdot CAOMGO + 0.013176 \cdot \\ CA\_TREATMENT\_T - 0.449208 \cdot C\_REQUIRED + 0.17427 \cdot E\_SCRAP + \\ 0.005 \cdot GAS + 0.0241847 \cdot GRADE\_CHANGING\_T + 0.003 \cdot \\ OXYGEN\_MELTING - 0.003858 \cdot OXYGEN\_REFINING + 0.011 \cdot \\ PEAK\_TARIFFE\_T + 0.056 \cdot REFINING\_T + 0.198 \cdot SCRAP\_BLUE + 0.195 \cdot \\ SCRAP\_PACK + 0.00297 \cdot C_1{}^3 \cdot E\_SCRAP \cdot SCRAP\_PACK + C\_1(0.738316 + \\ 0.000198 \cdot CAOMGO + 0.000792 \cdot GAS + 0.007 \cdot OTHER\_T - 0.000594 \cdot \\ OXYGEN\_REFININIG + 0.004 \cdot OTHER_T \cdot SCRAP\_PACK) + C(0.004954 + \\ 0.000792 \cdot C_1{}^2 \cdot C\_REQUIRED \cdot SCRAP\_PACK + C\_1(0.002376 + 0.000044 \cdot \\ C\_REQUIRED \cdot SCRAP\_PACK)).\end{gathered} \tag{3}$$

The average and the maximal relative deviation from the experimental data was 3.31% and 41.21%, respectively. The calculated influences of the individual parameters (individual variables) on the electric energy consumption are presented in Figure 5. It is possible to conclude that the dolomite, E-type scrap, low-alloyed steel (moderate content of Cr), other technological delays, and coke injection during refining were the most influential factors. Note that the coke, lime, limestone, scrap charging, reparation of the linings with the dolomite or magnesite, electrode settings, chemical analysis delay, oxygen and temperature analysis delay, and the delay during tapping were not considered in the model (Equation (3)). Additionally, based on the genetically developed model, it was possible to calculate that the average electric energy consumption could be reduced by up to 1.16% in the case of the maintenance and other technological delays.

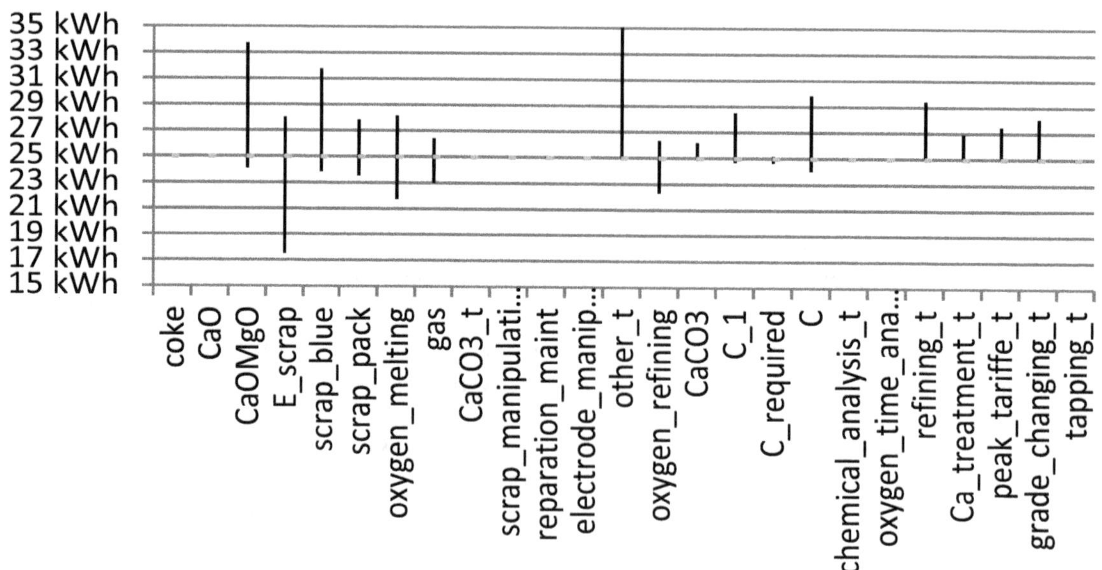

Figure 5. The calculated influences of the individual parameters on the electric energy consumption using a linear genetically developed model.

4. Validation of the Modeling Results

Additional data were gathered in January 2019 (278 batches) when the maintenance and technological delays were below 20 minutes. The average values and the standard deviation of the individual parameters are summarized in Table 2.

Table 2. The average values and the standard deviation of individual parameters from 3248 consecutively produced batches at the Štore Steel Ltd. in 2018.

Parameter	Average	Standard Deviation
Coke (kg)	800.58	68.56
Lime (kg)	989.06	105.04
Dolomite (kg)	696.31	59.42
E-type scrap (kg)	40.50	5.22
Low-alloyed steel (moderate content of cr) (kg)	7.41	4.95
Packets of scrap (kg)	7.67	2.96
Oxygen consumption (Nm^3)	1211.83	128.30
Natural gas consumption (Nm^3)	505.48	69.25
Lime addition (min)	0.13	1.01
Scrap charging (min)	0.21	0.42
Reparation of the linings with the dolomite or magnesite (min)	0.00	0.00
Electrode settings (min)	0.52	1.82
Other technological delays (min)	0.05	0.35
Oxygen consumption (Mm^3)	495.44	106.34
Limestone (kg)	96.40	210.69
Carbon content obtained by the first chemical composition analysis (%)	0.32	0.13
Required, final carbon content (%)	0.42	0.13
Coke (kg)	87.54	65.20
Chemical analysis delay (min)	4.22	2.62
Temperature and oxygen analysis delay (min)	0.51	1.44
Extended refining (min)	1.76	2.76
Delay due to Ca-treated steel production (min)	0.09	1.03
Delay due to waiting for lower electricity tariff (min)	0.01	0.08
Delay due to steel grade changing (min)	1.17	3.46
Delay during tapping (min)	1.25	1.65

The average relative deviation between the experimental data and the linear regression model was 3.65%, and that between the experimental data and the genetic programming model was 3.49%. This is in accordance with the average relative deviation from the data obtained in 2018. Consequently, we can conclude that the represented approach can be used as a precise EAF energy consumption tool that also considers the maintenance and technological delays.

5. Conclusions

The prediction of the electric energy consumption of the EAF operation at the Štore Steel company was presented. Twenty-five selected parameters from the individual production process steps in 2018 (3248 consecutively produced batches) were used for modeling. Two models were considered: the first was based on linear regression, and the second was based on the more accurate genetic programming. The average relative deviation of the models from the experimental data was 3.60% with the linear regression model, and 3.31% with the genetic programming model, respectively.

Based on the linear regression results, it was possible to conclude that 63.60% of the total variances could be explained by the variances of the independent variables. Based on the linear regression model, it was possible to calculate that the average electric energy consumption could be reduced by up to 1.04% in the case of maintenance and other technological delays, while on the other hand the time savings represented 24.89% of the average tapping time. Out of the 25 independent parameters, only lime, dolomite, scrap charging, chemical analysis delay, temperature and oxygen analysis delay, and delay during tapping were not significantly influential ($p > 0.05$).

An in-house genetic programming system, coded in AutoLISP, which is integrated into AutoCAD, was used to obtain 100 independent models for the prediction of the electric energy consumption during the EAF operation. A population size of 500 organisms was chosen. Out of these 500 organisms (computer programs), 50 were from the same developed linear regression model, and the remaining 450 organisms were randomly generated at the beginning of the simulated evolution. Afterward, the population was changed with the genetic operations (e.g., crossover) without introducing additionally developed linear regression models. Only the best ones were used for analysis. The most influential parameters (based on calculation) were dolomite, E-type scrap, low- alloyed steel (moderate content of Cr), other technological delays, and coke injection during refining. It must be emphasized that coke, lime, limestone, scrap charging, reparation of the linings with the dolomite or magnesite, electrode settings, chemical analysis delay, oxygen and temperature analysis delay, and delay during tapping were not considered in the genetically developed model.

Both models were also validated by using the data from 278 batches produced in 2019, when the maintenance and the technological delays were below 20 minutes per batch. The average relative deviation of the linear regression and genetic programming model prediction from the experimental data were 3.56% and 3.49%, respectively. This was in accordance with the average relative deviations from the data obtained in 2018.

The following points represent the highlights of our work:

- For modeling the EAF electric energy consumption, 25 parameters were used.
- Parameters involved melting (e.g., coke, dolomite, quantity), refining and tapping (e.g., injected oxygen, carbon, and limestone quantity), maintenance, and technological delays.
- The data from 3248 consecutively produced batches in 2018 were used.
- For modeling, linear regression and genetic programming were used.
- Both developed models were validated by using the data from 278 batches produced in 2019.
- Both models showed that the electric energy consumption could be reduced by up to 1.16% with the reduction of the maintenance and other technological delays.

In the future, a detailed analysis of charging and melting operation steps will be conducted including the time-dependent electric energy, natural gas, oxygen, and coke consumption. The represented approach is, with only slight modifications, practically applicable in a spectra of different EAFs as well as in other steelmaking process steps.

Author Contributions: M.K.: conceptualization, methodology, investigation, data analysis, software, writing, visualization; K.S.: conceptualization, investigation, data analysis, writing, editing, visualization; R.V.: software, data mining, data analysis, review and editing; B.Š.: project management, data analysis, review and editing.

References

1. Stopar, K.; Kovačič, M.; Kitak, P.; Pihler, J. Electric arc modeling of the EAF using differential evolution algorithm. *Mater. Manuf. Process.* **2017**, *32*, 1189–1200. [CrossRef]
2. Toulouevski, Y.N.; Zinurov, I.Y. Modern Steelmaking in Electric Arc Furnaces: History and Development. In *Innovation in Electric Arc Furnaces*; Springer: Berlin/Heidelberg, Germany, 2013; pp. 1–24.
3. Toulouevski, Y.N.; Zinurov, I.Y. EAF in Global Steel Production. In *Energy and Productivity Problems*; Springer: Berlin, Germany, 2017; pp. 1–6.
4. Tunc, M.; Camdali, U.; Arasil, G. Energy Analysis of the Operation of an Electric-Arc Furnace at a Steel Company in Turkey. *Metallurgist* **2015**, *59*, 489–497. [CrossRef]
5. Damiani, L.; Revetria, R.; Giribone, P.; Schenone, M. Energy Requirements Estimation Models for Iron and Steel Industry Applied to Electric Steelworks. In *Transactions on Engineering Technologies*; Springer: Singapore, 2019; pp. 13–29.
6. Shyamal, S.; Swartz, C.L.E. Real-time energy management for electric arc furnace operation. *J. Process Control* **2019**, *74*, 50–62. [CrossRef]

7. Gajic, D.; Savic-Gajic, I.; Savic, I.; Georgieva, O.; Di Gennaro, S. Modelling of electrical energy consumption in an electric arc furnace using artificial neural networks. *Energy* **2016**, *108*, 132–139. [CrossRef]
8. Zhao, S.; Grossmann, I.E.; Tang, L. Integrated scheduling of rolling sector in steel production with consideration of energy consumption under time-of-use electricity prices. *Comput. Chem. Eng.* **2018**, *111*, 55–65. [CrossRef]
9. Klemen, S.; Kovačič, M.; Peter, K.; Jože, P. Electric-arc-furnace productivity optimization. *Mater. Tehnol.* **2014**, *48*, 3–7.
10. Marchi, B.; Zanoni, S.; Mazzoldi, L.; Reboldi, R. Product-service System for Sustainable EAF Transformers: Real Operation Conditions and Maintenance Impacts on the Life-cycle Cost. *Procedia CIRP* **2016**, *47*, 72–77. [CrossRef]
11. Marchi, B.; Zanoni, S.; Mazzoldi, L.; Reboldi, R. Energy Efficient EAF Transformer—A Holistic Life Cycle Cost Approach. *Procedia CIRP* **2016**, *48*, 319–324. [CrossRef]
12. Belkovskii, A.G.; Kats, Y.L. Effect of the Mass of the Liquid Residue on the Performance Characteristics of an Eaf. *Metallurgist* **2015**, *58*, 950–958. [CrossRef]
13. Wei, G.; Zhu, R.; Dong, K.; Ma, G.; Cheng, T. Research and Analysis on the Physical and Chemical Properties of Molten Bath with Bottom-Blowing in EAF Steelmaking Process. *Metall. Mater. Trans. B* **2016**, *47*, 3066–3079. [CrossRef]
14. Wieczorek, T.; Blachnik, M.; Mączka, K. Building a Model for Time Reduction of Steel Scrap Meltdown in the Electric Arc Furnace (EAF): General Strategy with a Comparison of Feature Selection Methods. In *Artificial Intelligence and Soft Computing—ICAISC 2008*; Springer: Berlin/Heidelberg, Germany, 2008; pp. 1149–1159.
15. Malfa, E.; Nyssen, P.; Filippini, E.; Dettmer, B.; Unamuno, I.; Gustafsson, A.; Sandberg, E.; Kleimt, B. Cost and Energy Effective Management of EAF with Flexible Charge Material Mix. *BHM Berg und Hüttenmännische Monatshefte* **2013**, *158*, 3–12. [CrossRef]
16. Sandberg, E.; Lennox, B.; Undvall, P. Scrap management by statistical evaluation of EAF process data. *Control Eng. Pract.* **2007**, *15*, 1063–1075. [CrossRef]
17. Lee, B.; Sohn, I. Review of Innovative Energy Savings Technology for the Electric Arc Furnace. *JOM* **2014**, *66*, 1581–1594. [CrossRef]
18. Hocine, L.; Yacine, D.; Kamel, B.; Samira, K.M. Improvement of electrical arc furnace operation with an appropriate model. *Energy* **2009**, *34*, 1207–1214. [CrossRef]
19. Feng, L.; Mao, Z.; Yuan, P.; Zhang, B. Multi-objective particle swarm optimization with preference information and its application in electric arc furnace steelmaking process. *Struct. Multidiscip. Optim.* **2015**, *52*, 1013–1022. [CrossRef]
20. Moghadasian, M.; Alenasser, E. Modelling and Artificial Intelligence-Based Control of Electrode System for an Electric Arc Furnace. *J. Electromagn. Anal. Appl.* **2011**, *3*, 47–55. [CrossRef]
21. Mapelli, C.; Baragiola, S. Evaluation of energy and exergy performances in EAF during melting and refining period. *Ironmak. Steelmak.* **2006**, *33*, 379–388. [CrossRef]
22. Kim, D.S.; Jung, H.J.; Kim, Y.H.; Yang, S.H.; You, B.D. Optimisation of oxygen injection in shaft EAF through fluid flow simulation and practical evaluation. *Ironmak. Steelmak.* **2014**, *41*, 321–328. [CrossRef]
23. Cantacuzene, S.; Grant, M.; Boussard, P.; Devaux, M.; Carreno, R.; Laurence, O.; Dworatzek, C. Advanced EAF oxygen usage at Saint-Saulve steelworks. *Ironmak. Steelmak.* **2005**, *32*, 203–207. [CrossRef]
24. Makarov, A.N. Change in Arc Efficiency During Melting in Steel-Melting Arc Furnaces. *Metallurgist* **2017**, *61*, 298–302. [CrossRef]
25. Oda, J.; Akimoto, K.; Tomoda, T.; Nagashima, M.; Wada, K.; Sano, F. International comparisons of energy efficiency in power, steel, and cement industries. *Energy Policy* **2012**, *44*, 118–129. [CrossRef]
26. Balan, R.; Hancu, O.; Lupu, E. Modeling and adaptive control of an electric arc furnace. *IFAC Proc. Vol.* **2007**, *40*, 163–168. [CrossRef]
27. MacRosty, R.D.M.; Swartz, C.L.E. Dynamic Modeling of an Industrial Electric Arc Furnace. *Ind. Eng. Chem. Res.* **2005**, *44*, 8067–8083. [CrossRef]
28. Coetzee, L.C.; Craig, I.K.; Rathaba, L.P. Mpc control of the refining stage of an electric arc furnace. *IFAC Proc. Vol.* **2005**, *38*, 151–156. [CrossRef]
29. Mesa Fernández, J.M.; Cabal, V.Á.; Montequin, V.R.; Balsera, J.V. Online estimation of electric arc furnace tap temperature by using fuzzy neural networks. *Eng. Appl. Artif. Intell.* **2008**, *21*, 1001–1012. [CrossRef]

30. Hanoglu, U.; Šarler, B. Multi-pass hot-rolling simulation using a meshless method. *Comput. Struct.* **2018**, *194*, 1–14. [CrossRef]
31. Koza, J.R. *The Genetic Programming Paradigm: Genetically Breeding Populations of Computer Programs to Solve Problems*; MIT Press: Cambridge, MA, USA, 1992; pp. 203–321.
32. Koza, J.R. *Genetic Programming II: Automatic Discovery of Reusable Programs*; MIT Press: Cambridge, MA, USA, 1994; ISBN 0-262-11189-6.
33. Kovačič, M.; Jager, R. Modeling of occurrence of surface defects of C45 steel with genetic programming. *Mater. Tehnol.* **2015**, *49*, 857–863. [CrossRef]
34. Kovačič, M.; Šarler, B. Genetic programming prediction of the natural gas consumption in a steel plant. *Energy* **2014**, *66*, 273–284. [CrossRef]
35. Kovacic, M.; Brezocnik, M. Reduction of Surface Defects and Optimization of Continuous Casting of 70MnVS4 Steel. *Int. J. Simul. Model.* **2018**, *17*, 667–676. [CrossRef]

8

Investigation of Start-Up Characteristics of Thermosyphons Modified with Different Hydrophilic and Hydrophobic Inner Surfaces

Xiaolong Ma, Zhongchao Zhao *, Pengpeng Jiang, Shan Yang, Shilin Li and Xudong Chen

School of Energy and Power, Jiangsu University of Science and Technology, Zhenjiang, Jiangsu 212000, China; marlon@stu.just.edu.cn (X.M.); pengpengjiang@stu.just.edu.cn (P.J.); shanyang33@stu.just.edu.cn (S.Y.); shilinli@stu.just.edu.cn (S.L.); xudongchen@stu.just.edu.cn (X.C.)

* Correspondence: zhongchaozhao@just.edu.cn

Abstract: In this paper, the influence of wettability properties on the start-up characteristics of two-phase closed thermosyphons (TPCTs) is investigated. Chemical coating and etching techniques are performed to prepare the surfaces with different wettabilities that is quantified in the form of the contact angle (CA). The 12 TPCTs are processed including the same CA and a different CA combination on the inner surfaces inside both the evaporator and the condenser sections. For TPCTs with the same wettability properties, the introduction of hydrophilic properties inside the evaporator section not only significantly reduces the start-up time but also decreases the start-up temperature. For example, the start-up time of a TPCT with CA = 28° at 40 W, 60 W and 80 W is 46%, 50% and 55% shorter than that of a TPCT with a smooth surface and the wall superheat degrees is 55%, 39% and 28% lower, respectively. For TPCTs with combined hydrophilic and hydrophobic properties, the start-up time spent on the evaporator section with hydrophilic properties is shorter than that of the hydrophobic evaporator section and the smaller CA on the condenser section shows better results. The start-up time of a TPCT with CA = 28° on the evaporator section and CA = 105° on the condenser section has the best start-up process at 40 W, 60 W and 80 W which is 14%, 22% and 26% shorter than that of a TPCT with smooth surface. Thus, the hydrophilic and hydrophobic modifications play a significant role in promoting the start-up process of a TPCT.

Keywords: Thermosyphon; start-up characteristics; hydrophilic and hydrophobic; contact angle

---

1. Introduction

As a two-phase passive device, the thermosyphon has a wide-range of various industrial applications, for instance, electronic equipment [1], heat-recovery systems [2], solar water heater systems [3] and space applications [4] due to the simple structure, reliability, high efficiency and low cost. The basic concept of heat pipes was first proposed by Gaugler in 1944 [5]. A thermosyphon is a gravity-assisted heat pipe without wicks that depends on phase-change heat transfer in both the evaporator and condenser sections to transfer large amounts of heat with relatively small temperature difference [6] and low thermal resistance [7]. Figure 1 shows the schematic diagram of the cross section and working principle of a two-phase closed thermosyphon (TPCT). A TPCT is composed of evaporator, adiabatic and condenser sections. The operating process begins with a certain volume of working fluid in the evaporator section, which is then heated by a source of heat, such as a heating element or a thermal bath. The heating converts the saturated liquid into vapor that rises to the condenser section. Afterwards, the vapor condenses into liquid, which flows back down to the evaporator section by gravity, in the process transferring heat to the heat sink, such as cold water [8].

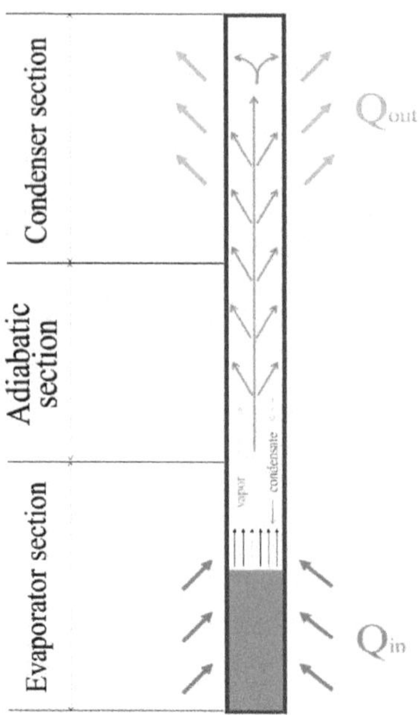

Figure 1. Cross section and working principle of a two-phase closed thermosyphon [9].

At present, the investigation of thermosyphons mainly includes the analysis of the thermal performance and start-up characteristics. In recent years, the thermal performance of thermosyphons in the aspects of filling ratio and surface modifications have been experimentally studied. Lataoui and Jemni [10] conducted an experimental study on a stainless steel thermosyphon to assess the influence of filling ratio, input power and the temperature of the cooling fluid on its thermal performance. Alireza Moradikazerouni et al. [11,12] investigated the effects of surface modifications on the heat sink, and the results found that different structural shapes on the heat surface have various heat transfer mechanisms. Rahimi et al. [13] modified the condenser and evaporator of a thermosyphon and compared the heat performance and resistance of a modified thermosyphon at different input powers with a flat thermosyphon. They found that the average thermal performance at tested heat loads was increased by 15.27% and the average thermal resistance of the thermosyphon was decreased by 2.35 times. Singh et al. [14] investigated the effect of surface modification on the thermal performance in an evaporator and condenser for flat thermosyphons with and without an anodized surface. Solomon et al. [15] studied the heat performance of thermosyphons with surface modifications at diverse inclination angles and input powers. The surface modifications significantly reduce the wall temperature of the evaporator and increase the heat-transfer coefficient. Solomon et al. [16] also analyzed the thermal performance of an anodized surface with a porous structure and observed a 15% reduction in the thermal resistance and 15% increase in the heat-transfer coefficient of the evaporator.

Additionally, Noie [17] investigated the effects of aspect ratio on the thermal performance of a thermosyphon and achieved heat performances of 60%, 90% and 30% for aspect ratios of 11.8, 7.45 and 9.8, respectively. Gedik [18] reported the influence of various operating conditions, such as heat input, inclination angle and the flow rate of cooling water on the heat-transfer characteristics of a TPCT. Moreover, the method whereby a nanofluid was used as working fluid in a thermosyphon has been theoretically and experimentally studied. Ma et al. [19] found that the heat-transfer rate of a nanofluid can rise to 3.11 times in an inclined square enclosure which indicates that the nanofluid is a potential choice as working fluid. Besides, the nanoadditives of various shapes on the fluid flow and heat transfer aspects of a nanofluid have different influences [20]. Hence, the parameters of surface modifications, operating conditions, working fluid and filling ratio have a great effect on the thermal performance of thermosyphons. Similarly, the start-up performance of the thermosyphon will also be affected by these factors.

The reliable operation of thermosyphons requires good start-up performance. The start-up of thermosyphon is a complex, transient process that is affected by several parameters. Sun et al. [21] studied the effects of filling ratio and heat input levels on the start-up characteristics of micro-oscillating heat pipes and observed two different start-up behaviors, start-up processes with and without bubble nucleation, depending principally on the spatial distribution of slugs/plugs in the micro-oscillating heat pipes. Guo et al. [22,23] found that the inclination angle is one of the factors that affect the start-up characteristics of thermosyphon. Then the influence of evaporator length on the start-up performance of a sodium-potassium alloy heat pipe was tested and obtained a uniform temperature distribution by increasing the evaporator section length. The Na-K heat pipe had excellent start-up performance, and the increase of inclination angle raised the temperature of the condenser. Wang et al. [24] analyzed the influence of inclination angle, heat input and flow rate of cooling water on the start-up properties of a thermosyphon with small diameter. Huang et al. [25] introduced the non-condensable gas used for regenerative building heating exchangers in a gravity loop thermosyphon and investigated its effect on the start-up time. They found that the non-condensable gas extended the start-up time of the thermosyphon, with a higher level corresponding to a longer time. Joung et al. [26] observed that a large amount of heat leakage increased the operating temperature and the start-up time of a loop heat pipe. In addition, Singh et al. [27] studied the start-up characteristics of a loop heat pipe and found that the start-up time increased with decreasing the applied heat load. Ji et al. [28] designed a loop heat pipe with composite porous wicks, and studied its heat transfer and start-up characteristics. Huang et al. [29] experimentally and mathematically analyzed the start-up process of a loop heat pipe. They concluded that the start-up process was closely subject to the structural parameters and environment of the loop heat pipe.

Although the thermal performance of a thermosyphon has been well studied from different aspects mentioned above, the start-up characteristics of a thermosyphon have been rarely investigated due to its complex and transient process. The start-up characteristics of a thermosyphon is in an unstable state. How to shorten the start-up time and make the thermosyphon quickly reach a stable state has an important impact on the operation of some equipment. Furthermore, the combination of hydrophilic and hydrophobic properties on the inner surfaces of a thermosyphon is barely investigated. As a high-efficiency heat-transfer device, the start-up characteristics of a thermosyphon is an important index to measure the reliability of the thermosyphon, which must be completed quickly and smoothly. Therefore, it is of great significance to investigate the start-up characteristics of a thermosyphon. In this paper, the effect of wettability properties on the start-up characteristics of TPCTs is fully investigated. Chemical coating and etching techniques are employed to manufacture surface wettability with different contact angles (CAs) at the inner wall of the thermosyphon. The influence of the surface with different CAs on the start-up time and wall superheat degrees of the evaporator section under different input power was compared and analyzed.

2. Methodology

2.1. Experimental System

Figure 2 shows the schematics of the experimental apparatus, while Figures 3 and 4 show the real thermosyphon and the experimental system, respectively. The experimental system is composed of a thermosyphon, a heat supply unit, a cooling unit and a data acquisition unit. The thermosyphon is made of copper with the lengths of evaporator, adiabatic and condenser sections shown in Figure 2 designed to be 100 mm, 50 mm and 100 mm, and the internal and external diameters of 8.32 and 9.52 mm, respectively. Deionized water of 3.2592 g is used as the working fluid and the filling ratio was 24%. The heat supply unit included an electrical resistor, a digital power meter and a voltage-regulating transformer. The evaporator and the adiabatic sections are wrapped with a polytetrafluoroethylene nanoparticle insulation in the inner layer and aluminum foil in the outermost layer for the purpose of reducing the heat loss as shown in Figure 4. The cooling system consisted of a refrigerating unit, a cooling water jacket, a rotameter of 6 $\sim$ 60 L/h and a number of pipelines. The jacket is wrapped with thermal insulation rubber outside. The data acquisition unit is made up of a data logger, a computer and 10 Pt100 thermocouples. The arrangement of 10 thermocouples is shown in Figure 2. In addition, a vacuum pump system consisting of burette, pressure gauge and vacuum pump is used to provide a vacuum in the thermosyphon, and the vacuum degree of each thermosyphon is 10^{-3} Pa. In addition, the boundary conditions of the experiment are shown in Table 1. The 12 TPCTs with different wettability properties on the start-up characteristics are fully investigated under these conditions.

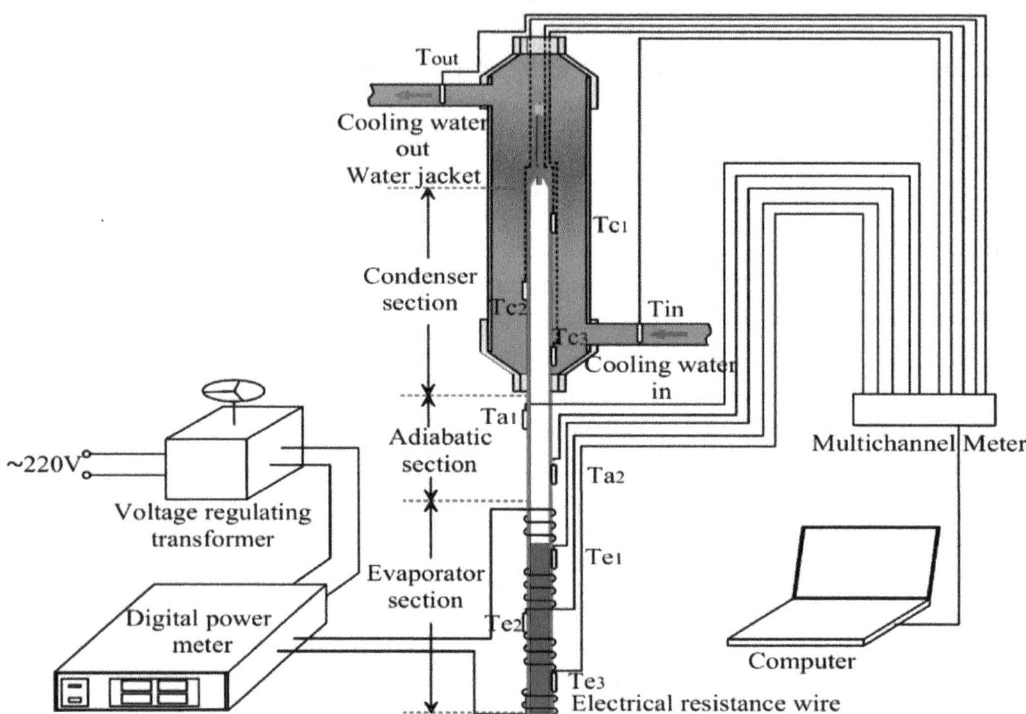

Figure 2. Schematic diagram of experimental apparatus.

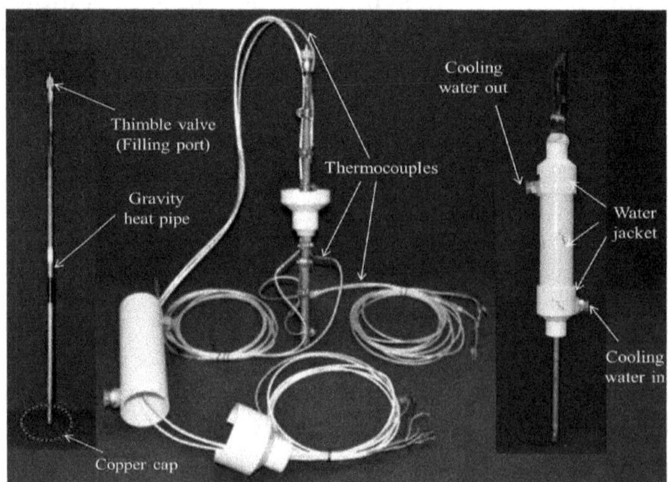

Figure 3. Image of the real thermosyphon used in the experiments.

Table 1. The boundary conditions of the experiment.

Input Power (W)	Temperature of Cooling Water (°C)	Flow Rate of Cooling Water (L/h)
40	18	20
60	18	30
80	18	40

Figure 4. Image of the experimental system.

2.2. Data Reduction and Problem Description

The heat transfer between the condenser section and the cooling water under various operating conditions (e.g., working fluid, applied heating power input and mass flow rate of cooling water) is determined using Equation (1):

$$Q_c = m_w c_p (T_{w,out} - T_{w,in}) \tag{1}$$

where m_w is the mass flow, $T_{w,in}$ is the temperature of cooling water at inlet of the condenser, $T_{w,out}$ is the temperature at outlet and c_p is the specific heat values of water.

During the experiment, strict insulation measures are taken on the outside of the evaporator section, the adiabatic section and the cooling water jacket in order to ensure minimum heat loss. The heat-balance method [9] is used to determine the heat loss of the system. The power relative error is defined as the ratio of the difference between the input heat Q_e in the evaporator section and the released heat Q_c in the condenser section to the input heat Q_e, and calculated using Equation (2) as:

$$\eta = \frac{Q_e - Q_c}{Q_e} \tag{2}$$

where

$$Q_e = Q_{in} = VI \tag{3}$$

Q_{in} is the heating power input on the evaporator section of the thermosyphon, while V and I are the voltage and the current monitored by the digital power meter. The measurement error of thermocouples was ± 0.5 °C, and the cooling water had a flow rate error of ± 2.5%. In the experiment, the maximum power relative error was 7.3%.

The objective of the paper is to investigate the effect of wettability properties on the start-up characteristics of TPCTs. The measurement includes the average start-up time and wall superheat degree of the evaporator section of the thermosyphon under different input power. The main problem in this study is the machining and preparation of different wettability properties on the inner surface. The values of different CAs need distinct process technology [30,31]. Besides, the cylindrical shape of the inner wall of the thermosyphon leads to a difficulty of processing and long-term stability. For the preparation of inner wettability surfaces, the techniques are chemical etching, electrochemical deposition, composite coating, anodizing, etc. Chemical coating and etching techniques are used to manufacture surface wettability with different CAs at the inner wall of the thermosyphon after a considerable number of experiments.

2.3. Surface Modification

In order to prepare the surfaces with various wettabilities in terms of different CAs, chemical techniques are performed using various materials. NaOH and $(NH_4)_2S_2O_8$ are used to etch the hydrophilic surface with a CA of approximately 28°. By coating various ratios of materials, such as N-butyl, stearic acid, xylenes and acetone, the CAs of surfaces are approximately 61°, 79°, 105°, 117° and 142°, respectively. Detailed surface modification methods have been described in the authors' previous work [9].

The 12 thermosyphons correspondingly produced test samples to verify the coating temperature resistance. The coated samples are put on the thermostatic magnetic stirrer for a high-temperature test. The automatic contact angle meter (Kino-SL150E) with the measuring error of ± 2° is used to measure the CA after high-temperature and long-time testing. It is found that the change in the CA is small and the maximum CA error is 4.2% in the testing temperature from 20 °C to 100 °C.

In order to ensure the accuracy and reproducibility of the CA of the wet surface, 4 μL of deionized water is titrated on each surface 5 times. Finally, the average value of the CA is taken. Figure 5a–f show the low- and high-magnification scanning electron microscopy (SEM) images of the surfaces with CAs of 28°, 61°, 79°, 105°, 117° and 142°, respectively, while Figure 5g shows the SEM image of a smooth surface. Figure 5d–f demonstrate the hydrophobic surfaces on which 4μL of water is dropped, with the static CAs of 105°, 117° and 142°, respectively, illustrating weak interactions between water drops and hydrophobic surfaces. Water drops are dispersed over the surfaces with the CAs of 28°, 61° and 79°, subject to high adhesive force between water and coated copper in the evaporator as shown in Figure 5a–c, respectively. Figure 5g shows that the TPCT7 remained smooth, without any resurface work.

According to different wettability properties, the thermosyphons are classified into 12 different TPCTs from TPCT1 to TPCT12 as listed in Table 2. From TPCT1 to TPCT6, each of their three sections has the same CA on the inner surfaces. A combination of hydrophilic and hydrophobic properties with different CAs is adopted to modify the inner surfaces of the evaporator and condenser of TPCTs from TPCT8 to TPCT12, while the inner wall of the three sections of TPCT7 is smooth surface without fabrication.

Table 2. Different two-phase closed thermosyphons (TPCTs) on inner modified surfaces with hydrophilic and hydrophobic properties.

TPCTs	Evaporator	Adiabatic	Condenser
TPCT1	28°	28°	28°
TPCT2	61°	61°	61°
TPCT3	79°	79°	79°
TPCT4	105°	105°	105°
TPCT5	117°	117°	117°
TPCT6	142°	142°	142°
TPCT7	smooth surface	smooth surface	smooth surface
TPCT8	28°	smooth surface	105°
TPCT9	28°	smooth surface	117°
TPCT10	28°	smooth surface	142°
TPCT11	105°	smooth surface	28°
TPCT12	142°	smooth surface	28°

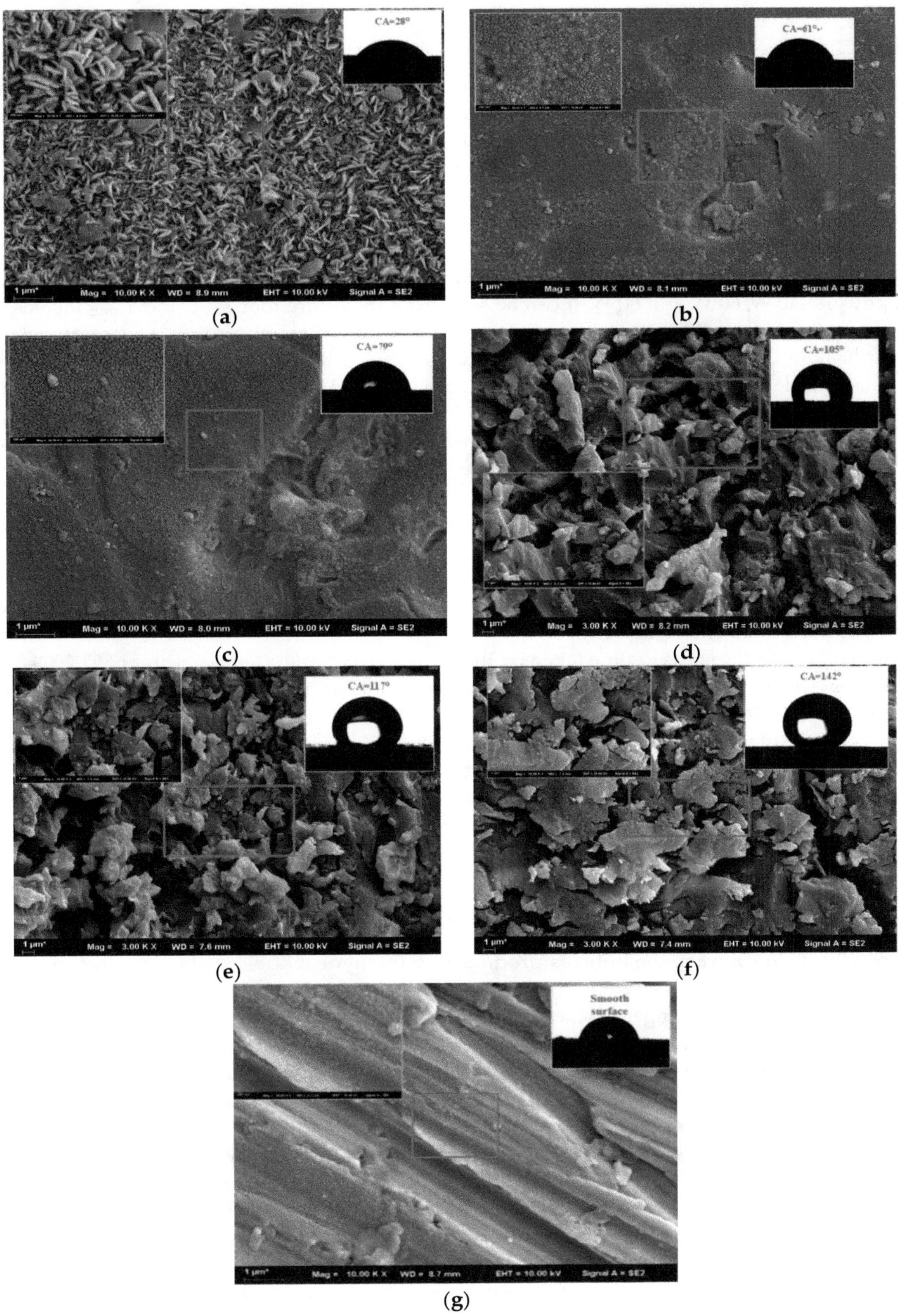

Figure 5. Scanning electron microscopy (SEM) images of: (**a**–**f**) hydrophilic and hydrophobic surfaces with contact angles (CAs) of 28°, 61°, 79°, 105°, 117° and 142°, respectively, and (**g**) smooth surface (inset: 4 μL of water droplet with static CA).

3. Results

The start-up performance is a crucial indicator of the thermosyphon operation. To enhance the overall performance of the thermosyphon, the start-up process in which the evaporator is heated to a steady state must be completed quickly and stably. In order to evaluate the influence of wettability on the start-up performance of the thermosyphon, the start-up times and two typical temperatures, i.e., the average temperatures of evaporator $T_{e.ave}$ (average of points Te1-Te3) and condenser $T_{c.ave}$ (average of points Tc1-Tc3), are selected and measured by thermocouples. In addition, all data are recorded every 5s through a multichannel meter and saved in the computer through Monitor and Control Generated System (MCGS) software.

Figure 6 shows the start-up processes of thermosyphons from TPCT1 to TPCT7 at different input powers: 40 W, 60 W and 80 W. It can be seen from the figure that $T_{e.ave}$ and $T_{c.ave}$ increase first and then maintain two different stable conditions, and all thermosyphons had a successful start-up. After the heat input power is imposed, $T_{e.ave}$ rapidly increases with extended heating process before the start-up is completed, implying that the working fluid is heated at the initial moment, then boiled and evaporated, after which the vapor reaches the condenser section and releases the latent heat. However, $T_{c.ave}$ does not rise at the early stage and the increase in the temperature on the condenser section is slower in the start-up process due to the heat-transfer delay. As $T_{c.ave}$ rises, the process whereby the working fluid is boiled into steam is accelerated by increasing the input power. For example, at the input powers of 40 W, 60 W and 80 W, the $T_{c.ave}$ values of TPCT1–TPCT7 increase sharply at 125 s, 70 s and 45 s, respectively. The generated vapor from the evaporator section is not condensed in the condenser section due to the heat is not taken away rapidly by the external environment and that is why the temperature of the condenser keeps increasing continuously. Once the working fluid cycle is completed, the system enters a steady state.

Figure 7 shows the start-up time taken by the average evaporator temperatures of TPCT1–TPCT7 to stabilize at different input powers with different CAs on the evaporator sections. At the same input power, the TPCT with hydrophobic properties need longer start-up time than those with hydrophilic properties do at different input powers. The start-up times of the evaporator sections of TPCT4–TPCT6 with hydrophobic properties are all longer than those of TPCT1-TPCT3 with hydrophilic properties at 40 W, 60 W and 80 W. The bubbles on both hydrophilic and hydrophobic surfaces go through the entire process of bubble formation, growth, coalescence and separation which reflects the complete cycle inside the thermosyphons. The hydrophilic surface has a more complex microstructure than the hydrophobic surface as shown in Figure 5. The smaller bubble diameters are generated on the hydrophilic surface due to the rapid replenishment of fresh liquid backflow on the hydrophilic surface being faster than that of the hydrophobic surface, which further promotes the heat transfer of the thermosyphon. In contrast, the bubbles produced by the hydrophobic surface could form the gas film with the other bubbles before leaving the hydrophobic surface. Thus, the evaporator with the hydrophobic surface takes longer to start up.

The TPCT1 responds more quickly than TPCT2–TPCT7 do when the thermosyphons enter the steady state at different input powers. Thus, TPCT1 has the best start-up for thermosyphons with the same wettabilities on the evaporator and the condenser sections among TPCT1–TPCT6, while the start-up time of TPCT1 at 40 W, 60 W and 80 W is 46%, 50% and 55% faster than that of TPCT7 (smooth surface). The reason is that the hydrophilic surfaces have more compact structures with stronger tension forces between pore and water, which is conducive to a smaller bubble diameter speeding up the departure frequency of the bubbles [32] and promotes the heat transfer of nucleate pool boiling. Conversely, the bubbles join together to form an air blanket in a very limited time before fleeing the hydrophobic surface [33,34]. The SEM images with different CAs presented in Figure 5 indicate that the pore diameter enlarges with increasing CA. Moreover, the relationship between the CA and the pore diameter is the same as the previous research result [35]. The short knife-like nanostructure coated on the surface with CA of 28° (Figure 5a) increase the heat-transfer area of phase transition and nucleation sites. Meanwhile, the nanostructure above enhances the hydrophilicity characteristics and improved the heat transfer of pool boiling [36], producing smaller bubbles. In consequence, the process of the heat transfer of pool boiling is significantly disturbed prompting more vapor to reach the condenser section to release more latent heat. Thus, the start-up performance of the thermosyphon is enhanced.

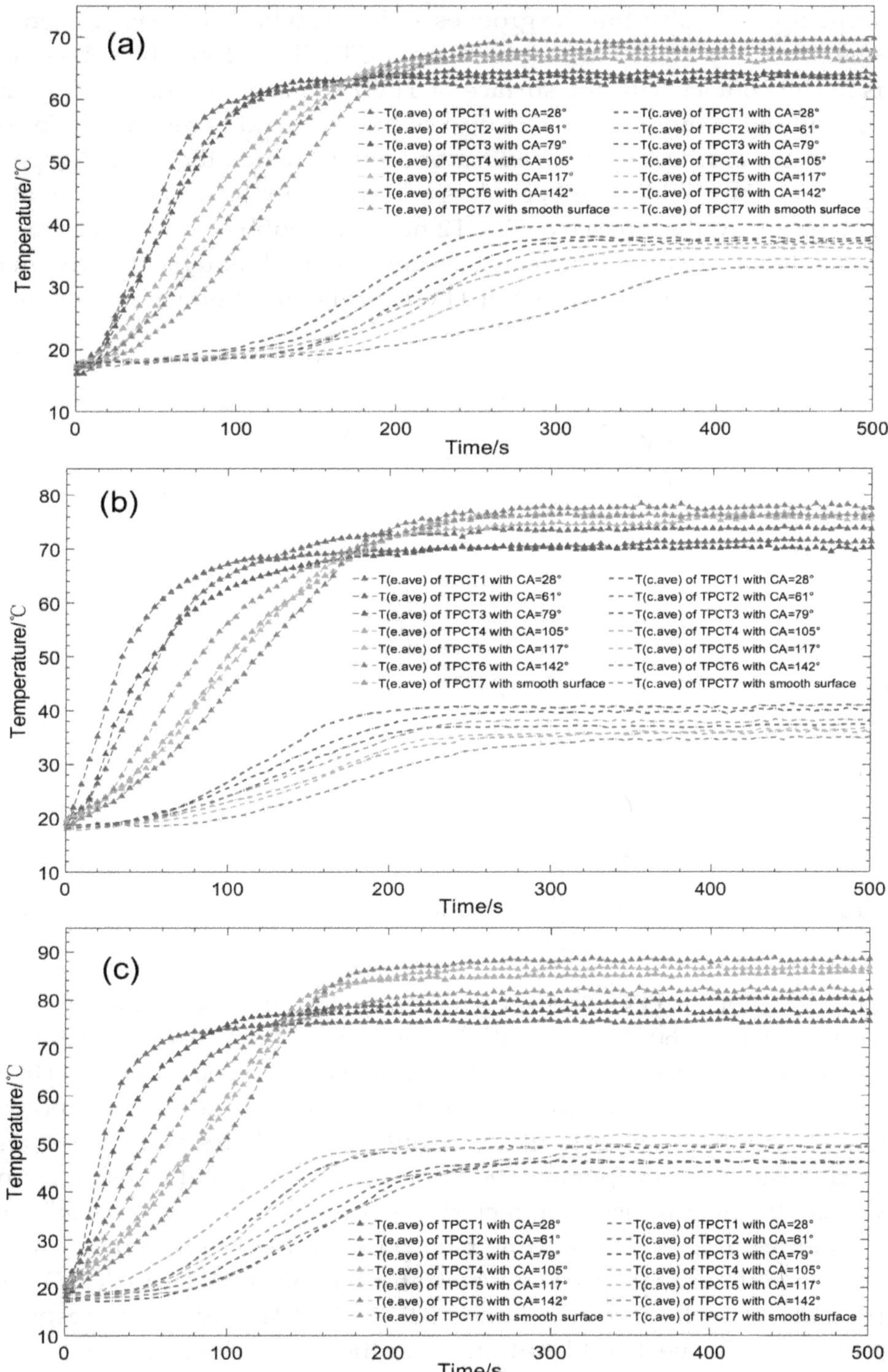

Figure 6. Start-up performances of TPCT1–TPCT7 at different input powers: (**a**) Q_{in} = 40 W; (**b**) Q_{in} = 60 W; (**c**) Q_{in} = 80 W

Generally, the start-up time is prolonged as the CA on the evaporator sections of TPCT1–TPCT6 increase at the same input power. The time of TPCT2 is shorter than that of TPCT3 at the input power of 40 W. However, the time of TPCT2 takes longer at the input powers of 60 W and 80 W. One of the reasons is that the accuracy of CA can affect the experimental results. The CA is the average value that may slightly change in the process of heat transfer. The calculation results show that the CA error of samples is within 4.2%, causing the start-up time of TPCT3 to be shorter than that of TPCT2. In addition, both the wettability's and the roughness of the thermosyphon surface are different, which could lead to different processes of bubble generation, growth and departure. It can be observed from Figure 5b,c that the surface structure is similar, but the grooves and gaps where it is easy to generate nucleation sites density on the surface of TPCT3 are lower than TPCT2. Therefore, the sub-cooled water in the thermosyphon is in full contact with the surface of TPCT3. There are more nucleation sites on the surface of TPCT2, resulting in more bubbles. Adjacent bubbles tend to merge and form large bubbles, which are trapped on the surface of the evaporator section. Thus, the start-up speed of TPCT2 is slower. At low heat flux, the bubble number is less and could not lead to merging of a large number of bubbles. At high heat flux, the surface of TPCT2 has more bubbles that makes merging easier for bubbles. The bubble departure diameter increases and the bubble departure frequencies decrease, so the start-up time of TPCT3 is less than that of TPCT2 at the input power of 60 W and 80 W.

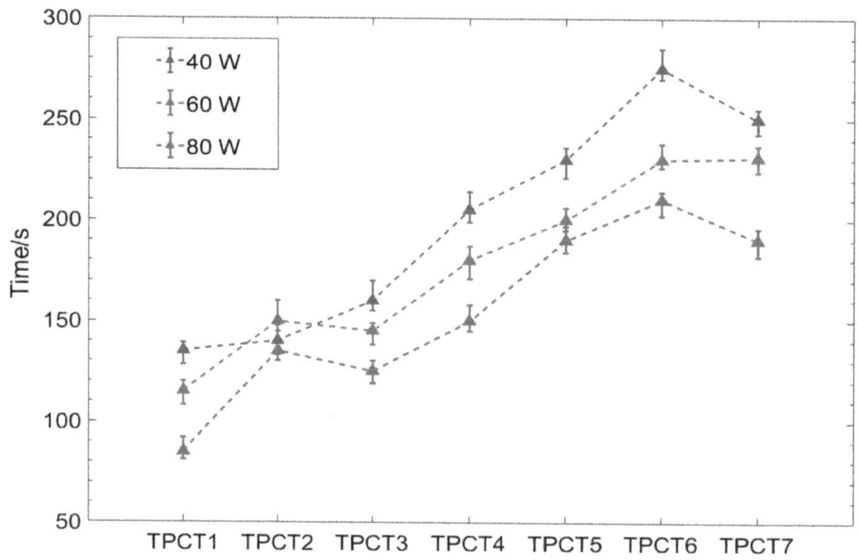

Figure 7. Times taken by the average evaporator temperatures of TPCT1–TPCT7 to stabilize at different input powers with various wettabilities on evaporator sections.

Figure 8 shows the plots of TPCTs with various wettabilities on evaporator sections versus the wall superheat degree at 40 W, 60 W and 80 W. The thermosyphons with hydrophilic surfaces (CA $< 90^{\circ}$) have not only a short start-up time, but also a low wall superheat degree. The wall superheat degree ($\Delta t = t_w - t_{sat}$, t_w: wall temperature, t_{sat}: saturated temperature) of evaporator sections for thermosyphons is an important factor to quantify the characteristics of thermosyphons. Since the heat exchange mechanism of the evaporator section is pool boiling heat transfer, the most important parameter to evaluate the heat-transfer characteristics of pool boiling is the wall superheat degree. The lower wall superheat degree at the same heat power means higher heat-exchange efficiency. As a result, the wall superheat degrees Δt of evaporator sections for TPCT1–TPCT7 were compared. The saturated pressure of each thermosyphon is 0.017212 MPa, and the corresponding saturated temperature is about 57 °C. The TPCT1 not only significantly reduces the start-up time, but also decreases the evaporator wall superheat degrees Δt compared with those of other TPCTs at the same input power. The wall superheat degrees Δt of TPCT1 at 40 W, 60 W and 80 W are 55%, 39% and 28% lower than that of TPCT7. Similarly, it has previously been reported that the modified surfaces with different CAs influence the pool boiling in the thermosyphon [37]. In the process of boiling heat transfer, the surfaces with hydrophobic properties produce a large bubble at low heat flux [38]. There are larger and deeper pores on the hydrophobic surfaces (Figure 5d–f) than those on the hydrophilic surfaces (Figure 5a–c). Thus, some air and gas film exist, which increase the thermal resistance and inhibit the heat transfer. Besides, the hydrophobic surface is close to the gas and then the bubble is easy to polymerize into a gas film, which prevents the liquid from replenishing to the heating wall. Therefore, the heat transfer is impeded and the boiling heat transfer performance begins to deteriorate. As a result, the superheat degrees of hydrophobic surfaces (CAs: 105°, 117° and 142°) surpass those of hydrophilic surfaces (CAs: 28°, 61°and 79°) and a smooth surface.

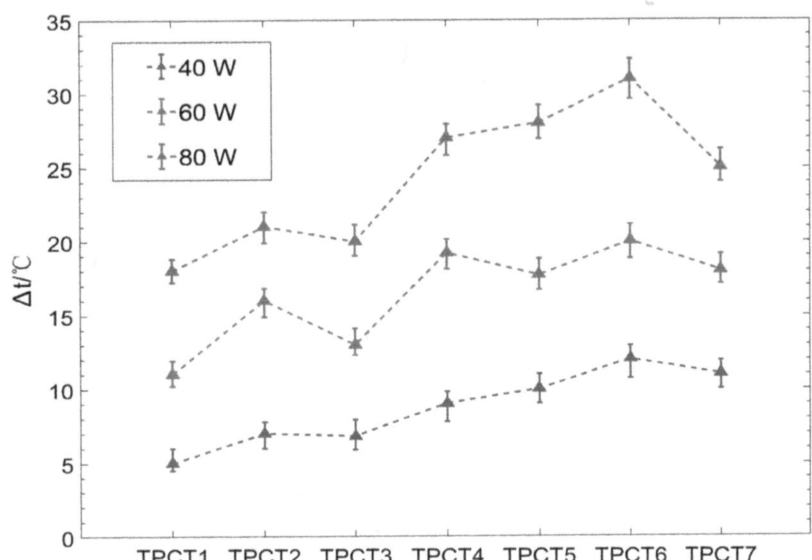

Figure 8. Relationships between wall superheat degree $\Delta t/°C$ on evaporator section and different TPCTs with various wettabilities.

Furthermore, Figure 8 demonstrates that the wall superheat degree Δt gradually increases with rising CA under the same power. As the input power increased, Δt of the hydrophobic evaporator section rises more evidently than that of the hydrophilic evaporator section. For example, the wall superheat degree of TCPT1 increases 13 °C from the input power of 40 W to 80 W while TCPT6 increases 19 °C. Since the hydrophobic surface produces more gas film as the power increases, the resulting large bubbles limit the heat transfer to the working fluid [39]. Thus, the superheat degree is augmented for hydrophobic surface. In contrast, the bubbles on the hydrophilic surface are smaller and quickly departs from the heating surface. Once the bubble departed, the surrounding working fluid quickly fills the remaining area and prevents the formation of large gas films with high thermal resistance. However, Δt of the evaporator of TPCT2 is higher than that of the evaporator of TPCT3 which is consistent with the start-up time due to the inhomogeneous structures as shown in Figure 5b,c. This gives a higher wall superheat degree to the evaporator of TPCT2 than that of the evaporator of TPCT3.

Figure 9 illustrates the start-up processes of TPCT7-TPCT12 modified with a combination of hydrophilic and hydrophobic properties at the input powers of 40 W, 60 W and 80 W. For TPCT8, TPCT9 and TPCT10 with hydrophilic inner surfaces on the evaporator sections, the start-up time is shorter than that of TPCT11 and TPCT12 with hydrophobic inner surfaces, which is consistent with the above discussion. With rising CA for TPCT7–TPCT9 on condenser sections, the start-up process on the evaporator section of TPCT8 is faster than those of TPCT9 and TPCT10. For TPCT8, the difference between the start-up time of the evaporator and the time reaching the stable state of the condenser is gradually decreased from 200 s to 150 s with the increase from 40 W to 80 W, which enhances the evaporator-condenser phenomenon induced by combined hydrophilic and hydrophilic properties. The addition of a hydrophilic surface to the surface of the evaporator section can quickly induce the sub-cooled water after the formation of bubbles, which can quickly take away the heat of the surface of the evaporator section. Thus, the stable temperature reduces. The addition of a hydrophobic surface on the surface of the condenser section rapidly reduces the water droplets in the condenser section, which indirectly speeds up the recycle of working fluid inside the thermosyphon. As the input power increased, TPCT8 and TPCT9 both have almost the same start-up times of 180 s and 200 s at input power of 60 W and 80 W, respectively. However, TPCT8 has a higher average temperature of 2 °C on the evaporator than that of TPCT9. Furthermore, the average temperature on the condenser section of TPCT9 (CA: 117°) gradually approaches that on the condenser section of TPCT8 (CA: 105 °C) with increasing input power at an average temperature of 54 °C.

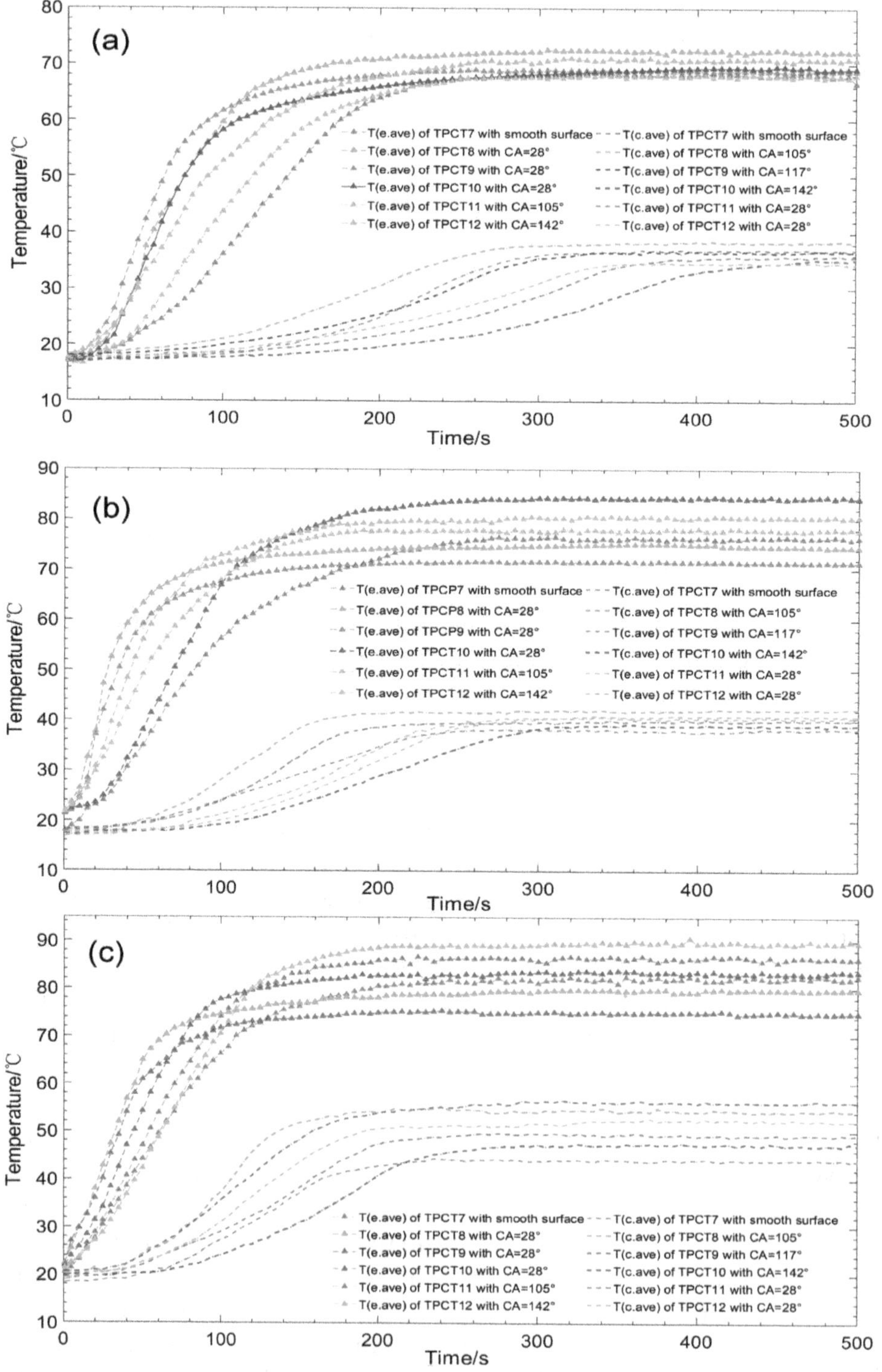

Figure 9. Start-up performances of TPCT7 (smooth surface) and TPCT8–TPCT12 with combined hydrophilic and hydrophobic properties at different heat input powers: (**a**) Q_{in} = 40 W;(**b**) Q_{in} = 60 W; (**c**) Q_{in} = 80 W.

For TPCT11 and TPCT12 with hydrophilic inner surfaces on the condenser sections (both with CA 28°), the condenser sections require shorter times to stabilize than that of TPCT7 with a smooth surface with increasing input power. Although the evaporator sections are hydrophobically modified, the steam still sufficiently releases heat on the inner surface of hydrophilically modified condenser sections, and the condensate fully contacts with the hydrophilic surface to release latent and sensible heats. Furthermore, it is found that the temperature difference between the evaporator (CA: 28°) and condenser (CA: 105 °C) of TPCT8 gradually decreases with increase in the input power, thus surpassing those of TPCT9–TPCT12 concerning isothermal properties. As a result, TPCT8 has the best start-up characteristics among TPCT8–TPCT12 under the same input power. It is postulated that the interaction between the liquid drops and the wall is attenuated with an increase in the CA, and the condensate falls down more easily as smaller drops and fast reflux to the evaporator section, which may suppress the release of latent and sensible heats in the condenser.

4. Conclusions

In this study, the influence of wettability properties inside the inner surface of thermosyphons on the start-up characteristics is fully investigated under different input powers. Chemical techniques are performed to fabricate the surfaces with different wettabilities that is quantified in the form of the CA inside the evaporator and the condenser sections. The experimental results demonstrate that different CAs not only significantly affect the start-up time but also influence the temperature variations and distributions of the outer walls of the evaporator and the condenser sections. Detailed conclusions can be drawn as follows:

(1) For thermosyphons with the same wettabilities on the evaporator and the condenser sections among TPCT1–TPCT6, the introduction of hydrophilic properties inside the evaporator section not only significantly shortens the start-up time but also decreases the start-up temperature. At the same input power, the start-up time of a thermosyphon with CA < 90° is shorter than that with CA > 90°. The start-up time of TPCT with CA = 28° has the shortest start-up time, while under the input powers of 40 W, 60 W and 80 W it is 46%, 50% and 55% shorter than that of TPCT with a smooth surface, respectively. The start-up time becomes longer with the increase of CA of the evaporator sections.

(2) As the CAs on the evaporator sections of TPCT1–TPCT6 increase, the wall superheat degree gradually increases. The TPCT with CA = 28° has a minimum superheat degree at input powers of 40 W, 60 W and 80 W, and the wall superheat degree is 55%, 39% and 28% lower than that of TPCT with smooth surfaces, respectively. In addition, the superheat degree of the hydrophobic evaporator section increases more obviously than that of the hydrophilic with increasing input power. In the experiment, the TPCT with CA = 142° has the highest superheat degree among TPCT1–TPCT6.

(3) For thermosyphons with combined hydrophilic and hydrophobic properties, the start-up time of the evaporator section with CA < 90° and the condenser section with CA > 90° is less than the evaporator section with CA > 90° and the condenser section with CA < 90°. With the increase of CA on the condenser sections of TPCT8–TPCT10, the start-up process of TPCT8 is faster than those of TPCT9 and TPCT10 with the same CA = 28° on the evaporator section. The experimental results and data analysis demonstrate that the start-up time of the TPCT8 with CA = 28° on the evaporator section and CA = 105° on the condenser section is the shortest among TPCT7–TPCT12.

(4) In this paper, the surfaces with CAs of 28°, 61°, 79°, 105°, 117° and 142° are fabricated inside the evaporator and the condenser sections of the thermosyphons. The experimental results show that the TPCT with CA = 28° on both the evaporator and the condenser section has the best start-up characteristics considering start-up time and wall superheat degree, which reflect the optimal wettabilitiy inside the inner surface of thermosyphons for industrial reference. Further research directions should extend to the preparation and investigation of superhydrophobic and superhydrophilic surfaces in the analysis of the thermal performance and start-up characteristics.

Author Contributions: Project administration, supervision, writing-review and editing, Z.Z.; methodology and writing-original draft preparation, X.M. and P.J.; investigation, data curation, formal analysis and validation, X.M., P.J., S.Y., S.L. and X.C. All authors have read and agreed to the published version of the manuscript.

Nomenclature

c_p	Specific heat values (J/kg·K)
m	Mass flow(kg/s)
Q	Heat load (W)
T	Temperature (°C)
I	Current (A)
V	Voltage (V)

Greek symbols

Δt	Wall superheat degree (°C)
η	Efficiency of the thermosyphon

Subscripts

a	Adiabatic section
ave	Average
c	Condenser section
e	Evaporator section
in	Cooling water inlet/ Input power
out	Cooling water outlet
sat	Saturated
w	Wall/water

Acronyms

CA	Contact angle
MCGS	Monitor and Control Generated System
SEM	Scanning electron microscopy
TPCP	Two-phase closed thermosyphon

References

1. Chang, Y.; Cheng, C.; Wang, J.; Chen, S. Heat pipe for cooling of electronic equipment. *Energy Convers. Manag.* **2008**, *49*, 3398–3404. [CrossRef]
2. Jafari, D.; Franco, A.; Filippeschi, S.; Di Marco, P. Two-phase closed thermosyphons: A review of studies and solar applications. *Renew. Sustain. Energy Rev.* **2016**, *53*, 575–593. [CrossRef]
3. Ziapour, B.M.; Khalili, M.B. PVT type of the two-phase loop mini tube thermosyphon solar water heater. *Energy Convers. Manag.* **2016**, *129*, 54–61. [CrossRef]
4. Jouhara, H.; Merchant, H. Experimental investigation of a thermosyphon based heat exchanger used in energy efficient air handling units. *Energy* **2012**, *39*, 82–89. [CrossRef]
5. Gaugler, R.S. Heat Transfer Device. U.S. Patent 2,350,348, 6 June 1994.
6. Faghri, A. *Heat Pipe Science and Technology*; Taylor & Francis: Philadelphia, PA, USA, 1995.
7. Shabgard, H.; Allen, M.J.; Sharifi, N.; Benn, S.P.; Faghri, A.; Bergman, T.L. Heat pipe heat exchangers and heat sinks: Opportunities, challenges, applications, analysis, and state of the art. *Int. J. Heat Mass Transf.* **2015**, *89*, 138–158. [CrossRef]
8. Amatachaya, P.; Srimuang, W. Comparative heat transfer characteristics of a flat two-phase closed thermosyphon (FTPCT) and a conventional two-phase closed thermosyphon (CTPCT). *Int. Commun. Heat Mass Transfer.* **2010**, *37*, 293–298. [CrossRef]
9. Zhao, Z.; Jiang, P.; Zhou, Y.; Zhang, Y.; Zhang, Y. Heat transfer characteristics of two-phase closed

thermosyphons modified with inner surfaces of various wettabilities. *Int. Commun. Heat Mass Transfer.* **2019**, *103*, 100–109. [CrossRef]
10. Lataoui, Z.; Jemni, A. Experimental investigation of a stainless steel two-phase closed thermosyphon. *Appl. Therm. Eng.* **2017**, *121*, 721–727. [CrossRef]
11. Moradikazerouni, A.; Afrand, M.; Alsarraf, J.; Mahian, O.; Wongwises, S.; Tran, M.-D. Comparison of the effect of five different entrance channel shapes of a micro-channel heat sink in forced convection with application to cooling a supercomputer circuit board. *Appl. Therm. Eng.* **2019**, *150*, 1078–1089. [CrossRef]
12. Moradikazerouni, A.; Afrand, M.; Alsarraf, J.; Wongwises, S.; Asadi, A.; Nguyen, T.K. Investigation of a computer CPU heat sink under laminar forced convection using a structural stability method. *Int. J. Heat Mass Transf.* **2019**, *134*, 1218–1226. [CrossRef]
13. Rahimi, M.; Asgary, K.; Jesri, S. Thermal characteristics of a resurfaced condenser and evaporator closed two-phase thermosyphon. *Int. Commun. Heat Mass Transf.* **2010**, *37*, 703–710. [CrossRef]
14. Singh, R.R.; Selladurai, V.; Ponkarthik, P.; Solomon, A.B. Effect of anodization on the heat transfer performance of flat thermosyphon. *Exp. Therm. Fluid Sci.* **2015**, *68*, 574–581. [CrossRef]
15. Solomon, A.B.; Daniel, V.A.; Ramachandran, K.; Pillai, B.; Singh, R.R.; Sharifpur, M.; Meyer, J. Performance enhancement of a two-phase closed thermosiphon with a thin porous copper coating. *Int. Commun. Heat Mass Transf.* **2017**, *82*, 9–19. [CrossRef]
16. Solomon, A.B.; Mathew, A.; Ramachandran, K.; Pillai, B.; Karthikeyan, V. Thermal performance of anodized two phase closed thermosyphon (TPCT). *Exp. Therm. Fluid Sci.* **2013**, *48*, 49–57. [CrossRef]
17. Noie, S. Heat transfer characteristics of a two-phase closed thermosyphon. *Appl. Therm. Eng.* **2005**, *25*, 495–506. [CrossRef]
18. Gedik, E. Experimental investigation of the thermal performance of a two-phase closed thermosyphon at different operating conditions. *Energy Build.* **2016**, *127*, 1096–1107. [CrossRef]
19. Ma, Y.; Shahsavar, A.; Moradi, I.; Rostami, S.; Moradikazerouni, A.; Yarmand, H.; Zulkifli, N.W.B.M. Using finite volume method for simulating the natural convective heat transfer of nano-fluid flow inside an inclined enclosure with conductive walls in the presence of a constant temperature heat source. *Phys. A Stat. Mech. Appl.* **2019**, 123035. [CrossRef]
20. Vo, D.D.; Alsarraf, J.; Moradikazerouni, A.; Afrand, M.; Salehipour, H.; Qi, C. Numerical investigation of γ-AlOOH nano-fluid convection performance in a wavy channel considering various shapes of nanoadditives. *Powder Technol.* **2019**, *345*, 649–657. [CrossRef]
21. Sun, Q.; Qu, J.; Yuan, J.; Wang, H. Start-up characteristics of MEMS-based micro oscillating heat pipe with and without bubble nucleation. *Int. J. Heat Mass Transf.* **2018**, *122*, 515–528. [CrossRef]
22. Guo, Q.; Guo, H.; Yan, X.; Wang, X.; Ye, F.; Ma, C. Experimental study of start-up performance of sodium-potassium heat pipe. *J. Eng. Thermophys.* **2014**, *35*, 2508–2512.
23. Guo, Q.; Guo, H.; Yan, X.; Wang, X.; Ye, F.; Ma, C. Effect of the evaporator length on start-up performance fo sodium-potassium alloy heat pipe. *J. Eng. Thermophys.* **2016**, *37*, 1717–1720.
24. Wang, X.; Xin, G.; Tian, F.; Cheng, L. Start-up behavior of gravity heat pipe with small diameter. *CIESC J.* **2012**, *63*, 94–98.
25. Huang, J.; Wang, L.; Shen, J.; Liu, C. Effect of non-condensable gas on the start-up of a gravity loop thermosyphon with gas–liquid separator. *Exp. Therm. Fluid Sci.* **2016**, *72*, 161–170. [CrossRef]
26. Joung, W.; Yu, T.; Lee, J. Experimental study on the loop heat pipe with a planar bifacial wick structure. *Int. J. Heat Mass Transf.* **2008**, *51*, 1573–1581. [CrossRef]
27. Singh, R.; Akbarzadeh, A.; Mochizuki, M. Operational characteristics of a miniature loop heat pipe with flat evaporator. *Int. J. Therm. Sci.* **2008**, *47*, 1504–1515. [CrossRef]
28. Ji, X.; Wang, Y.; Xu, J.; Huang, Y. Experimental study of heat transfer and start-up of loop heat pipe with multiscale porous wicks. *Appl. Therm. Eng.* **2017**, *117*, 782–798. [CrossRef]
29. Huang, B.; Huang, H.; Liang, T. System dynamics model and startup behavior of loop heat pipe. *Appl. Therm. Eng.* **2009**, *29*, 2999–3005. [CrossRef]
30. Dai, C.; Yang, L.; Xie, J.; Wang, T. Nutrient diffusion control of fertilizer granules coated with a gradient hydrophobic film. *Colloids Surf. A Physicochem. Eng. Asp.* **2020**, *588*, 124361. [CrossRef]
31. Anosov, A.; Smirnova, E.Y.; Sharakshane, A.; Nikolayeva, E.; Zhdankina, Y.S. Increase in the current variance in bilayer lipid membranes near phase transition as a result of the occurrence of hydrophobic defects. *Biochim. Biophys. Acta (BBA)-Biomembr.* **2020**, *1862*, 183147. [CrossRef]

32. Li, Y.; Liu, Z.; Wang, G. A predictive model of nucleate pool boiling on heated hydrophilic surfaces. *Int. J. Heat Mass Transfer.* **2013**, *65*, 789–797. [CrossRef]
33. Phan, H.T.; Caney, N.; Marty, P.; Colasson, S.; Gavillet, J. Surface wettability control by nanocoating: The effects on pool boiling heat transfer and nucleation mechanism. *Int. J. Heat Mass Transfer.* **2009**, *52*, 5459–5471. [CrossRef]
34. Hsu, C.; Chen, P. Surface wettability effects on critical heat flux of boiling heat transfer using nanopartical coatings. *Int. J. Heat Mass Transfer.* **2012**, *55*, 3713–3719. [CrossRef]
35. Leese, H.; Bhurtun, V.; Lee, K.P.; Mattia, D. Wetting behavior of hydrophilic and hydrophobic nanostructured porous anodic alumina. *Colloids Surf. A Physicochem. Eng. Asp.* **2013**, *420*, 53–58. [CrossRef]
36. Shi, B.; Wang, Y.; Chen, K. Pool boiling heat transfer enhancement with copper nanowire arrays. *Appl. Therm. Eng.* **2015**, *75*, 115–121. [CrossRef]
37. Bankoff, S. Ebullition from solid surfaces in the absence of a pre-existing gaseous phase. *Trans. Am. Mech. Eng.* **1957**, *79*, 735–740.
38. Zheng, X.; Ji, X.; Wang, Y.; Xu, J. Study of the pool boiling heat transfer on the superhydrophilic and Superhydrophobic surface. *Prog. Chem. Ind.* **2016**, *12*, 3793–3798.
39. Thiagarajan, S.J.; Yang, R.; King, C.; Narumanchi, S. Bubble dynamics and nucleate pool boiling heat transfer on microporous copper surfaces. *Int. J. Heat Mass Transfer* **2015**, *89*, 1297–1315. [CrossRef]

BFC-POD-ROM Aided Fast Thermal Scheme Determination for China's Secondary Dong-Lin Crude Pipeline with Oils Batching Transportation

Dongxu Han [1], Qing Yuan [2], Bo Yu [1,*], Danfu Cao [3] and Gaoping Zhang [3]

1 School of Mechanical Engineering, Beijing Key Laboratory of Pipeline Critical Technology and Equipment for Deepwater Oil & Gas Development, Beijing Institute of Petrochemical Technology, Beijing 102617, China; handongxubox@bipt.edu.cn
2 National Engineering Laboratory for Pipeline Safety, Beijing Key Laboratory of Urban Oil and Gas Distribution Technology, China University of Petroleum, Beijing 102249, China; 2015314026@student.cup.edu.cn
3 Storage and Transportation Company, Sinopec Group, Xuzhou 221000, China; caodf.gdcy@sinopec.com (D.C.); zhanggp.gdcy@sinopec.com (G.Z.)
* Correspondence: yubobox@bipt.edu.cn

Abstract: Since the transportation task of China's Secondary Dong-Lin crude pipeline has been changed from Shengli oil to both Shengli and Oman oils, its transportation scheme had to be changed to "batch transportation". To determine the details of batch transportation, large amounts of simulations should be performed, but massive simulation times could be costly (they can take hundreds of days with 10 computers) using the finite volume method (FVM). To reduce the intolerable time consumption, the present paper adopts a "body-fitted coordinate-based proper orthogonal decomposition reduced-order model" (BFC-POD-ROM) to obtain faster simulations. Compared with the FVM, the adopted method reduces the time cost of thermal simulations to 2.2 days from 264 days. Subsequently, the details of batch transportation are determined based on these simulations. The Dong-Lin crude oil pipeline has been safely operating for more than two years using the determined scheme. It is found that the field data are well predicted by the POD reduced-order model with an acceptable error in crude oil engineering.

Keywords: fast thermal simulation; crude oil pipeline; batch transportation; body-fitted coordinate-based proper orthogonal decomposition reduced-order model (BFC-POD-ROM); transport scheme determination

---

1. Introduction

This paper focuses on the fast thermal scheme determination for oils batching transportation in China's Secondary Dong-Lin crude pipeline, which is obtained by the proper orthogonal decomposition based reduced-order model (POD-ROM) method. Thus, in this section, the Secondary Dong-Lin crude oil pipeline and the thermal simulations for batch transportation are briefly reviewed first. The POD reduced-order model and applications on the crude pipeline's thermal simulation are reviewed subsequently.

The Secondary Dong-Lin pipeline (also called Dongying-Linyi parallel pipeline), owned by the Sinopec Company, is an important crude oil transportation pipeline across the Shandong province in China. The pipeline is designed to transport the Shengli (SL) crude oil to Linyi, which is produced in the Shengli oilfields in Dongying. Since the fluidity of SL oil is poor, the heated transportation process was adopted before October 2015 [1].

The imported Oman oil (OM), however, with a low condensation point (0 °C) and good fluidity, was planned to be transported in the Secondary Dong-Lin pipeline by October 2015. Thus, its task turned to the transportation of both the imported OM oil and produced SL oil, which totally deviated from the original design. The "batch transportation with different oil temperatures" must be adopted and tremendous corresponding thermal simulations were required.

Different from the batch transportation of petroleum products [2], the thermal characteristics of crude oils' batch transportation are very complex because the different crude oils must be transported in different temperatures due to their large fluidity differences. Batch pipelining with different oil temperatures was first applied to the Pacific Pipeline System, situated in California USA. and commissioned in 1999 [3,4]. Unfortunately, few technique reports are available for its thermal characteristics. Recently, some studies have been performed to discover the thermal behaviors of batch pipelining of crude oils. Cui et al. [5] studied the thermal periodic characteristics for crude oils' batch transportation. Wang et al. [6] gave a report on the thermal and hydraulic behaviors. Yuan [7] studied the thermal characteristics of crude oil batch pipelining with inconstant flow rates.

All the thermal simulations in the above references, however, use the finite volume method (FVM), which is not very suitable for thermal scheme determinations of real engineering pipelines, such as the Secondary Dong-Lin pipeline in the present paper. Moreover, to obtain a proper operational scheme, thousands of thermal simulations should be done using FVM, consuming hundreds of days of simulation even with 10 computers' parallel computing. Thus, to overcome this problem, this paper adopts the POD reduced-order model to significantly improve the simulation speed.

To describe a physical problem with a reduced-order model, the first step is to obtain a series of basis functions, which can express accurately the problem with a small degree of freedom. Normally, the basis function is extracted from a large amount of data by mathematical methods, such as POD [8,9], empirical mode decomposition (EMD) [10], or dynamic mode decomposition (DMD) [11]. POD is adopted in this paper for model reduction. The reduced-order model (ROM), based on POD, can not only describe the problems, but also accelerate the calculations. Thus, this technique is studied extensively for heat transfer and is widely used in engineering.

Regarding the field of heat transfer, Banerjee et al. [12] established a POD-Galerkin ROM for heat transfer based on a finite element method, and Raghupathy et al. [13] established a boundary condition-independent ROM by combining a POD-Galerkin method with a finite volume method. The research of POD-Galerkin ROM are becoming increasingly mature, thus, it is widely applied to engineering [14–18].

The POD reduced-order models above, however, cannot be applied to the thermal simulation of the Secondary Dong-Lin pipeline. To the author's knowledge, the POD-based ROMs in the relevant literature are established for a fixed physical domain, although the boundary conditions and initial fields might vary, while, for the Secondary Dong-Lin crude oil pipeline, the physical domains vary along the pipeline since its diameter and buried depth is different from place to place. To solve this problem, the current research group first proposes a "body-fitted coordinate-based proper orthogonal decomposition reduced-order model" (BFC-POD-ROM) for the heat transfer problem [19,20], in which physical domains with different shapes or sizes can be mapped to the same computational domain.

Therefore, in this paper, BFC-POD-ROM is adopted to obtain the fast thermal simulation of China's Secondary Dong-Lin crude pipeline. Then, the detailed thermal scheme is determined based on the simulations. Finally, the predicted oil temperature distributions are verified through the field data of the Secondary Dong-Lin crude pipeline.

2. Oil Transportation Scheme and Thermal State of Secondary Dong-Lin Pipeline

2.1. Basic Situation

The Secondary Dong-Lin pipeline (constructed in 1999) was originally designed as a supplementary crude pipeline for the Old Dong-Lin pipeline (constructed in 1979), since the old one could not accomplish the crude transportation task by itself.

Prior to October 2015, the main task of the Secondary Dong-Lin pipeline was to transport the SL oil from the Sheng-Li oilfield in Dongying to the consumers in Linyi. Additionally, it transported the SL oil produced in the Bin-Nan oil production factory (also called Bin-Nan oil) through injection (See Figure 1), while the old one was arranged to transport the imported OM oil.

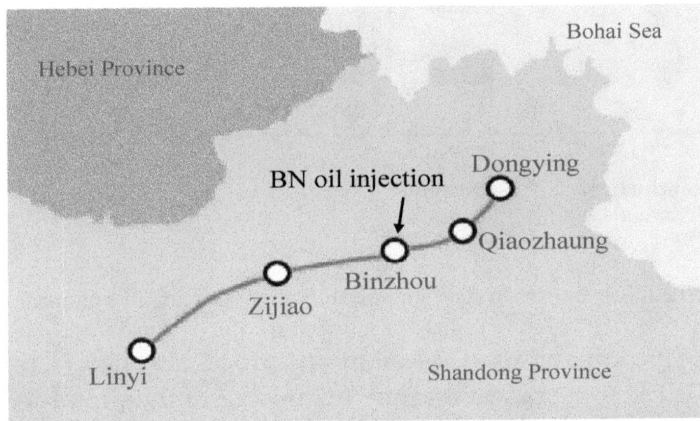

Figure 1. Route of the Secondary Dong-Lin crude pipeline.

The whole secondary pipeline is located in the Shandong Province of China. The route of the Secondary Dong-Lin pipeline is drawn in the map (See Figure 1) and its sketch map with details is found in Figure 2.

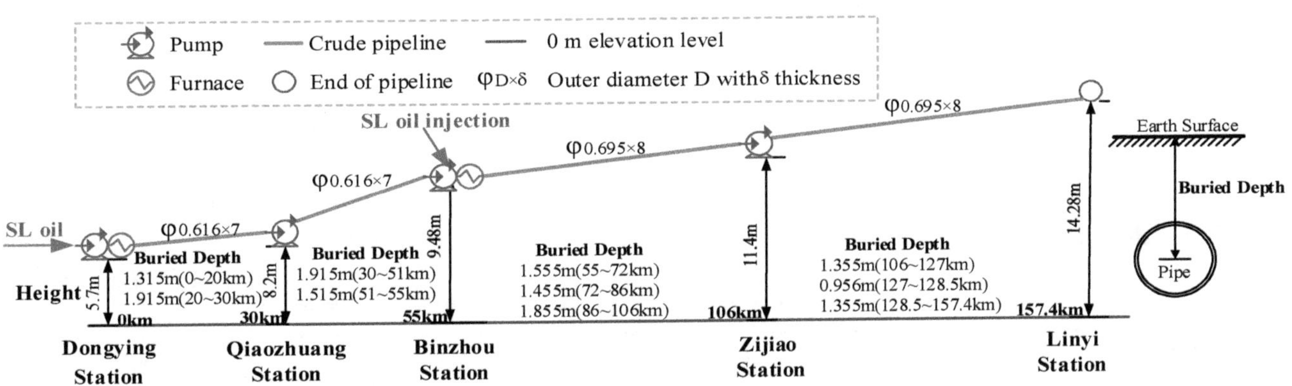

Figure 2. Sketch map of the Secondary Dong-Lin crude pipeline.

Figure 2 shows the total length of the pipeline is 157.4 km with a slight elevation change. Four pump stations (Dongying, Qiaozhuang, Binzhou, and Zijiao stations) are separately located in the positions of 0 km, 30 km, 55 km, and 106 km along the pipeline. Among them, the Dongying Station and Binzhou Station are equipped with furnaces, which means the crude oil can be heated in these two stations. The outer diameter of the pipeline is 0.616 m with a wall thickness of 7 mm before Binzhou station and becoming 0.695 m and 8 mm after Binzhou Station. The buried depth of the pipeline varies along the pipeline, with the minimum depth at 0.956 m and the maximum at 1.915 m.

Since the old Dong-Lin pipeline is too old and should be decommissioned as per regulations, the Secondary Dong-Lin pipeline was planned, by Sinopec Company, to take over the transportation tasks of the Old Dong-Lin pipeline after November 2015. Thus, the SL and OM oils would be transported in the same pipeline.

As demanded by the downstream consumer, the imported OM oil is of much better chemical quality and should not be mixed with SL oil. Therefore, the batching transportation is the only choice for the Sinopec Company (See Figure 3). Additionally, to decrease the viscosity of SL oil, the company decided to blend some OM oil into the SL oil at the beginning of the pipeline. The blended oil is named SL_{OM} oil in this paper. The optimal amount of SL_{OM} compositions also needs to be determined by thermal simulation. The rough scheme of batch transportation is shown in Figure 3.

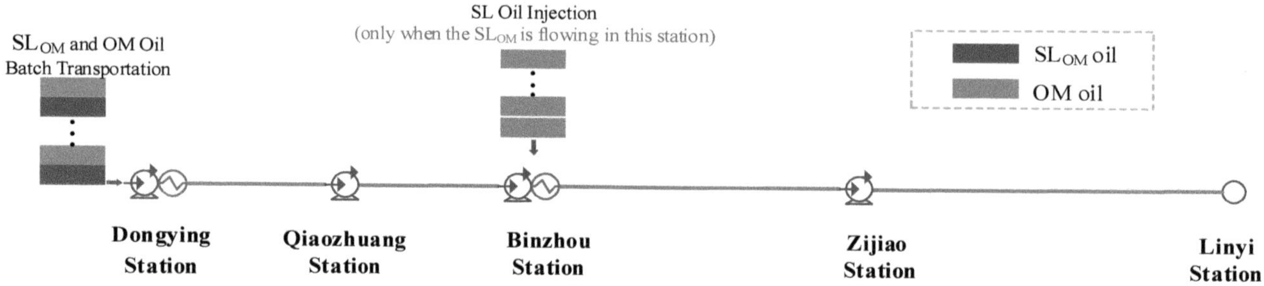

Figure 3. Sketch map of the batch transportation scheme.

Figure 3 shows the SL_{OM} oil and the OM oil are pumped alternately into the pipeline from the Dongying station, which is the first station and the beginning of the pipeline. To avoid degrading the quality of OM oil, the SL oil can only be injected when the oil flowing in Binzhou station is SL_{OM} oil.

Even though the rough scheme is easy to be pictured, to determine the safe and economic detailed transport scheme, many thermal analyses should be performed. The thermal state before and with batch transportation are introduced as follows.

2.2. *Thermal State of the Dong-Lin oil Pipeline*

2.2.1. Thermal State before the Batch Transportation

Prior to the batch transportation, the crude oils in the Secondary Dong-Lin oil pipeline were SL oil. Since the condensation point of SL is 11 °C, as shown in Table 1, it might be gelled in the pipeline in the winter when the environmental temperature is lower than the condensation point [21]. This could lead to scrapping the pipeline, which is unacceptable for the Sinopec Company.

Table 1. Basic physical properties of Shengli (SL) and Oman (OM) oil.

Oil	ρ_o (kg/m^3)	$c_{p,o}$ (J/kg·°C)	θ_{cp} (°C)	μ (Pa·s)
SL oil	937	2000	11	Polynomial $P_n(T)$
OM oil	868	2100	0	See Figure 4
SL_{OM} oil	Can be predicted through properties of SL oil and OM oil by equations in Ref. [22]. Sinopec also did extensive testing on the SL_{OM} oil with different ratios of SL and OM oil.			

Thus, before October 2015, the heating process was adopted in the Secondary Dong-Lin pipeline to ensure the safety of the pipeline. Additionally, the heat process can also lower the viscosity and improve the fluidity of SL oil, which can reduce the power cost at pumps. Since the temperature of oil flowing out of Dongying Station and Binzhou station was kept at a certain value, the thermal state was approximately steady. Figure 5 gives the oil temperature distribution along the pipeline in October 2015.

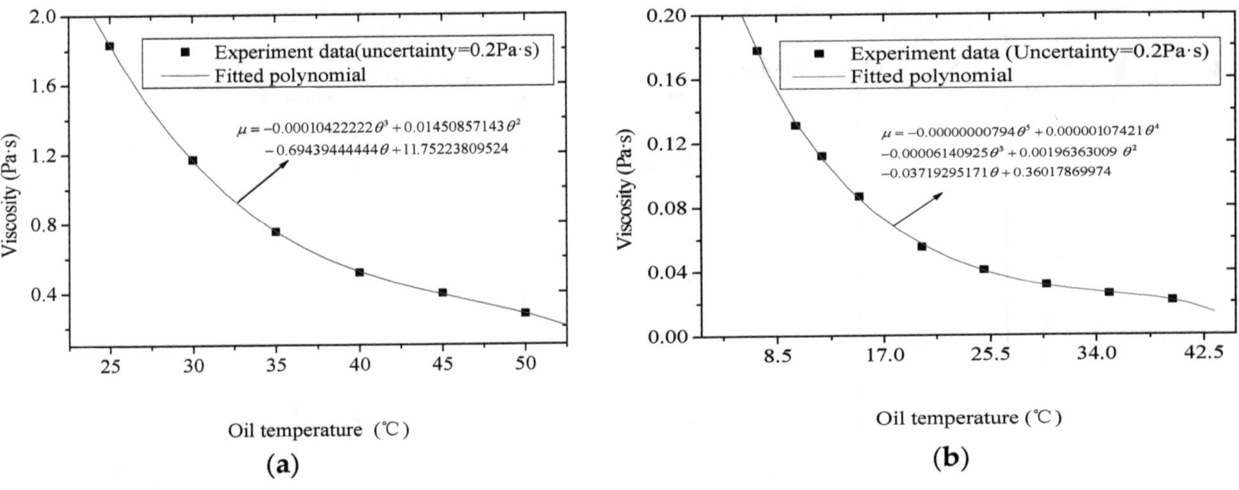

Figure 4. Viscosity-temperature curves and fitted equations of: (**a**) Shengli (SL) oil; and (**b**) Oman (OM) oil.

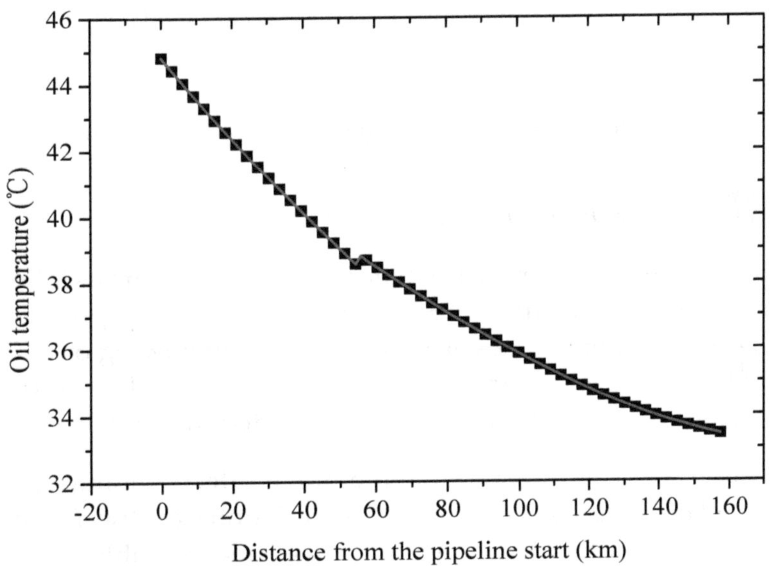

Figure 5. Oil temperature distribution along the pipeline in October 2015.

2.2.2. Thermal State of the Batch Transportation

Table 1 and Figure 4 show the fluidity differences of the SL_{OM} and OM oils are very significant at the same temperature. To save the energy consumed by furnaces, the SL_{OM} and OM oil are transported at different temperatures (See Figure 6). The SL_{OM} oil with high temperature is also called "hot oil" while the OM oil is called "cool oil". Figure 6 gives the typical oil temperature distribution flowing out of the Dongying station during a certain time frame.

Since the oil temperature varies, the thermal state of the whole pipeline is unsteady and changes dramatically when the SL_{OM} oil and OM oil alternate. Additionally, the thermal behaviors are so complicated that the thermal analysis to determine the detailed transportation scheme must be done by numerical simulations. Considering the changing environmental temperature from month to month, to clarify the thermal behavior of a certain case, the current authors must simulate the pipeline's thermal behavior from the beginning to the coldest month. This can be very time consuming.

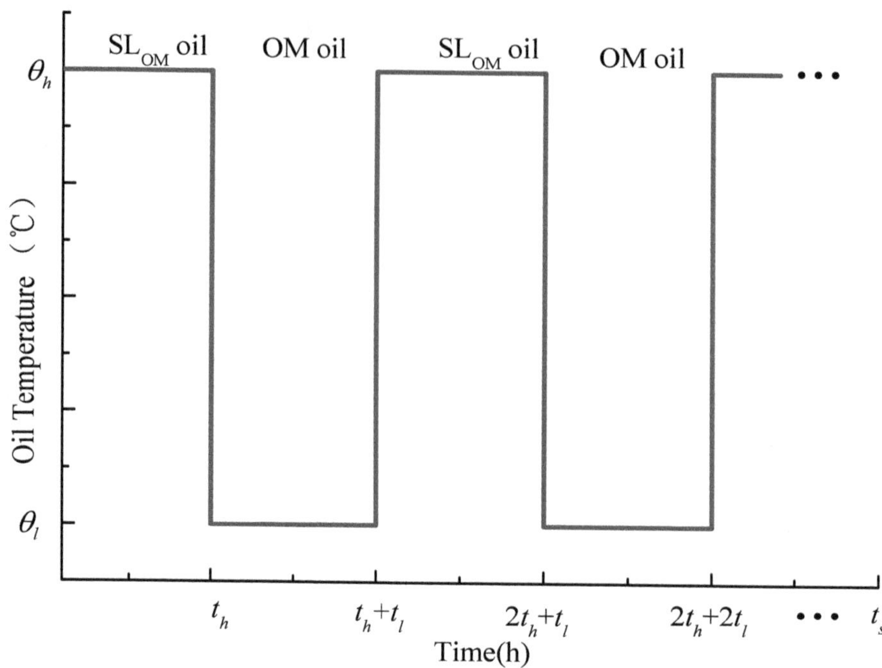

Figure 6. Oil temperature flowing out of Dongying station versus time in a certain time frame.

2.3. *Detailed Batch Transportation Scheme to be Determined*

The thermal-related operating parameters in batch transportation are as follows: The compositions of SL_{OM} (critical to its fluidity), flow fluxes of SL_{OM} and OM oils, transportation time of each patch (value of t_h and t_l in Figure 6), temperatures of SL_{OM} and OM oil flowing out of the Dongying station (value of θ_h and θ_l in Figure 6), how much and when the oil should be heated by the furnaces in Binzhou station, and the temperature and flux of SL oil injected in Binzhou station.

Thus, thousands of cases should be simulated to find a safe, economical, and relatively optimal transportation scheme. The time consumption of simulations by the frequently-used FVM can be hundreds of days on a personal computer, which is unacceptable for engineering practices. To quickly determine the thermal scheme for this paper, the current authors applied the body-fitted coordinate-based POD reduced-order model developed by their research group to the pipeline thermal simulation. The method is elaborately introduced in Section 4.

3. Physical and Mathematical Model for Secondary Dong-Lin Pipeline

Regarding the batch transportation of the Secondary Dong-Lin pipeline, as shown in Figure 3, sometimes the oil temperature can be higher than that of the surrounding pipe wall and soil, while the opposite is true at other times. This leads to the complicated, unsteady thermal state of the pipeline. Since the pipeline is 154.7 km long, the full 3-D simulation is impractical for the unsteady heat transfer in the pipeline. Thus, the problem is simplified by splitting the pipeline into a series of thermal elements, as shown in Figure 7.

Figure 7 shows the thermal element consists of a pipeline cross-section and a segment of the axial pipeline. The physical and mathematical model of the pipeline cross-section and the axial pipeline (namely the oil stream) are introduced in Sections 3.1 and 3.2, respectively.

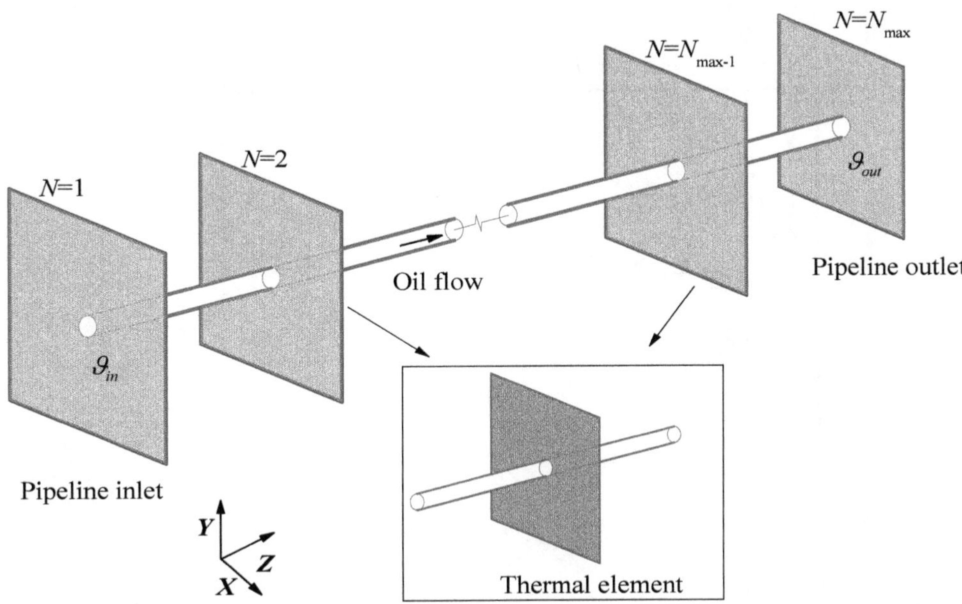

Figure 7. The simplification of the whole 3-D pipeline.

3.1. Physical and Mathematical Model of the Pipeline Cross-section

The sketch map of the pipeline cross-section is shown in Figure 8a. Since an oil pipe is buried in a certain place, it will influence the local temperature field in the soil. The further the soil from the pipe the smaller the influence can be, so it can be neglected for soil very far from the pipe. Thus, to simplify the simulation, it is believed there is a "thermal influence region" (See Figure 8a) of the oil pipe. It is assumed the oil pipe has no impact on the soil temperature field outside the influence region. According to the literature [6] and engineering experience, the thermal influence region of the hot crude oil pipeline is within 10 m, which means $L = 10$ m and $H = 10$ m, as shown in Figure 8a.

Considering the symmetry of the pipeline section (See Figure 8a), the physical model is obtained and is shown in Figure 8b. The physical domain is governed by heat conduction and the whole boundary can be divided into six parts, which are Lines O-A, A-B, B-C, C-D′-D, F-F′-O and semicircle D-E-F (See Figure 8b).

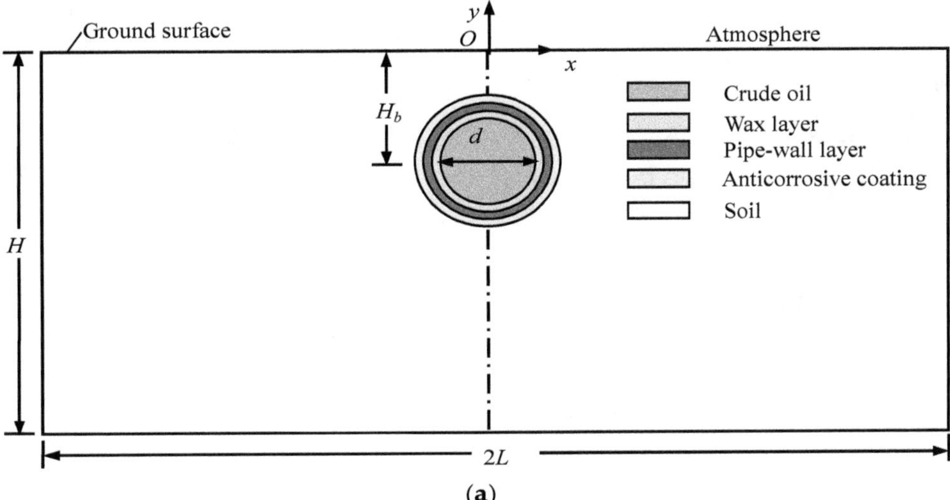

(**a**)

Figure 8. *Cont.*

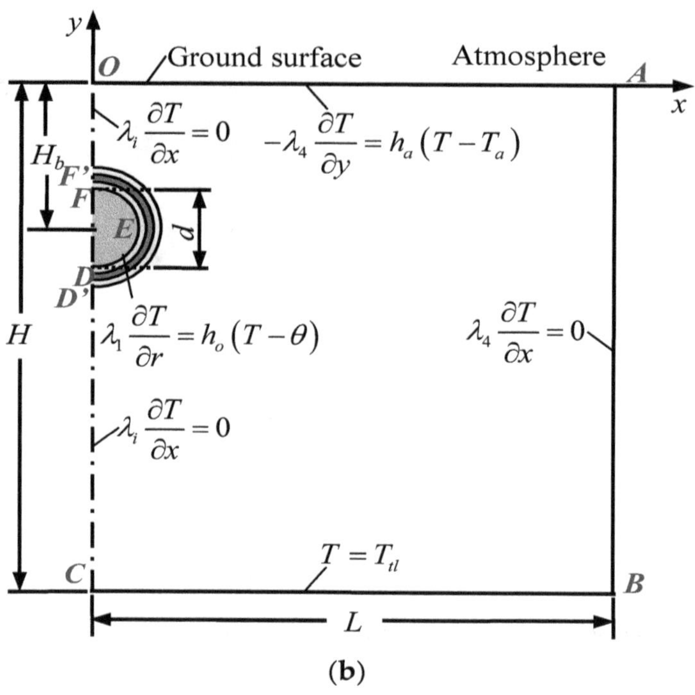

Figure 8. Thermal influence region and physical model of the pipeline cross-section on Cartesian coordinates: (**a**) Thermal influence region; (**b**) physical model.

Considering line O-A and semicircle D-E-F, they are convective heat transfer boundaries that describe the heat convection between the thermal region and the outside (air and oil, respectively, for Line O-A and semicircle D-E-M). Regarding Line A-B, it is the thermal adiabatic boundary, which means the oil pipe has no influence on the soil outside the thermal region. Looking at Line B-C, the temperature has a fixed value, which is the temperature of the thermostat soil layer. Considering Line C-D′-D and Line F-F′-O, they are symmetric boundary conditions.

Using Cartesian coordinates, the governing equation is shown in Equation (1):

$$\frac{\partial(\rho_i c_{p,i} T)}{\partial t} = \frac{\partial}{\partial x}(\lambda_i \frac{\partial T}{\partial x}) + \frac{\partial}{\partial y}(\lambda_i \frac{\partial T}{\partial y}) \tag{1}$$

where T stands for the temperature field in the cross-section. The subscript, $i = 1, 2, 3, 4$, stands for the wax layer [23], pipe-wall layer, anticorrosive layer (also called "the three layers"), and soil region around the pipe, respectively.

To take advantage of the BFC-based POD reduced-order model stated in the "Introduction", all the simulations in this paper are on body-fitted coordinates. To obtain the mathematical model on body-fitted coordinates, the physical domain is mapped to the calculation domain (See Figure 9) on body-fitted coordinates.

Correspondingly, Equation (1) is mapped to body-fitted coordinates and the governing equation for BFC is obtained as Equation (2) [24]:

$$J\frac{\partial(\rho_i c_{p,i} T)}{\partial t} = \frac{\partial}{\partial \xi}\left[\frac{\lambda_i}{J}\left(\alpha\frac{\partial T}{\partial \xi} - \beta\frac{\partial T}{\partial \eta}\right)\right] + \frac{\partial}{\partial \eta}\left[\frac{\lambda_i}{J}\left(\gamma\frac{\partial T}{\partial \eta} - \beta\frac{\partial T}{\partial \xi}\right)\right] \tag{2}$$

where, $\alpha = x_\eta^2 + y_\eta^2, \beta = x_\xi x_\eta + y_\xi y_\eta, \gamma = x_\xi^2 + y_\xi^2, J = x_\xi y_\eta - x_\eta y_\xi$.

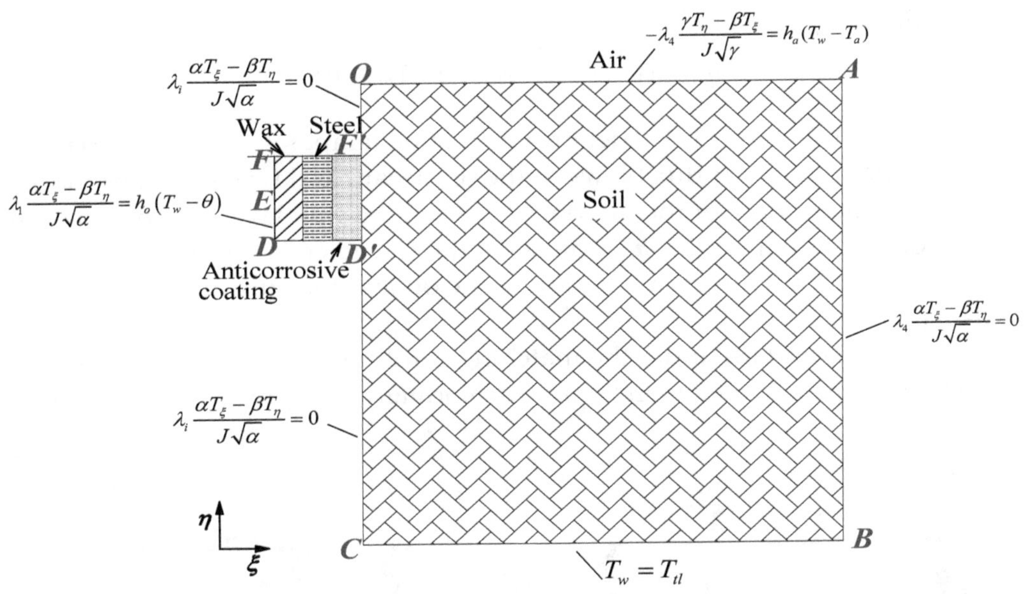

Figure 9. Physical model of pipeline cross-section on body-fitted coordinate.

Similarly, the boundary conditions on BFC also are obtained by mapping the boundary conditions on Cartesian coordinates, as shown in Figure 8b. The gained boundary conditions are as follows (See Figure 9):

$$\text{The boundary O-A}: \quad -\lambda_4 \frac{\gamma T_\eta - \beta T_\xi}{J\sqrt{\gamma}} = h_a(T_w - T_a) \tag{3}$$

$$\text{The boundary A-B}: \quad -\lambda_4 \frac{\alpha T_\xi - \beta T_\eta}{J\sqrt{\alpha}} = 0 \tag{4}$$

$$\text{The boundary B-C}: \quad T_w = T_{tl} \tag{5}$$

$$\text{The boundary C-D'-D and F-F'-O}: \quad \lambda_i \frac{\alpha T_\xi - \beta T_\eta}{J\sqrt{\alpha}} = 0 i = 1,2,3,4 \tag{6}$$

$$\text{The boundary D-E-F}: \quad \lambda_1 \frac{\alpha T_\xi - \beta T_\eta}{J\sqrt{\alpha}} = h_o(T_w - \theta) \tag{7}$$

Used in Equations (3)–(7), the subscript, w, stands for the wall in contact with the oil stream; subscripts, o and a, stand for oil and air, respectively; and the subscript, tl, stands for the thermostat layer.

Normally, the governing equation, Equation (2), is solved by FVM under the boundary conditions using Equations (3)–(7). Compared with the reduced order method introduced in Section 4, this currently presented method is called the "full order method".

3.2. Physical and mathematical model of the oil stream

The pipeline thermal simulation is the coupling of the cross-section and oil stream simulations. Since the physical and mathematical models for the cross-sections are already given above, the physical and mathematical model for the oil stream are shown in Equations (8)–(10):

Mass conservation equation:

$$\frac{\partial}{\partial t}(\rho_o A) + \frac{\partial}{\partial z}(\rho_o v A) = 0 \tag{8}$$

Energy conservation equation:

$$C_{p,o}\left(\frac{\partial \theta}{\partial t} + v\frac{\partial \theta}{\partial z}\right) - \frac{f v^3}{2d} = -\frac{4q}{\rho_o d} \tag{9}$$

Matching condition:

$$q = \lambda_1 \frac{\alpha T_\xi - \beta T_\eta}{J\sqrt{\alpha}} = h_o(T_w - \theta) \quad (10)$$

Among Equations (8), (9), and (10), θ denotes the hot oil temperature, v denotes the flow velocity of the oil stream, q denotes the heat flux between the oil stream and wax layer around the oil stream, and T_w stands for the temperature of the interface between the wax layer and oil flow, and its values are calculated by solving Equation (2). h_o represents the forced convection heat transfer coefficient of the oil flow and wax layer, which is a function of the oil temperature and must be determined by experimental data [22].

To solve Equations (7) and (8) under the matching conditions of Equation (9), the characteristics method is applied under the grid shown in Figure 10. Normally, the grid size along the pipe is between 0.5 km and 2 km.

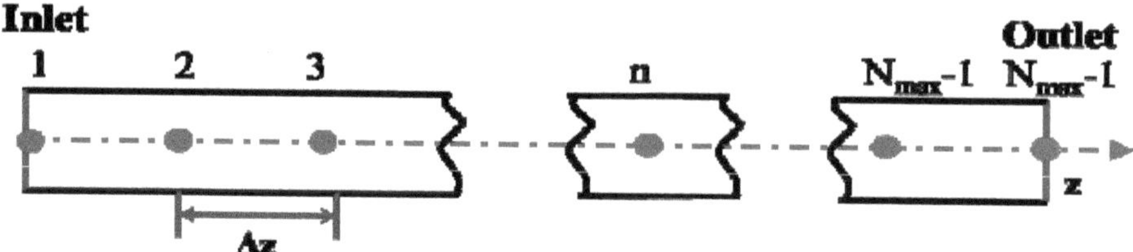

Figure 10. The grid along the pipe.

The final discretization of Equations (8) and (9) is shown in Equation (11):

$$\theta_i^n = \frac{\frac{f^{n-1}(v_i^{n-1})^3}{2d_i} - \frac{4q_{i-1}^{n-1}}{\rho d_i} - C_{p,o}\frac{\theta_i^{n-1} - \theta_{i-1}^{n-1}}{\Delta z} v_i^{n-1} + \frac{C_{p,o}\theta_i^{n-1}}{\Delta t}}{C_{p,o}/\Delta t} \quad (11)$$

Equation (11) shows the superscript, n, means the present value and $n-1$ stands for the value in the previous time step. q_{i-1}^{n-1} in Equation (11) is the key variable, which binds the pipeline cross-sections' thermal simulation with the oil temperature calculation along the pipe. Using Equation (10), the value of q_{i-1}^{n-1} can be calculated after the temperature field in the cross-section is obtained by solving Equation (2).

4. Model Reduction for Pipeline's Cross-section Thermal Simulation

To realize the thermal simulation of the crude oil pipeline, Equation (2) and Equations (8)–(9) must be solved. During the thermal simulation of the crude pipeline, the most time-consuming part is its cross-sections' temperature field calculation (solving of Equation (2)), which can occupy more than 99.99% of the whole process. The authors of this paper try to adopt the POD reduced-order model to significantly improve the calculation speed.

Using the POD reduced-order model [19], the temperature can be written as Equation (12):

$$T(\xi, \eta, t) = \sum_{k=1}^{M} a_k(t)\phi_k(\xi, \eta) \quad (12)$$

where, $\phi_k(\xi, \eta)$ $(k = 1, 2, \ldots M)$ are the POD basis functions which are dependent on space (ξ, η) and independent of time, t. The basis functions can be obtained by analyzing the sampling data by POD. $a_k(t)$ $(k = 1, 2, \ldots M)$ are the amplitudes, which are dependent on time and independent of space. $a_k(t)$ $(k = 1, 2, \ldots M)$ are the unknowns and their equations are called the "reduced-order model". M is the order of basis functions to describe the temperature field and the dimensions of the unknowns, $a_k(t)$.

Equation (12) shows there are two key points in the POD reduced-order model, the POD basis functions and the reduced-order model. The POD basis function and reduced-order models are introduced in Sections 4.1 and 4.2, respectively. The standard POD reduced-order model implementation procedure is introduced briefly in Section 4.3.

4.1. POD Basis Function

POD is a powerful mathematical method. Using the analysis on a set of simulation data (sampling matrix) obtained by full-order simulations, POD can extract a series of basis functions, which capture the dominant information of the physical problems. Regarding the unsteady-state heat transfer problem in this paper, the main process is as follows:

Suppose a two-dimensional unsteady-state heat transfer problem (such as the problem in this paper) of some specific conditions, which has N_ξ control points in the ξ direction and N_η control points in the η direction. The sampling matrix of this physical problem would be constructed.

The POD reduced-order model is applied to solve the unsteady problem with different conditions (boundary conditions and geometric shapes of the simulated domain) in this paper. The temperature fields at representative instances under various sampling conditions should be put into the sampling matrix. Suppose L kinds of sampling conditions are sampled and, for every condition, the temperature field at each time instance is obtained with a total number, K. Considering the ith condition, if the temperature fields in each time instance are sampled, the sampling matrix for the ith condition can be constructed as Equation (13). Similarly, for each condition, a sampling matrix for it can be obtained.

$$\mathbf{T}_i = \begin{bmatrix} T(\xi_1,\eta_1,t_1) & T(\xi_1,\eta_1,t_2) & \cdots & T(\xi_1,\eta_1,t_{K-1}) & T(\xi_1,\eta_1,t_K) \\ \vdots & \vdots & \cdots & \vdots & \vdots \\ T\left(\xi_{N_\xi},\eta_1,t_1\right) & T\left(\xi_{N_\xi},\eta_1,t_2\right) & \cdots & T\left(\xi_{N_\xi},\eta_1,t_{K-1}\right) & T\left(\xi_{N_\xi},\eta_1,t_K\right) \\ T(\xi_1,\eta_2,t_1) & T(\xi_1,\eta_2,t_2) & \cdots & T(\xi_1,\eta_2,t_{K-1}) & T(\xi_1,\eta_2,t_K) \\ \vdots & \vdots & \cdots & \vdots & \vdots \\ T\left(\xi_{N_\xi},\eta_2,t_1\right) & T\left(\xi_{N_\xi},\eta_2,t_2\right) & \cdots & T\left(\xi_{N_\xi},\eta_2,t_{K-1}\right) & T\left(\xi_{N_\xi},\eta_2,t_K\right) \\ \vdots & \vdots & \cdots & \vdots & \vdots \\ \vdots & \vdots & \cdots & \vdots & \vdots \\ T\left(\xi_1,\eta_{N_\eta},t_1\right) & T\left(\xi_1,\eta_{N_\eta},t_2\right) & \cdots & T\left(\xi_1,\eta_{N_\eta},t_{K-1}\right) & T(\xi_1,\eta_1,t_K) \\ \vdots & \vdots & \cdots & \vdots & \vdots \\ T\left(\xi_{N_\xi},\eta_{N_\eta},t_1\right) & T\left(\xi_{N_\xi},\eta_{N_\eta},t_2\right) & \cdots & T\left(\xi_{N_\xi},\eta_{N_\eta},t_{K-1}\right) & T\left(\xi_{N_\xi},\eta_1,t_K\right) \end{bmatrix} \tag{13}$$

What needs to be noted is the temperature of all the moments is not necessarily put into the sampling matrix, if the dominant information has been contained in the sampling matrix. Subsequent to a sampling matrix for each condition being obtained, all the matrices are combined to produce a larger sampling matrix, $\mathbf{S} \in \mathbb{R}^{m\times n}$, shown in Equation (14), where $m = N_\xi \times N_\eta$ and $n = \sum_{i=1}^{L} K_i$ (K_i is column number of $\mathbf{T}_i$). Thus, the matrix, $\mathbf{S}$, contains the information of the temperature evolution at time-bearing different conditions.

$$\mathbf{S}=\begin{bmatrix} \mathbf{T}_1 & \mathbf{T}_2 & \cdots & \mathbf{T}_L \end{bmatrix} \tag{14}$$

Using the "snapshot method" or "singular value decomposition (SVD) method" [9], the basis functions matrix can be obtained. Usually, to save the time consumption, the SVD method is adopted as $m < n$ and the "snapshot method" is used as $m > n$. Both methods are proper for $m = n$. In this paper, the SVD method is applied in Section 5.2. Thus, the SVD method is introduced as follows:

Consider the sampling matrix, $\mathbf{S} \in \mathbb{R}^{m\times n}$, with rank, $d \le min(m,n)$. Normally, in most engineering problems, there are no two same samplings put into a matrix, $\mathbf{S}$, $d = min(m,n)$. Since the SVD method is only used as $m \leq n$, d is equal to m. The SVD method guarantees real numbers, $\sigma_1 \geq \sigma_2 \geq \ldots \geq \sigma_d > 0$, orthogonal matrices, $\mathbf{U} \in \mathbb{R}^{m \times m}$ and $\mathbf{V} \in \mathbb{R}^{n \times n}$, which satisfy Equation (15):

$$\mathbf{U}^{\mathrm{T}}\mathbf{S}\mathbf{V} = \begin{pmatrix} \mathbf{D} & \mathbf{0} \\ \mathbf{0} & \mathbf{0} \end{pmatrix} \tag{15}$$

where, $\mathbf{D} = diag(\sigma_1, \ldots, \sigma_d) \in \mathbb{R}^{d \times d}$,U, and $\mathbf{V}$ are eigenvectors of $\mathbf{SS}^{\mathrm{T}}$ and $\mathbf{S}^{\mathrm{T}}\mathbf{S}$, respectively. **U** and **V** are also called left singular vectors and right singular vectors. The first d columns of **U** and **V** are eigenvectors with eigenvalues, $\lambda_i = \sigma_i^2$, and the other columns are eigenvectors with eigenvalues, $\lambda_i = 0$. The first d columns in **U** are the POD basis functions required in the POD reduced-order model. Thus, the basis functions matrix, $\boldsymbol{\psi}$, can be written as Equation (16):

$$\boldsymbol{\psi} = \begin{bmatrix} \boldsymbol{\phi}_1 & \boldsymbol{\phi}_2 & \cdots & \boldsymbol{\phi}_\mathrm{d} \end{bmatrix} \tag{16}$$

where, $\boldsymbol{\phi}_k$ is the kth columns of **U**, and $\boldsymbol{\phi}_k$ can be expressed as Equation (17):

$$\boldsymbol{\phi}_k = \left[\phi_k(\xi_1, \eta_1), \ldots, \phi_k(\xi_{N_\xi}, \eta_1), \phi_k(\xi_1, \eta_2), \ldots, \phi_k(\xi_{N_\xi}, \eta_2), \ldots, \ldots, \phi_k(\xi_{N_\xi}, \eta_{N_\eta})\right]^{\mathrm{T}} \tag{17}$$

4.2. Reduced-order model (equations of $a_k(t)$)

The reduced order model is established by projecting the governing equation, Equation (2), onto the space spanned by the first M basis functions. Following a series of deductions, the current research group established the BFC-based POD-Galerkin reduced-order model for the heat conduction problem (find the details in Ref. [20]), shown in Equation (18):

$$\sum_{k=1}^{M} \frac{\mathrm{d}a_k}{\mathrm{d}t} G_{ik} = -\left(\oint \sqrt{\alpha} q^{(\xi)} \phi_i d\eta - \sqrt{\gamma} q^{(\eta)} \phi_i d\xi\right) - \sum_{k=1}^{M} a_k H_{ik} i = 1, 2 \ldots M \tag{18}$$

where,

$G_{ik} = \int_\Omega J \rho c_p \phi_k \phi_i d\Omega.$
$H_{ik} = \int_\Omega \left[\frac{\lambda}{J}\left(\alpha \frac{\partial \phi_k}{\partial \xi} - \beta \frac{\partial \phi_k}{\partial \eta}\right) \frac{\partial \phi_i}{\partial \xi} + \frac{\lambda}{J}\left(\gamma \frac{\partial \phi_k}{\partial \eta} - \beta \frac{\partial \phi_k}{\partial \xi}\right) \frac{\partial \phi_i}{\partial \eta}\right] d\Omega$

Where the domain, Ω, here stands for the calculation domain on BFC shown in Figure 9. Equation (17) is a system of linear equations with $a_k (k = 1, 2, \ldots M)$ as unknowns, which can be solved by LU decomposition. Usually, for heat conduction problems, the value of M is below 30, which means the number of unknowns is below 30, while, in a full-order model, the unknowns can be thousands (3761 in the problem of this paper) or more, depending on the number of grids. Thus, the reduced-order model can reduce the order from thousands to less than 30, which makes the simulation speed increase significantly.

The boundary conditions are the same with the full-order model given above. The discretization of them in the POD reduced-order model can be found in Ref. [19].

4.3. The standard POD reduced-order model implementation procedure

Figure 11 gives the standard implementing procedure of the POD reduced-order model, which includes sampling, basis function extraction, reduced-order model solving, and physical field reconstructing. The details can be found in Reference [19].

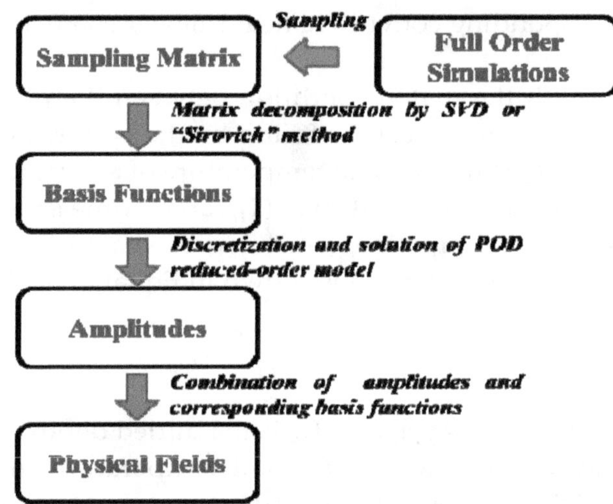

Figure 11. Procedure of proper orthogonal decomposition (POD) reduced-order model implementation.

5. Application and Discussions

The process of BFC-POD-ROM-aided fast thermal scheme determination for the Secondary Dong-Lin crude pipeline is elaborately introduced. First, the property and boundary variables of the pipeline are given. Subsequently, the implementation process and performance of the POD reduced-order model are illustrated. Finally, the thermal scheme for the Secondary Dong-Lin crude pipeline is determined and the results of the POD reduced-order model are compared with the field data.

5.1. Property and Boundary Variables for Secondary Dong-Lin crude Pipeline

To simulate the thermal behavior of the Secondary Dong-Lin pipeline, aside from the parameters given in Section 2, the property and boundary parameters should be given as well.

(1) Property parameters

Figure 8 shows there are several different-property domains in the physical mode of the pipeline, which are the oil stream, three layers, and soil region. Since the properties of the oils have been given in Section 2, only the property parameters of the soil and the three layers are offered here (see Table 2.).

Table 2. Properties of the soil and the three layers.

Soil and the three layers	ρ_i (kg/m^3)	$C_{p,i}$ (J/kg·°C)	λ_i (W/m·°C)
Soil (0 km–56 km)	2235	1.67	943
Soil (56 km–157.4 km)	2235	1.35	943
Anticorrosive Layer	1000	0.4	1670
Pipe-wall Layer	7850	50	460
Wax Layer	1000	0.15	2000

As shown in Table 2, the soil thermal conductivity from 0 km to 55 km of the pipeline is 1.67 W/m·°C, while the soil thermal conductivity becomes 1.35 W/m·°C after 56km of the pipeline. The heat conductivities are obtained by inverse calculation through the operational data of the Sinopec Company, which is a frequently-used method in oil pipeline thermal simulations [7].

(2) Boundary parameters

As shown in Figure 8, the boundary parameters of the pipeline cross-section include the temperature of the oil stream, θ, and the corresponding convective heat transfer coefficient, h_o, the temperature of air, T_a, the corresponding convective heat transfer coefficient, h_a, and the temperature of the thermostat layer, T_{tl}.

θ can be determined through the match condition between the pipeline cross-section and oil stream, as shown in Equation (10), and h_o can be found by the empirical formula in Ref. [22] using the above boundary parameters. T_a is not the real temperature of air, rather a pseudo air temperature interpolated by the temperature of the thermostat layer (T_{tl}) and the measured value of soil temperature, T_b, in the buried depth of the crude oil pipeline, which has been a typical method in engineering for easy simulation [22]. The interpolation equation is shown in Equation (19):

$$T_a = T_b + \frac{T_b - T_{tl}}{H} H_b \tag{19}$$

where, H is the depth of the thermostat layer, and H_b is the buried depth of the oil pipe (See Figure 8).

Thus, the thermostat layer temperature, T_{tl}, soil temperature in a buried depth, T_b, and convective heat transfer coefficient, h_a, are the boundary parameters needed (see Table 3). Using field measurements for this study, T_{tl} is 12.3 °C and h_a is 20 (W/m^2·°C). T_b varies from day to day and mile to mile. Table 3 gives a series of field data in five typical spots measured by the Sinopec Company. T_b for other days and other spots are interpolated by the data from Table 3.

Table 3. Boundary parameters.

	Soil Temperature in Buried Depth T_b (°C)				
Date	**0 km**	**30 km**	**55 km**	**106 km**	**157.4 km**
Oct. 31st 2015	22.67	22.44	20.28	20.75	24.76
Nov. 30th 2015	15.90	19.39	14.89	17.31	21.09
Dec. 31st 2015	7.39	13.39	9.94	12.71	15.68
Jan. 31st 2016	5.86	11.27	8.03	9.33	13.02
Feb. 29th 2016	5.23	10.14	7.24	8.32	11.31

5.2. POD-ROM-based fast thermal simulation for Secondary Dong-Lin crude pipeline

Figure 11 shows that to obtain the POD-ROM-based fast thermal simulation, there are two main steps. The first is sampling and basis function extracting. The second is reduced-order model solving and physical field reconstructing.

5.2.1. Sampling and Basis function

The quality of basis functions has a significant influence on the accuracy of POD reduced-order model-based fast thermal simulation. The acceptable basis functions should contain the main characters of the temperature field evolution of the Secondary Dong-Lin crude pipeline. This mainly depends on the sampling process, in which the samplings should be representative and, for sake of time consumption, as few as possible.

Figure 7 shows the whole pipeline is separated into a series of slices and the main differences among the slices are geometrics (see Figure 2) and boundary conditions (see Table 3). Thus, for this particular problem, the obtained temperature basis function must be capable of depicting the temperature field under different combinations of geometrics and boundary conditions. The main process for sampling and basis function extraction is as follows:

First, the sampling conditions are given in Tables 4–6. The sampling conditions are a thermal analogy of the real conditions. Take "Sampling 1" as an example: Sampling 1 is set as an analog of the thermal situation of the 0–20 km part in the Secondary Dong-Lin crude pipeline. To obtain such an analog, the geometry and heat conductivity of Sampling 1 are set the same within the 0–20km part in the Secondary Dong-Lin crude pipeline. To save time in the samplings' calculation, the samplings' simulation times are set at 50 days, which is much shorter than the real conditions (five months). The alternate frequency (1.5 d/3.5 d one time) is also much higher than the real conditions. The real

T_b (soil temperature in buried depth) changes from month to month so are analogized by changing it every 10 days (see Table 6).

Table 4. The samplings.

Sampling No.	Geometry (See Table 5)	T_b (°C)	λ_4 (W/m·°C)	θ_h (°C)	θ_l (°C)	t_h/t_l (d/d)	T_{initial} (°C)	t_s (d)
1	Geo1	See Table 6	1.67	23	38	1.5/3.5	0	50
2	Geo2			20	38		45	
3	Geo3			15	35		75	
4	Geo4		1.35	14	36	0.5/1.5	105	
5	Geo5			13	35		157	

Note: T_{initial} (0) stands for the temperature field in the 0 km zone of the crude pipeline before the commission of the batching transportation scheme. T_{initial} (45) stands for the temperature field at 45 km. The others follow in kind.

Table 5. Geometry parameters of Geo1–Geo5.

Geo No.	d (m)	H_b (m)	δ_w (m)	δ_{ac} (m)
Geo 1	φ616 × 7	1.315	0.003	0.007
Geo2	φ616 × 7	1.915	0.003	0.007
Geo3	φ695 × 8	1.3555	0.003	0.007
Geo4	φ695 × 8	1.5555	0.003	0.007
Geo5	φ695 × 8	1.8555	0.003	0.007

Table 6. T_b at different times.

Time	0 d–10 d	10 d–20 d	20 d–30 d	30 d–40 d	40 d–50 d
T_b (°C)	22.67	15.9	7.39	5.86	5.23

Subsequently, using FVM, the temperature field in each slice is calculated with time steps at 600s on body-fitted grids (grid number is 3761), as shown in Figure 12.

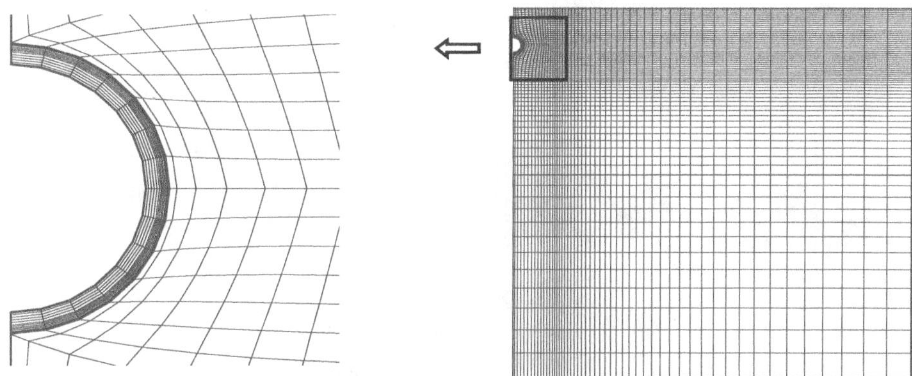

Figure 12. Body-fitted coordinate-based (BFC) grid of the pipeline cross-section.

Based on the temperature fields obtained by the FVM, the sampling matrix shown in Equation (14) can be constructed. To reduce the dimension of the sampling matrix (which can cause problems for the matrix decomposition) and avoid unnecessary information noise, sampling is dense when the temperature field changes quickly. Quite the reverse, it is sparse when the temperature field is changing slowly.

Thus, for this particular problem, the temperature field in every time step is put into the sampling matrix in time intervals [0 h, 5 h] after the alternates of θ_h and θ_l. One temperature field is adopted every five and 10 time steps in the time interval [5 h, 10 h] and [10 h, t_h or t_l], respectively. To summarize, the sampling matrix, **S**, shown in Equation (14) consists of 8330 temperature fields, which makes **S** a matrix with 3761 rows and 8330 columns.

Finally, the basis functions are extracted from the sampling matrix, **S**, by SVD. The energy distribution of basis functions is shown in Figure 13. The first 23 basis functions, $M = 23$ in Equation (12), are applied into the POD reduced-order model for fast simulation. Figure 14 gives contours of four typical basis functions based on Geo 1.

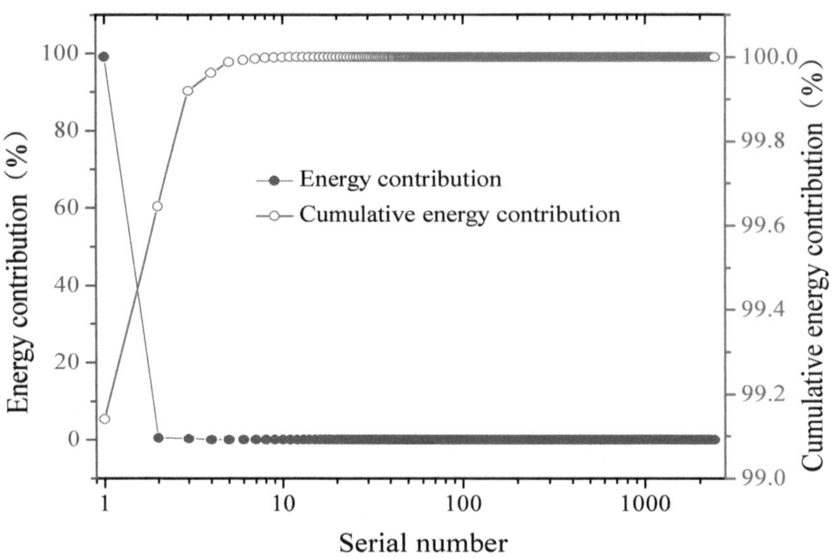

Figure 13. Energy distribution of basis functions.

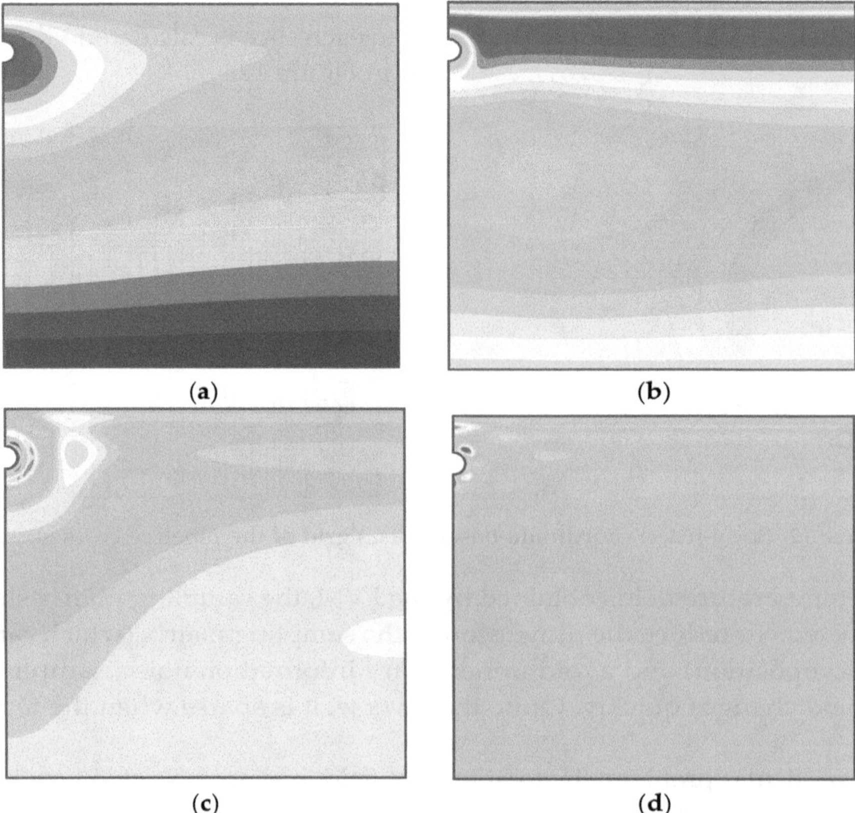

Figure 14. Contours of four typical basis functions based on Geo 1: (**a**) 1st; (**b**) 5th; (**c**) 15th; (**d**) 23rd.

5.2.2. POD Reduced-order model validation and thermal characteristics of batch transportation

To test the accuracy of the POD reduced-order model-based fast simulation, the first three potential schemes (see Table 7) given by Sinopec are simulated by a POD reduced-order model as well as the FVM.

Table 7. Parameters for Schemes 1–3.

Scheme No.	Q_h^D/Q_l^D (m^3h^{-1}/m^3h^{-1})	Q_h^{BI} (m^3h^{-1})	θ_h/θ_l (°C/°C)	t_h/t_l (d/d)	r_{SL}/r_{OM}
1	2161/2470			7.56/2.44	90:36
2	2322/2321	172	See Table 8	8.11/2.83	90:26
3	1872/2470			7.56/2.44	90:20

Table 8. Values of θ_h and θ_l in different months.

Time	θ_l (°C)	θ_h (°C)			Furnaces
		Scheme 1	Scheme 2	Scheme 3	
Oct. 2015	23.20	37.03	38.22	39.04	Both furnaces in Dongying and Binzhou are closed
Nov. 2015	20.00	36.37	37.78	38.75	
Dec. 2015	14.88	33.77	35.39	36.51	
Jan. 2016	13.60	33.89	35.64	36.84	
Feb. 2016	12.90	33.69	35.48	36.71	

Regarding the Secondary Dong-Lin crude pipeline, the coldest month each year is February. The environmental temperature decreased progressively from October 2015 to February 2016, as shown in Table 3, which made February 2016 the riskiest month. The oil temperature distribution has great significance for engineering. Considering Scheme 1, the oil temperature distributions during the last cycle (the last 2.44 d/7.56 d) in February 2016 are shown in Figure 15.

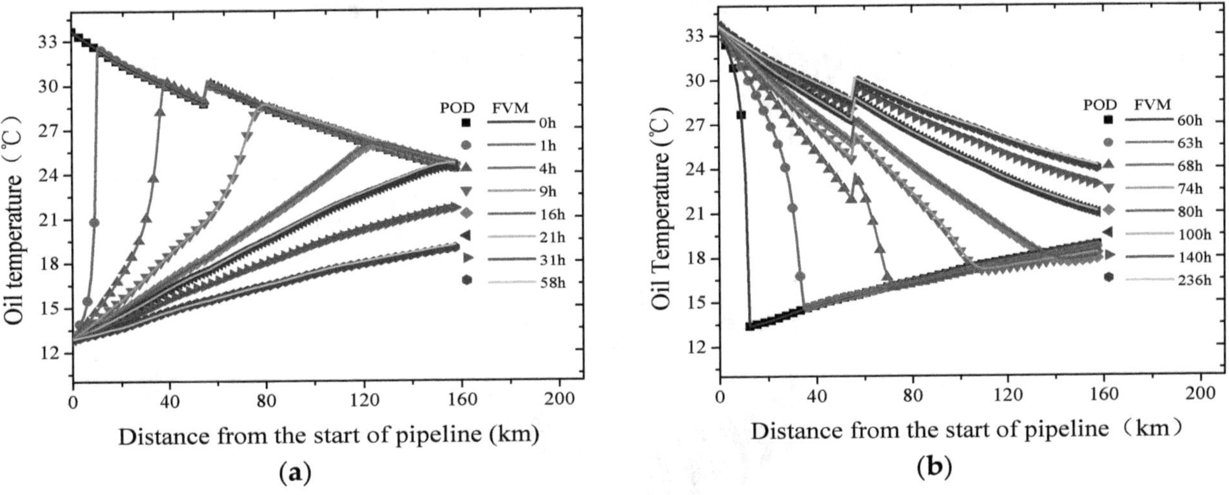

Figure 15. The oil temperature along the pipeline during the last cycle of Scheme 1: (**a**) 0 d–2.44 d; (**b**) 2.44 d–10 d.

The curves in Figure 15 are explained as follows: Beginning with the last cycle, $t = 0$ h in Figure 15a, the whole pipeline is filled with SL_{OM} oil. Table 8 shows the temperature of the oil in $z = 0$ km is 33.69 °C and, due to the heat loss to the environment, the temperature decreases along the pipeline from 0 km to 55 km. Since the injected SL_{OM} oil in $z = 55$ km is at 50 °C (higher than the upstream SL oil's temperature of 28.73 °C), the mixed oil becomes 30.18 °C, which shows a temperature jump at $z = 55$ km (See Figure 15a). The temperature after $z = 55$ km decreases for the earlier stated reason before $z = 55$ km.

Then, the OM oil, at 12.9 °C, is pumped into the pipeline, which is at a lower temperature than the SL oil's temperature. Thus, the OM oil is warmed when it flows along the pipeline, which can be shown by the curves at t = 1 h, 4 h, and 9 h. What should be noted is that the oil pipeline is not occupied fully by the OM oil until t = 16 h. Thus, the curves of t = 1 h, 4 h, and 9h show an up-and-down trend. The "uptrend-curves" part of the pipeline is filled with OM oil, which absorbs energy from the environment. The "downtrend-curves" part of the pipeline is filled with SL_{OM} oil, which releases energy to the environment. When at t = 16 h, the SL oil is completely driven out of the pipeline by the OM oil behind.

While the OM oil keeps absorbing the energy from the environment, the temperature of the environment decreases, leading to the OM oil temperature decreasing with time, as shown in Figure 15a. The "reduction" lasts until the oils alternate at the beginning of the pipeline. Figure 15b shows the temperature curves after the alternation. It illustrates the opposite thermal characteristics of Figure 15a, when SL_{OM} oil with a higher temperature than OM oil is pumped into and fully occupies the pipeline.

Figure 15a,b have the symbols and lines representing the results of the POD reduced-order model and the FVM, respectively. Even though the thermal characteristics of the Dong-Lin crude pipeline are very complicated, as stated above, the results of the POD reduced-order model agree well with those of the FVM. The main thermal characteristics along the pipeline of Schemes 2 and 3 are similar to those of Scheme 1, given in Figure 15, and the differences are just the values. Aside from the oil temperature distribution along the pipeline, the engineers are also concerned about the oil temperature flowing out of the pipeline. Thus, for the three schemes, Figure 16 gives the outflow oil temperature (in the end of the pipeline) versus time curves.

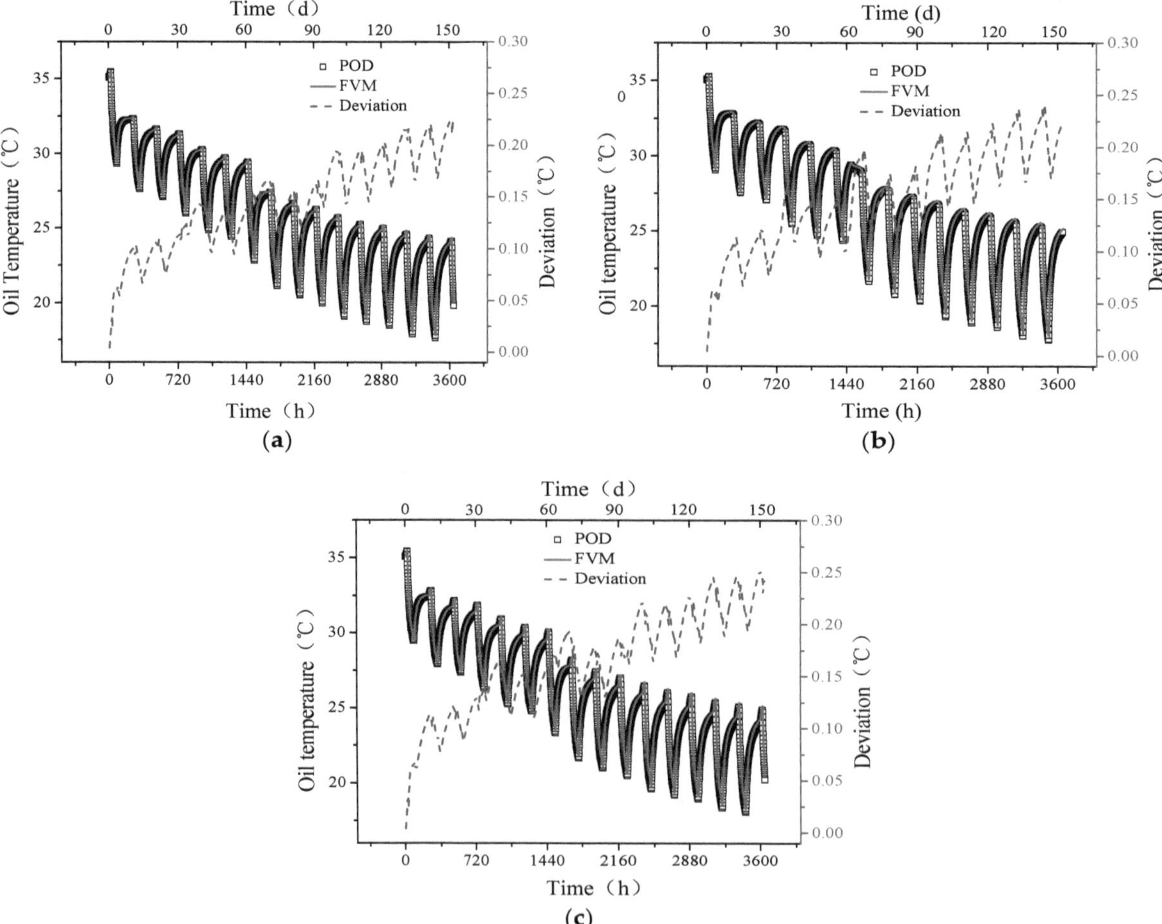

Figure 16. The oil temperature flowing out of the pipeline in Linyi Station: (**a**) Scheme 1; (**b**) Scheme 2; and (**c**) Scheme 3.

Figure 16 shows the oil temperature in the end of the pipeline periodically declines from October 2015 to February 2016. The periodic alternating of the oils pumped into the pipeline and the increasingly colder weather are responsible for the two trends. Looking at Figure 16, it can be found that the POD reduced-order model has good accuracy. The largest errors (compared with the FVM) for Schemes 1, 2, and 3 are 0.22 °C, 0.24 °C, and 0.25 °C, respectively. The mean errors (compared with FVM) for Schemes 1, 2, and 3 are 0.15 °C, 0.15 °C, and 0.16 °C, respectively.

To illustrate the accuracy of the POD reduced-order model more vividly, for Scheme 1, Figures 17 and 18, respectively, give the temperature fields at the beginning and end of the pipeline. Figures 17 and 18 have dashed lines representing the POD reduced-order model results and solid lines representing the FVM. The results agree well with each other.

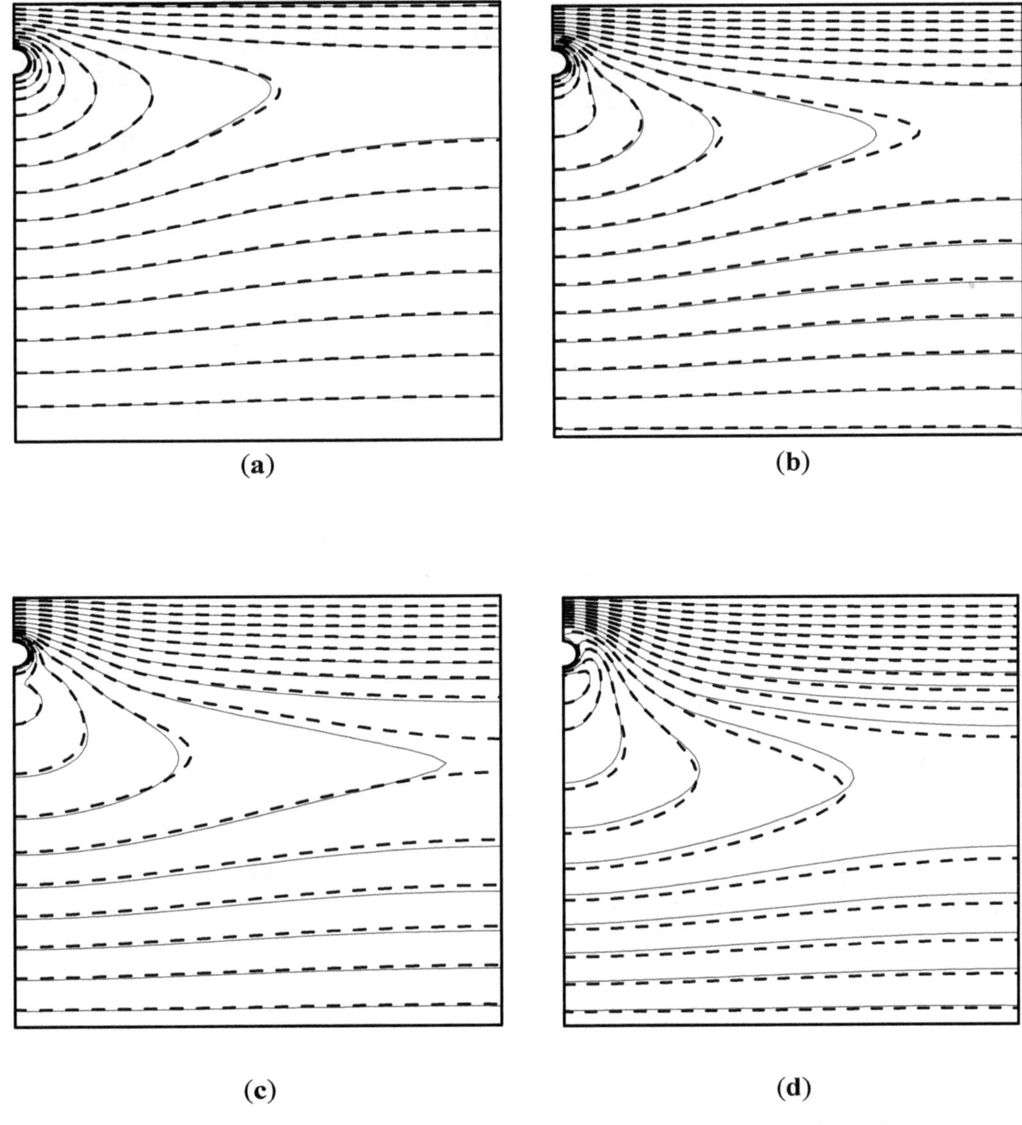

Figure 17. The cross-section's temperature field in the beginning of the pipeline (dashed lines: POD reduced-order model. Solid lines: Finite volume method (FVM)): (**a**) 38th day; (**b**) 76th day; (**c**) 114th day; (**d**) 152nd day.

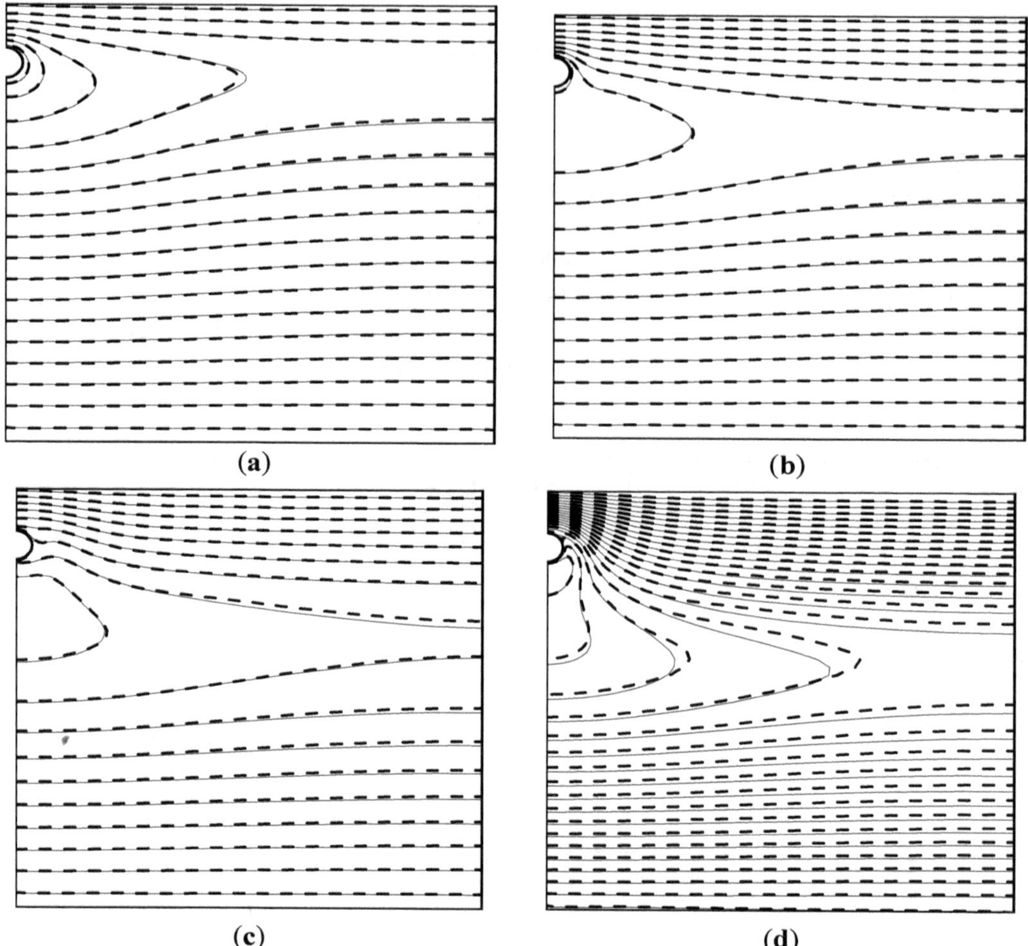

Figure 18. The cross-section's temperature field at the end of the pipeline (Dashed lines: POD reduced-order model. Solid lines: FVM): (**a**) 38th day; (**b**) 76th day; (**c**) 114th day; (**d**) 152nd day.

To illustrate the speed advantage of the POD reduced-order model, Table 9 shows the time consumption of the POD reduced-order model and the FVM. The simulation speed is more than 100 times faster than the FVM, which means much time is saved in the engineering.

Table 9. Time consumptions of the FVM and POD reduced-order models.

Scheme No.	FVM (h)	POD (h)	Acceleration Factor
Scheme 1	62.7	0.49	128
Scheme 2	64.5	0.52	124
Scheme 3	64.0	0.56	114

Considering the above analysis, it can be found that, for the thermal simulation of the Dong-Lin crude pipeline with oil batching transportation, the POD reduced-order model has good accuracy and significant efficiency. Thus, rather than the commonly-used FVM, the POD reduced-order model is adopted to determine the operational scheme in this paper.

5.3. Detail Batch Transportation Scheme Determination and Field Verification

Through the first three potential schemes given by Sinopec, the performance of the POD reduced-order model was verified. Following that, considering the capacity of oil stocks, blend ratios of SL_{OM} oils, behaviors of furnaces, and other factors, the Sinopec Company drew 1024 potential schemes in sum. All the schemes were simulated by the POD reduced-order model to find their thermal performance critical to the energy cost of the oil heated and oil pumped because oil fluidity is related to oil temperature (see Figure 4).

Considering the thermal and hydraulic consumption of each scheme given by the simulations, the Sinopec Company chose Scheme No. 124 (shown in Table 10) as the final operating scheme based on the considerations of power consumption and transportation risk. The risk means the restart-ability of the oil pipeline after being shut-down for an accident. The lower the temperature and the longer the shut-down time, the higher the risk is.

Table 10. Parameters for the determined scheme.

Scheme No.	Q_h^D/Q_l^D (m^3h^{-1}/m^3h^{-1})	Q_h^{BI} (m^3h^{-1})	θ_h/θ_l (°C/°C)	t_h/t_l (d/d)	r_{SL}/r_{OM}
124	1370/2064	172	See Table 11	4.58/1.17	9:2

Table 11. The determined values of θ_h and θ_l in different months.

Time	θ_l (°C)	θ_h (°C)	Furnace in Dongying	Furnace in Linyi
Oct. 2015 Nov. 2015 Dec. 2015 Jan. 2016 Feb. 2016	23.5 23 21 19 17	41.5 41.5 41.5 41.5 41.5	Close for OM oil. Open for SL_{OM} oil and keep heating it to 41.5 °C	Open for both OM and SL_{OM} oil Raise OM oil 8 °C Raise SL_{OM} oil 5 °C

To ensure the success of the new scheme commissioning, the Sinopec Company did much more preparation than was expected. Thus, the company began to execute the chosen scheme on November 4, 2015. Figure 19 gives the comparisons between the simulated results and the operational field data.

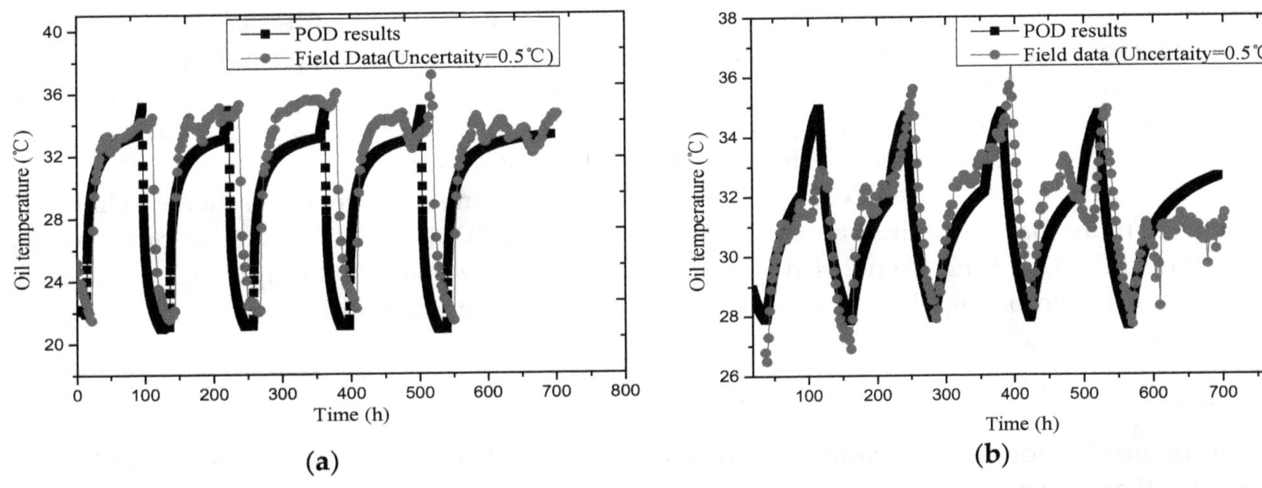

Figure 19. The comparisons between field data and the results predicted by the POD reduced-order model: (**a**) The oil temperature flowing into Binzhou station versus time on February 2016; (**b**) the oil temperature flowing into Linyi station versus time on February 2016.

Figure 19 demonstrates the thermal behavior under the chosen scheme can be predicted with an error acceptable to crude oil engineering. There are some spikes in the field data shown in Figure 19. The flow flux and the temperature are not controlled precisely during real-time engineering, which means the pipeline is operated around the chosen scheme, and not exactly on the scheme. To operate a crude oil pipeline exactly according to the planning scheme is impossible due to unpredictable factors (such as the flow shock of the upstream, and the temporary change of oil transportation task) existing in practical engineering.

The oil temperature flowing into other stations shows the same behavior. To properly describe the deviations between the field data and simulation results, the mean oil temperatures flowing into Binzhou and Linyi station are calculated. It should be noted that the mean temperature of OM oil and SL_{OM} oil are calculated independently (see Tables 12 and 13). The deviations are also shown in Tables 12 and 13.

Table 12. The mean temperature of OM oil in different months.

Oil	Data	Nov. 2015		Dec. 2015		Jan. 2016		Feb. 2016	
θ_{OM} in BZ	Field Data	28.89	(0.24, 0.8%)	26.78	(0.48, 1.8%)	25.66	(1.21, 4.7%)	23.54	(0.42, 1.8%)
	POD Results	29.13		26.30		24.45		23.12	
θ_{OM} in LY	Field Data	32.18	(1.27, 3.9%)	32.93	(0.15, 0.4%)	29.62	(0.48, 1.6%)	29.98	(0.32, 1.1%)
	POD Results	33.45		33.08		30.10		30.30	

Note: BZ is for Binzhou station and LY is for Linyi station. The values in the brackets are absolute and relative deviations, respectively.

Table 13. The mean temperature of SL_{OM} oil in different months.

Oil	Data	Nov. 2015		Dec. 2015		Jan. 2016		Feb. 2016	
$\theta_{SL_{OM}}$ in BZ	Field Data	36.02	(0.12, 0.3%)	35.66	(0.48, 1.3%)	33.89	(1.21, 3.6%)	33.64	(0.06, 0.2%)
	POD Results	36.14		36.35		34.63		33.58	
$\theta_{SL_{OM}}$ in LY	Field Data	33.99	(0.99, 2.9%)	34.48	(1.37, 4.0%)	31.29	(0.79, 2.5%)	31.72	(0.63, 2.0%)
	POD Results	34.98		35.85		32.08		32.35	

Note: BZ is for Binzhou station and LY is for Linyi station. The values in the brackets are absolute and relative deviations, respectively.

Tables 12 and 13 show the mean oil temperature errors are less than 1.27 °C and 1.37 °C for OM and SL_{OM} oil, respectively, which is acceptable for oil transportation engineering. There are three reasons to generate such a deviation. First, the physical model used in the current study is an approximate description and the POD reduced-order model itself is also an approximate mathematical model. Second, there are some inevitable errors in parameter values in the physical model, such as the heat capacity, density and viscosity of oil, the heat conductivity in the regions of the wax layer, pipe-wall layer, anticorrosive layer, and soil, and the forced convection heat transfer coefficient of the oil flow and the wax layer. Third, the pipeline is operated around the chosen scheme, not exactly on the scheme, as stated above, which is believed to be the biggest reason.

To summarize, the BFC-based POD-ROM is adopted to determine the detailed scheme to improve the efficiency more than a hundred times. Compared with the FVM, the POD reduced-order model reduces the simulation time from 264 days, using 10 computers' parallel computing, to 2.2 days. The Dong-Lin crude oil pipeline has been safely operating for more than two years using the determined scheme.

6. Conclusions

The determination of the crude pipeline's detailed batch transportation scheme could cost tremendous thermal simulation time, which is an unsolved problem in oil transportation engineering. To solve this problem for China's Secondary Dong-Lin crude pipeline, a fast scheme determination strategy was developed for the first time.

The main idea of the strategy was that, rather than the traditional FVM, the BFC-based POD reduced-order model was adopted to increase the speed of the thermal simulation. The whole strategy included three main steps, which are summarized as follows:

(1) Sampling matrix construction and basis function obtainment.

The quality of basis functions has significant influence on the accuracy of the POD reduced-order model-based fast thermal simulation. Thus, the corresponding samplings should be representative and, for the sake of time consumption, as few as possible. Regarding crude oil batch transportation, the following point is recommended: Design the sampling conditions by using the "analogy method", which means the geometry and heat conductivity should be the same within the pipeline and the boundary condition can be designed as an analogy of the real condition (see Tables 4–6). To reduce the dimensions of the sampling matrix, sampling can be dense as the temperature field changes quickly and sparse as the temperature field is slowly changing.

(2) The validation of the POD reduced-order model.

The POD reduced order model is an approximate method. Its accuracy is dependent on the quality of obtained basis functions, the number of basis functions adopted in the POD reduced-order model, and the complexity of the problem. Thus, the POD reduced-order model should be validated by the full order model (FVM in this paper) before being applied to the engineering.

Regarding the specific problem in this paper, through the comparisons between the POD reduced-order model and FVM, it was found that the POD reduced-order model had good accuracy with a mean error of 0.16 °C, which is acceptable for engineering. Thus, it is believed the POD reduced-order model can be applied to thermal simulations of crude oil batch transportation. The obtained basis function is feasible for China's Secondary Dong-Lin crude pipeline and the number of adopted basis functions is appropriate.

(3) The determination of the transportation scheme.

It was found that the POD reduced-order model can be more than one hundred times faster than the FVM. Therefore, to find a proper operational scheme, it is feasible to simulate hundreds, or thousands of schemes with reasonable time consumption. Moreover, this method can be combined with some optimization methods to find an optimized operating scheme.

Aided by the body-fitted coordinate-based POD reduced-order model, the details of the batch transportation scheme were determined and can be found in Tables 10 and 11. The Dong-Lin crude oil pipeline has been safely operating for more than two years using the determined scheme. Compared with the field data, the predicted results by the POD reduced-order model are of an acceptable accuracy for crude oil engineering. The mean oil temperature errors were less than 1.27 °C and 1.37 °C for OM and SL_{OM} oil, respectively.

Author Contributions: Conceptualization, B.Y.; Methodology, D.H.; Code, D.H.; Validation, Q.Y., D.C. and G.Z.; Writing—Original Draft Preparation, D.H.; Writing—Review & Editing, B.Y., D.H., and Q.Y.

Nomenclature

Roman Symbols

a_k	Amplitude of the kth POD basis function
A	Flowing area of the pipeline (m^2)
$c_{p,i}$	Specific heat capacity of region i (J/(kg·°C))
$c_{p,o}$	Specific heat capacity of oil (J/(kg·°C))
d	Diameter of the pipeline (mm)
f	The Darcy coefficient
h_o	Heat convection coefficient between the wax layer and oil stream (W/(m^2·°C))
h_a	Heat convection coefficient between the soil and air (W/(m^2·°C))
H	Thermal influence region on the vertical direction (m)
H_b	The buried depth of crude oil pipeline (m)
L	Thermal influence region on the horizontal direction (m)
q	The heat flux between the wax layer and oil stream (W/m^2)
Q_h^D, Q_l^D	Flow flux of hot and cool oil in Dongying station (m^3/h)
Q_h^{BI}	The flow flux injected in Binzhou station (m^3/h)
r_{SL}, r_{OM}	Ratio of SL oil and OM oil in SL_{OM} oil

S	The sampling matrix
t_h	Transportation time of hot oil during one period (h)
t_l	Transportation time of cool oil during one period (h)
t_s	The total transportation time of hot and cool oils (h)
T	Temperature of the pipeline's cross-section
T_a	Temperature of the air (°C)
T_c	Temperature of the soil thermostat layer (°C)
v	Flow velocity of the oil stream (m/s)
x, y	Cartesian coordinate in the cross-section of pipeline (m)
z	Coordinate along the cross-section of pipeline (m)
Greek Symbols	
δ_w	Thickness of the wax layer (m)
δ_{ac}	Thickness of the anticorrosive layer (m)
ϕ_k	The *k*th POD basis function
$\mathbf{\Phi}_k$	Vector of the *k*th POD basis function
λ_i	Heat conductivity coefficient of region *i* (W/(m·°C))
μ	Viscosity (Pa·s)
θ	Temperature of the oil (°C)
θ_{cp}	Condensation point of crude oil (°C)
θ_h	Temperature of the hot oil, namely SL_{OM} oil (°C)
θ_l	Temperature of the cool oil, namely OM oil (°C)
$\theta_{SL_{OM}}$	Temperature of SL_{OM} oil (°C)
θ_{OM}	Temperature of OM oil (°C)
ρ_i	Density of region *i* (W/(m·°C))
ρ_o	Density of oil (kg/m^3)
ξ, η	Body-fitted coordinate in the cross-section of pipeline (m)
Subscripts	
1,2,3,4	Regions of wax layer, pipe-wall layer, anticorrosive layer and soil respectively
ac	Anticorrosive layer
h	Hot oil
l	Cool oil, namely low temperature oil
o	oil
OM	Oman oil
SL	Shengli oil
SL_{OM}	Mixture of SL oil and OM oil
tl	Thermostat layer
w	Wax layer
ξ, η	Partial derivatives of the variable

References

1. Yu, Y.; Wu, C.; Xing, X.; Zuo, L. Energy saving for a Chinese crude oil pipeline. In Proceedings of the ASME 2014 Pressure Vessels and Piping Conference, Garden Grove, CA, USA, 20–24 July 2014. [CrossRef]
2. Zhang, H.R.; Liang, Y.T.; Xia, Q.; Wu, M.; Shao, Q. Supply-based optimal scheduling of oil product pipelines. *Petrol. Sci.* **2016**, *13*, 355–367. [CrossRef]
3. Shauers, D.; Sarkissian, H.; Decker, B. California line beats odds, begins moving viscous crude oil. *Oil Gas J.* **2000**, *98*, 54–66.
4. Mecham, T.; Wikerson, B.; Templeton, B. Full Integration of SCADA, field control systems and high speed hydraulic models-application Pacific Pipeline System. In Proceedings of the International Pipeline Conference, Calgary, AB, Canada, 1–5 October 2000. [CrossRef]
5. Cui, X.G.; Zhang, J.J. The research of heat transfer problem in process of batch transportation of cool and hot oil. *Oil Gas Stor. Trans.* **2013**, *23*, 15–19. [CrossRef]
6. Wang, K.; Zhang, J.J.; Yu, B.; Zhou, J.; Qian, J.H.; Qiu, D.P. Numerical simulation on the thermal and hydraulic behaviors of batch pipelining crude oils with different inlet temperatures. *Oil Gas Sci. Technol.* **2009**, *64*, 503–520. [CrossRef]

7. Yuan, Q.; Wu, C.C.; Yu, B.; Han, D.; Zhang, X.; Cai, L.; Sun, D. Study on the thermal characteristics of crude oil batch pipelining with differential outlet temperature and inconstant flow rate. *J. Petrol. Sci. Eng.* **2018**, *160*, 519–530. [CrossRef]
8. Lumley, J.L. *The Structure of Inhomogeneous Turbulent Flows in Atmospheric Turbulence and Radio Wave Propagation*; Nauka: Moscow, Russian, 1967.
9. Sirovich, L. Turbulence and dynamics of coherent structures, Part 1: Coherent structures. *Q. Appl. Math.* **1987**, *45*, 561–571. [CrossRef]
10. Huang, N.E.; Shen, Z.; Long, S.R.; Wu, M.L.; Shih, H.H.; Zheng, Q.; Yen, N.C.; Tung, C.C.; Liu, H.H. The empirical mode decomposition and Hilbert spectrum for nonlinear and nonstationary time series analysis. *Proc. Roy. Soc. London A* **1998**, *454*, 903–995. [CrossRef]
11. Schmid, P.J. Dynamic mode decomposition of numerical and experimental data. *J. Fluid Mech.* **2010**, *656*, 5–28. [CrossRef]
12. Banerjee, S.; Cole, J.V.; Jensen, K.F. Nonlinear model reduction strategies for rapid thermal processing systems. *IEEE Trans. Semiconduct. Manuf.* **1988**, *11*, 266–275. [CrossRef]
13. Raghupathy, A.P.; Ghia, U. Boundary-condition-independent reduced-order modeling of complex 2D objects by POD-Galerkin methodology. In Proceedings of the IEEE Semiconductor Thermal Measurement and Management Symposium, San Jose, CA, USA, 15–19 March 2009. [CrossRef]
14. Fogleman, M.; Lumley, J.; Rempfer, D.; Haworth, D. Application of the proper orthogonal decomposition to datasets of internal combustion engine flows. *J. Turbul.* **2004**, *5*, 1–18. [CrossRef]
15. Ding, P.; Tao, W.Q. Reduced order model based algorithm for inverse convection heat transfer problem. *J. Xi'an Jiaotong Univ.* **2009**, *43*, 14–16. [CrossRef]
16. Thomas, A.B.; Raymond, L.F.; Pau, G.A.; Thomas, J.; Breault, R.W. A reduced-order model for heat transfer in multiphase flow and practical aspects of the proper orthogonal decomposition. *Theor. Comp. Fluid Dyn.* **2012**, *43*, 68–80. [CrossRef]
17. Gaonkar, A.K.; Kulkarni, S.S. Application of multilevel scheme and two level discretization for POD based model order reduction of nonlinear transient heat transfer problems. *Comput. Mech.* **2015**, *55*, 179–191. [CrossRef]
18. Selimefendigil, F.; Öztop, F. Numerical study of natural convection in a ferrofluid-filled corrugated cavity with internal heat generation. *Numer. Heat Tr. A-Appl.* **2015**, *67*, 1136–1161. [CrossRef]
19. Han, D.; Yu, B.; Yu, G.; Zhao, Y.; Zhang, W. Study on a BFC-based POD-Galerkin ROM for the steady-state heat transfer problem. *Int. J. Heat Mass Tran.* **2014**, *69*, 1–5. [CrossRef]
20. Han, D.; Yu, B.; Zhang, X. Study on a BFC-Based POD-Galerkin Reduced-Order Model for the unsteady-state variable-property heat transfer problem. *Numer. Heat Tr. B-Fund.* **2014**, *65*, 256–281. [CrossRef]
21. Yu, G.; Yang, Q.; Dai, B.; Fu, Z.; Lin, D. Numerical study on the characteristic of temperature drop of crude oil in a model oil tanker subjected to oscillating motion. *Energies* **2018**, *11*, 1229. [CrossRef]
22. Yang, Y.H. *Design and Management of Oil Pipeline*, 1st ed.; Press of China Petroleum University: Qing Dao, China, 2006.
23. Cheng, Q.; Gan, Y.; Su, W.; Liu, Y.; Sun, W.; Xu, Y. Research on exergy flow composition and exergy loss mechanisms for waxy crude oil pipeline transport processes. *Energies* **2017**, *10*, 1956. [CrossRef]
24. Tao, W.Q. *Numerical Heat Transfer*; Xi'an Jiaotong University Press: Xi'an, China, 2001.

10

Visualization Study of Startup Modes and Operating States of a Flat Two-Phase Micro Thermosyphon

Liangyu Wu [1], Yingying Chen [1], Suchen Wu [2], Mengchen Zhang [2], Weibo Yang [1,*] and Fangping Tang [1]

[1] School of Hydraulic, Energy and Power Engineering, Yangzhou University, Yangzhou 225127, China; lywu@yzu.edu.cn (L.W.); yychen@microflows.net (Y.C.); tangfp@yzu.edu.cn (F.T.)
[2] Key Laboratory of Energy Thermal Conversion and Control of Ministry of Education, School of Energy and Environment, Southeast University, Nanjing 210096, China; scwu@microflows.net (S.W.); 220130421@seu.edu.cn (M.Z.)
* Correspondence: wbyang@yzu.edu.cn

Abstract: The flat two-phase thermosyphon has been recognized as a promising technique to realize uniform heat dissipation for high-heat-flux electronic devices. In this paper, a visualization experiment is designed and conducted to study the startup modes and operating states in a flat two-phase thermosyphon. The dynamic wall temperatures and gas–liquid interface evolution are observed and analyzed. From the results, the sudden startup and gradual startup modes and three quasi-steady operating states are identified. As the heat load increases, the continuous large-amplitude pulsation, alternate pulsation, and continuous small-amplitude pulsation states are experienced in sequence for the evaporator wall temperature. The alternate pulsation state can be divided into two types of alternate pulsation: lengthy single-large-amplitude-pulsation alternated with short multiple-small-amplitude-pulsation, and short single-large-amplitude-pulsation alternated with lengthy multiple-small-amplitude alternate pulsation state. During the continuous large-amplitude pulsation state, the bubbles were generated intermittently and the wall temperature fluctuated cyclically with a continuous large amplitude. In the alternate pulsation state, the duration of boiling became longer compared to the continuous large-amplitude pulsation state, and the wall temperature of the evaporator section exhibited small fluctuations. In addition, there was no large-amplitude wall temperature pulsation in the continuous small-amplitude pulsation state, and the boiling occurred continuously. The thermal performance of the alternate pulsation state in a flat two-phase thermosyphon is inferior to the continuous small-amplitude pulsation state but superior to the continuous large-amplitude pulsation state.

Keywords: thermosyphon; two-phase flow; startup; phase change; operating state; visualization

---

1. Introduction

The rapid development of microelectronic technology poses an important challenge for high-heat-flux electronic cooling [1]. A number of efficient cooling technologies, including microchannels [2–5], heat pipes [6], boiling [7], solid–liquid change [8], fractal surface [9,10] and liquid cooling [11,12], have been proposed and are used for electronic component cooling [13], spacecraft thermal control [14], microfluidic engineering [15] and battery thermal management [16]. Of these, the flat two-phase thermosyphon has been regarded as the preferred heat removal technique in a confined space [17,18]. Differing from conventional heat pipes, the evaporator and condenser section of the flat two-phase thermosyphon are replaced by two plates. Therefore, the flat two-phase thermosyphon can expand one-dimensional heat transfer into two-dimensional heat transfer on a plane, resulting in an efficient heat transfer and satisfactory temperature uniformity [19]. Utilized in a solar collector system, the two-phase thermosyphon combines good energy behavior with simplicity of manufacture [14–17]. In addition, the flat two-phase thermosyphon can be integrated with electronic devices [20]. Therefore, the flat two-phase thermosyphon has been introduced as an effective way to meet the challenges of heat dissipation and temperature uniformity for high-heat-flux electronic devices.

Unlike in unconfined spaces, the vapor-liquid phase change inside a flat two-phase thermosyphon involves a direct interaction between boiling and condensation accompanied by complex gas–liquid two-phase flow behaviors because of the narrow space [21]. For example, when the liquid level in the cavity is high, the liquid surface may contact the condenser surface because of the two-phase flow fluctuation, which forms a "liquid bridge" between the liquid surface and the condenser surface because of the surface tension. The formation of a liquid bridge may increase the thermal resistance of the condenser surface as the liquid film thickness increases [22]. In addition, Wu et al. [23] reported that, instead of gravity and buoyancy, the surface tension and the shear force at the gas–liquid interface are the dominant forces affecting the gas–liquid two-phase flow during boiling and condensation in the flat two-phase thermosyphon. Therefore, the coupled boiling–condensation process has a non-negligible effect on the gas–liquid two-phase flow behavior and heat transfer process [24] in flat two-phase thermosyphons.

Available experimental and theoretical studies of flat two-phase thermosyphons focused primarily on the steady-state thermal performance, such as temperature uniformity [25,26], equivalent thermal conductivity [27], thermal resistance [28], and maximum heat transfer capacity. In addition, a number of studies have been conducted to investigate the coupled boiling–condensation heat transfer in a confined space [17,21,29]. However, few studies have given attention to the thermal response and corresponding gas–liquid two-phase flow in the flat two-phase thermosyphon, specifically the boiling and condensation behaviors under the startup and quasi-steady processes. Visual representation of vapor–liquid two-phase flow is of significance to understand the coupled boiling–condensation phase change heat transfer inside flat two-phase thermosyphons. Therefore, a visualization experiment was conducted to investigate the gas–liquid phase change heat transfer. The startup modes and operating state in the flat two-phase thermosyphon are investigated and analyzed by the observed dynamic temperature variations of the evaporator and condenser surface as well as the gas–liquid two-phase interface evolution.

2. Description of Experiment

In order to visualize the vapor–liquid two-phase flow, a flat two-phase thermosyphon was manufactured with transparent sidewalls. The visualization experimental setup, as shown in Figure 1, includes the flat two-phase thermosyphon, a heating unit, a cooling unit, and a data acquisition unit. The gas–liquid two-phase behavior and the coupled evaporator–condenser heat transfer process are observed in the flat two-phase thermosyphon with this experimental setup.

The flat two-phase thermosyphon primarily comprises a quartz glass tube, an evaporator plate, a condenser plate, a sealing plate of heat sink, and a charging pipe, as shown in Figure 2. The various components of the flat two-phase thermosyphon are illustrated in Figure 3. The glass tube is tightly clamped between the evaporator and condenser plates, and a close cavity is formed where the working medium is filled. The outer diameter of glass tube is 50 mm and the inner diameter is 44 mm. For clear observation of the vapor–liquid two-phase flow in the flat two-phase thermosyphon, a height of 15 mm was used for the glass tube during the experiment. Annular grooves were milled on the evaporator and condenser plates and were filled with fluorine rubber O-rings for sealing the flat two-phase thermosyphon. The evaporator and condenser sections are square brass plates with 45-mm sides, as shown in Figure 3. The thickness of the evaporator plate is 5 mm, while that of the condenser plate is 10 mm. A cooling channel is set on the back of the condenser plate; therefore, a sealing plate is needed for the sealing of the cooling water. The working medium is fed into the cavity through the charging pipe. In this study, de-ionized water was used as the working medium, as shown in Table 1.

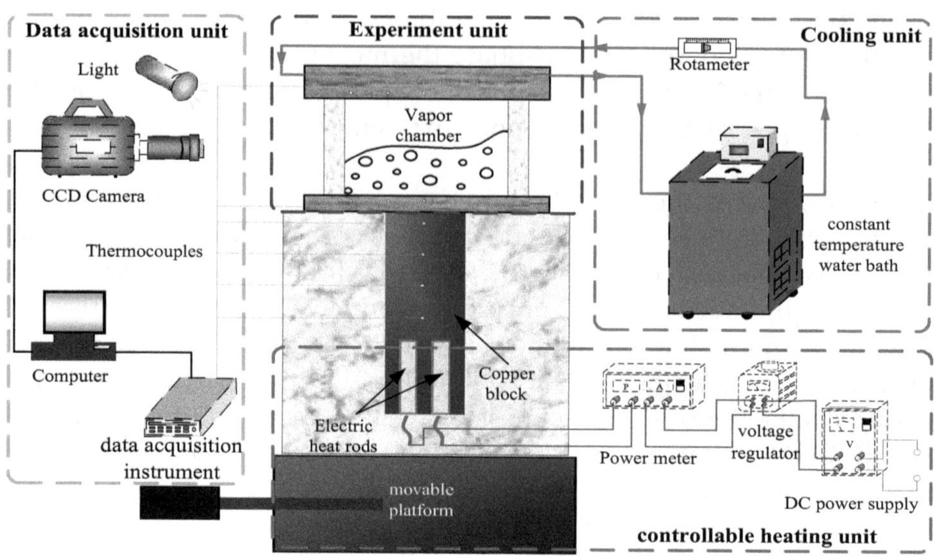

Figure 1. Schematic of experimental setup.

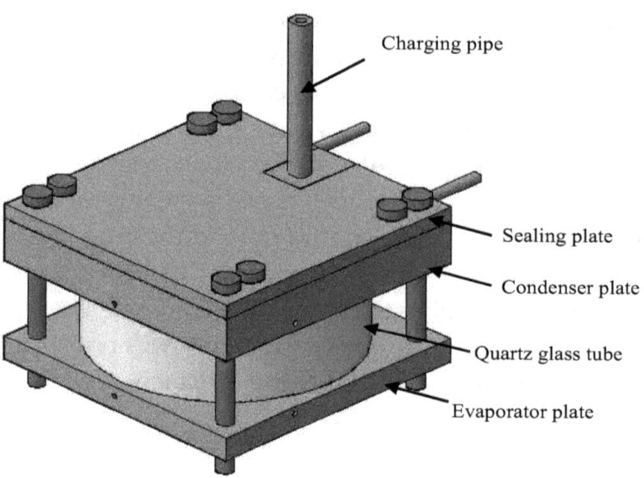

Figure 2. Schematic of flat two-phase thermosyphon.

Table 1. The thermophysical properties of de-ionized water (3.17 kPa, 25 °C).

Thermophysical Property	Value
Density (kg/m^3)	997
Enthalpy (kJ/kg)	104.67
latent heat (kJ/kg)	2435
thermal conductivity (W/K)	607
Specific isobar heat capacity (kJ/(kg·K)	4.182

The heating unit, which is used to provide and control a heat source for the flat two-phase thermosiphon, is supplied by a direct current power combined with a voltage regulator. Electric heating rods are embedded in the bottom of a copper block to heat the evaporator and a power meter is paralleled with the heating rods to measure the heating power, as shown in Figure 4. The electric heating rods are 6 mm in diameter and 50 mm in length. The 20-mm diameter copper block is clamped by a bracket comprising a number of polytetrafluoroethylene plates, so that the copper block can make close contact with the evaporator section of the flat two-phase thermosyphon. To obtain the accurate axial heat flux density of the copper block, four 0.5-mm diameter holes were drilled to a depth of 10 mm in the copper block. The holes were distributed along the axial direction of the copper block, and the distance from the holes to the top surface of the copper block are 5 mm, 20 mm, 35 mm, and 50 mm, as shown in Figure 5. During the experiment, heating power ranging from 20–90 W was applied to examine the effect of heat load on vapor–liquid two-phase flow.

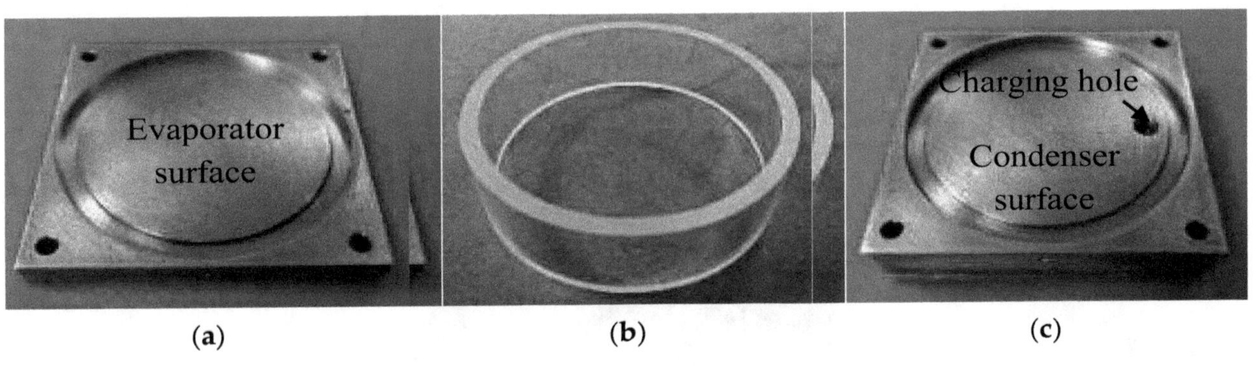

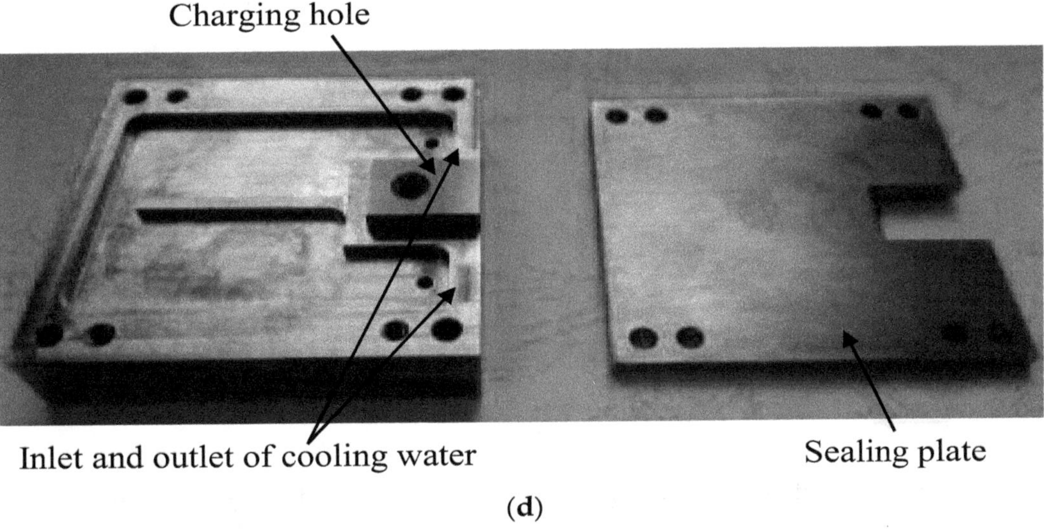

Figure 3. Structure of flat two-phase thermosyphon: (**a**) evaporator surface; (**b**) quartz glass tube; (**c**) condenser surface; and (**d**) cooling water tank at back of condenser section.

The cooling unit is connected to the condenser section of the flat two-phase thermosyphon, and comprises a constant temperature water bath and a glass rotameter. The constant temperature water bath provides a constant-temperature fluid circulation for the cooling of the condenser section, and the flow rate of the circulating water is measured by the glass rotameter. In the experiment, the water temperature was set to room temperature of 25 °C, and the flow rate of the circulating water was set to 80 mL/min. In order to measure the sensible heat gain of the circulating water, a number of thermocouples were installed at the inlet and outlet of the condenser section. The sensible heat gain is determined as the heat load of the flat two-phase thermosyphon. The cooling heat transfer rate is $Q_{cool} = \rho V C p \Delta T_{cool}$, where ρ and Cp are the density and specific heat of the cooling water, respectively, and ΔT_{cool} is the temperature difference between the cooling water at inlet and outlet.

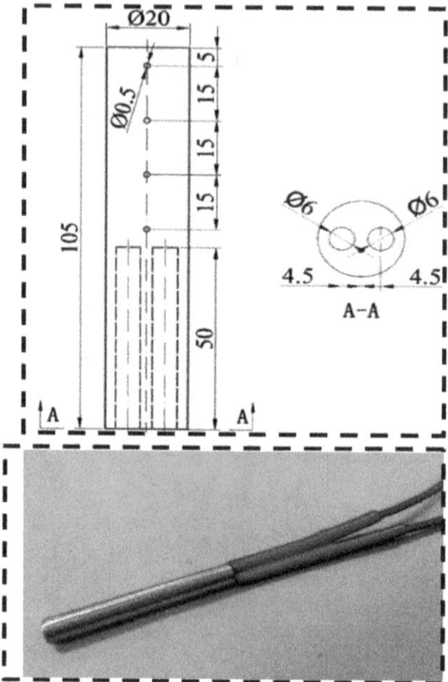

Figure 4. Structure of copper block and heating rods (unit: mm).

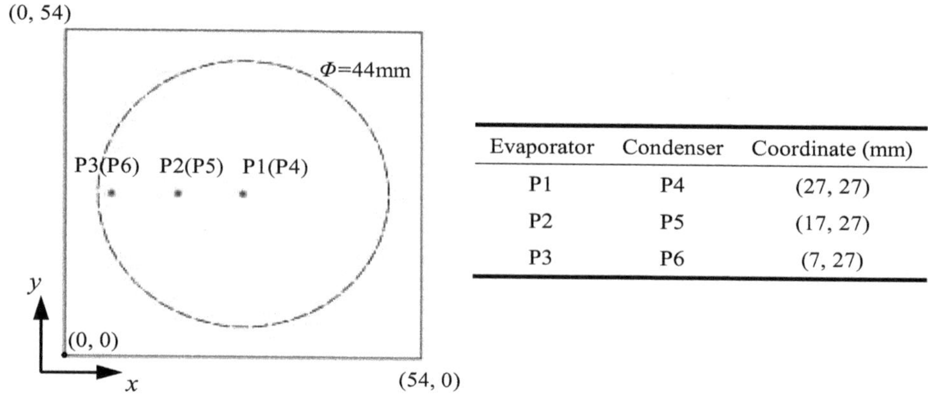

Evaporator	Condenser	Coordinate (mm)
P1	P4	(27, 27)
P2	P5	(17, 27)
P3	P6	(7, 27)

Figure 5. Position of measuring points on evaporator and condenser surface.

The data acquisition unit comprises a charge-coupled device (CCD) camera, a data collector (Agilent 34970A, Santa Clara, CA, USA), a computer, and a light. The vapor–liquid two-phase behavior in the confined cavity of the flat two-phase thermosyphon was monitored in real time and recording was started when the flat two-phase thermosyphon operated. The frame rate of CCD was set to 500 fps. To record the dynamic temperature variations of evaporator and condenser surfaces, three thermocouples (see Figure 5) were installed on the evaporator and condenser surfaces. The temperature was recorded by the data acquisition instrument. The sampling rate for the acquisition of the temperature is 2 Hz.

3. Results and Discussion

The vapor–liquid two-phase state inside the cavity is directly related to the wall temperature of the evaporator and condenser section, and, therefore, determines the phase change mechanisms and thermal performance of the flat two-phase thermosyphon [30]. An experiment was conducted to visually monitor the operating state of the working medium inside the cavity during the start-up and the quasi-steady processes. Based on the observed vapor–liquid two-phase state and the measured wall temperatures of the condenser and evaporator sections under different working conditions, it is possible to analyze the startup modes and operating states of a flat two-phase thermosyphon.

3.1. Startup Modes

When the evaporator section is heated, the flat two-phase thermosyphon initially goes through a start-up process and then attains a quasi-steady operating state. During the start-up process, the bubbles start to generate on the evaporator surface, and then gradually increase in size and rise from the surface. As a result, a complex two-phase vapor–liquid is formed close to the evaporator surface because of the dual effect of the natural convection and bubble disturbance. According to the dynamic temperature variations of the evaporator surface, two startup modes are identified for the flat two-phase thermosyphon: sudden startup mode and gradual startup mode.

The difference of two startup modes is mainly caused by the filling rate, i.e., the thermal response is largely dependent on the charging ratio. A larger degree of superheating is required to induce the startup of the flat two-phase thermosyphon when the charging ratio is higher. However, small superheating is required to induce the startup when the charging ratio is low. Therefore, different startup modes appear even though the heat input remains the same. The charging ratio φ is defined as the ration of the working fluid volume to the interior volume of thermosyphon.

3.1.1. Sudden Startup Mode

As shown in Figure 6a, in the sudden startup mode, the wall temperature of the evaporator section rises rapidly initially without temperature fluctuations, and the vapor–liquid two-phase working medium remains stationary in the early startup stage. Subsequently, the wall temperature continues to increase gradually, and no bubbles are generated inside the two-phase thermosyphon. During this process, a significant degree of superheating is required to induce the startup of the flat two-phase thermosyphon. The input heat flux is absorbed as sensible heat thorough the working medium and solid wall; therefore, the energy continues to be absorbed in the thermosyphon. As can be seen from the figure, at approximately 1000 s, the evaporator wall temperature suddenly decreases by approximately 10 °C (i.e., a temperature overshoot occurs in the startup process), and the condenser wall temperature suddenly rises by approximately 7 °C. Subsequently, there is a pronounced temperature pulsation for the condenser and evaporator walls.

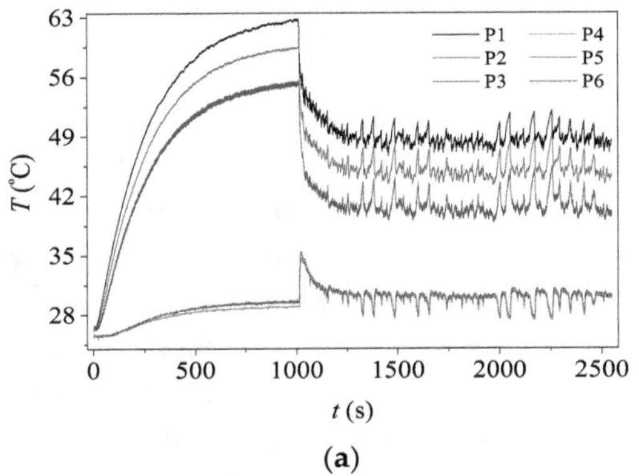

(a)

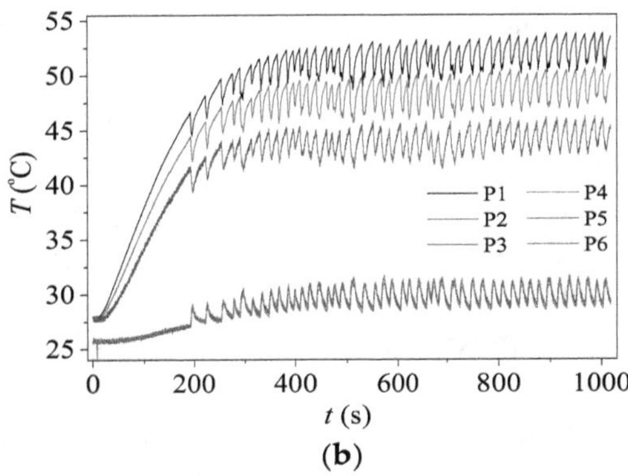

(b)

Figure 6. Dynamic wall temperature variations: (**a**) sudden startup mode (φ = 73%, q = 8.9 W/cm^2) and (**b**) gradual startup mode (φ = 47%, q = 8.5 W/cm^2).

This can be explained by the fact that, over time, the significant degree of superheating because of the continued accumulation of energy breaks the critical equilibrium state of the vapor–liquid two-phase working medium and bubbles begin to form on the evaporator surface and grow rapidly, and then leave the evaporator surface, causing a strong disturbance of the liquid surface [31]. When the vapor comes into contact with the condenser surface, it condenses to form condensate droplets and returns to the evaporation section because of the effect of gravity. The heat absorbed by the evaporator section is rapidly transmitted to the condensation section through the generation of bubbles and the condensation of vapor, and then the condenser and evaporator surface temperatures gradually attain an equilibrium state. Therefore, it can be concluded that the sudden startup is an overshoot startup process, in which the maximum surface temperature of evaporator section in the startup process is significantly greater than the value of steady-state wall temperature.

3.1.2. Gradual Startup Mode

In contrast to the sudden startup mode, there is no wall temperature overshoot in the gradual start-up mode, as shown in Figure 6b. The wall temperature of the evaporator section shows a significant pulsation after a short pre-heating period (approximately 200 s), and periodic variations of bubble formation, growth, and detachment from the evaporator surface can be observed inside the flat two-phase thermosyphon. Subsequently, increases in the departing frequency and the number of bubbles generated on the evaporator surface lead to a slow rise in the evaporator and condenser wall temperatures. Over time, the evaporator and condenser wall temperatures gradually become stable, and finally enter a steady state.

It can be seen from the figure that, during the entire startup process, the temperature change from the initial time to the steady state is gradual. This is because the bubbles increase in size shortly after the evaporator is heated in the gradual startup mode, i.e., a small degree of superheat is required for the formation of the critical bubble core. Therefore, the heat input into the evaporator section can be rapidly transmitted through the phase change process of the working fluid. As a result, a gradual increase of the evaporator section wall temperature (no overshoot) is observed in the gradual startup mode.

3.2. Operating States

The working fluid in the flat two-phase thermosyphon enters a quasi-steady state after the startup process. According to the wall temperature pulsation characteristics and visual images of the vapor–liquid two-phase fluid, three quasi-steady operating states are experienced in sequence with increasing heat load: continuous large-amplitude pulsation state (State A), alternate pulsation state (State B), and continuous small-amplitude pulsation state (State C).

3.2.1. Continuous Large-Amplitude Pulsation State (State A)

State A typically occurs when the evaporator section is supplied with a small heat load. As shown in Figure 7, during the first tens of seconds of a typical cycle, the liquid is superheated in the evaporator section and remains in the static status, while the vapor is cooled in the condenser section. When a critical degree of superheat is achieved, the liquid starts boiling and bubbles are generated. In the boiling process, the bubble generation only lasts for a few seconds, accompanied by a sudden large fluctuation in temperature; then the liquid returns to the static state and goes into the next cycle. The static status and boiling status can be clearly distinguished during this process. In this operating state, the wall temperature of the evaporator section fluctuates cyclically with a continuous large amplitude, where only a monotonic increase and decrease of the wall temperature are experienced in a typical cycle. Because of the small heat load, the energy absorption rate in the cavity is relatively slow, and bubbles are generated intermittently on the surface of the evaporator section. The following energy phases of the evaporation section are repeated cyclically until the end of the heat input: accumulation, release, re-accumulation, and re-release. In this process, the fluid inside a thermosyphon experiences boiling, stagnation, re-boiling, and re-stagnation.

In order to intuitively exhibit the features of a large-amplitude pulsation cycle, Figure 7b shows the wall temperature pulsation of the evaporator and condenser sections from 700 to 900 s, which includes four cycles of continuous large-amplitude pulsation. One pulsation cycle is from 700 to 769 s. Figure 7c shows the typical gas–liquid two-phase behavior for a pulsation cycle, for which the corresponding time points and wall temperatures are marked by point 1 to point 6 in

Figure 7b. During the cycle, the fluid remains in a static state from t = 731 s (point 1) to t = 765 s (point 2), and the wall temperature of the evaporator section gradually increases because of the energy accumulation. When the wall temperature increases to point 2, the degree of superheat reaches a critical value. As a result, the nucleation sites on the evaporation surface are activated and bubbles are generated continuously. Because of the bubble motion and vapor–liquid phase change, energy is rapidly transferred from the evaporator section to the condenser section, resulting in a significant decrease in the evaporator temperature and a rapid increase in the condenser temperature.

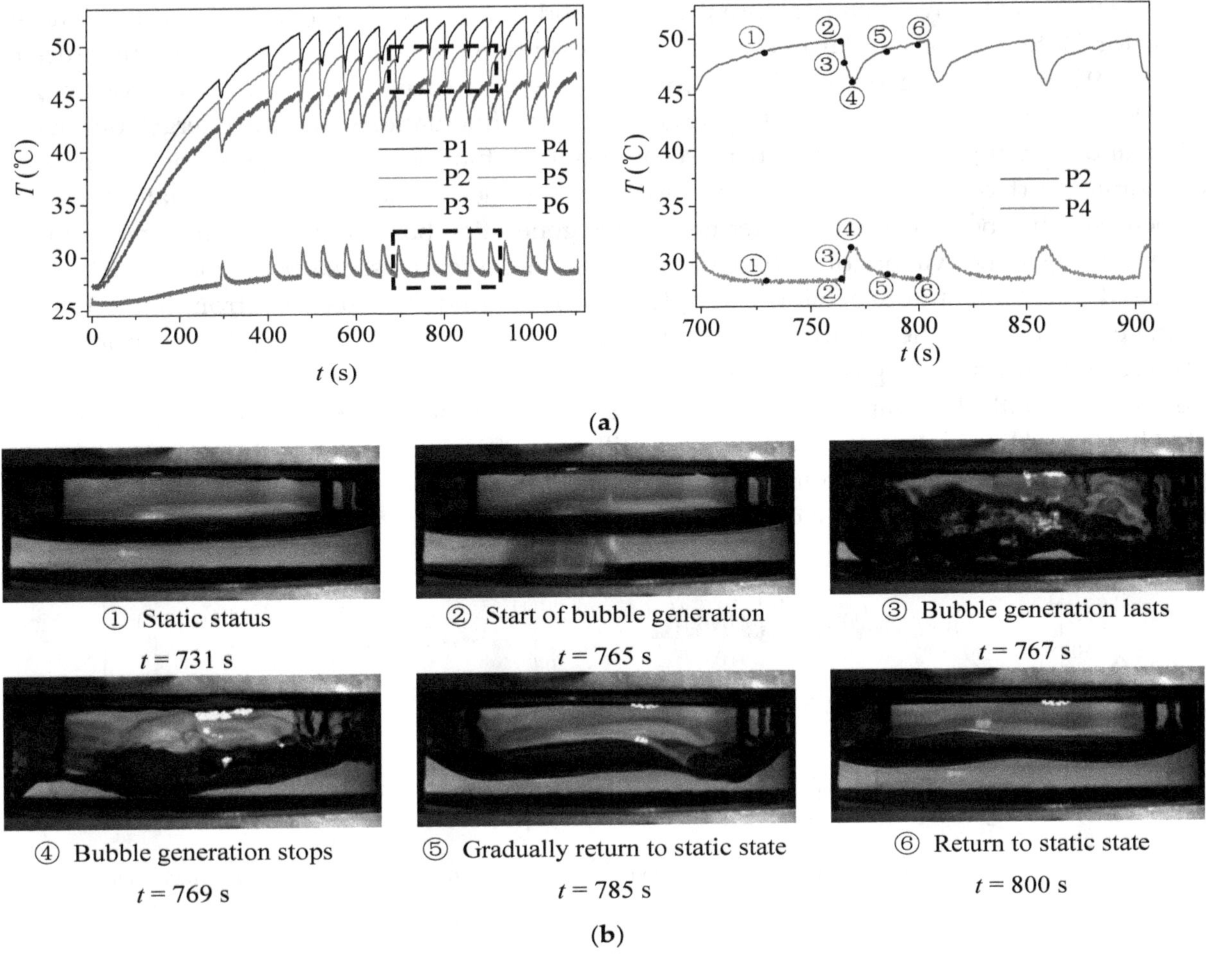

Figure 7. Continuous large-amplitude pulsation state (φ = 47%, q = 6.4 W/cm^2): (**a**) dynamic variation of wall temperature pulsation; (**b**) gas–liquid two-phase behavior in one cycle.

It should be noted that the continuous bubble generation (point 3) is of short duration and only lasts for approximately 4 s. After t = 769 s (point 4), the wall temperature of the evaporator section decreases; the degree of superheat is smaller than the critical value, causing bubble generation to cease, so the liquid gradually returns to the static state (point 5) and the energy absorption starts again in the next cycle (point 6). It can be seen that the heat accumulation in the evaporator section is of short duration in this operating state because of the small heat input, and the heat can be rapidly transferred to the condenser section through the motion of the vapor bubbles. Therefore, bubble generation is only observed for a short period.

3.2.2. Alternate Pulsation State (State B)

When compared with State A, the energy accumulation is accelerated with increasing heat input, so the duration of static status becomes shorter while the boiling status lasts longer in State B. As the bubble generation frequency increases, the pressure perturbation appears in the cavity, leading to the small amplitude fluctuation of wall temperature. In State B, the wall temperature of the evaporator section shows a fluctuation with small amplitude in addition to the large-amplitude fluctuations, and it exhibits a similar cyclical fluctuation. According to the pulsation amplitude and its duration for the wall temperature of evaporator section, State B can be divided into two alternate pulsation types. One is characterized by a long-duration single large-amplitude pulsation alternating with multiple short-duration small-amplitude pulsations (State B–1), and the other is characterized by a single short-duration large-amplitude pulsation alternating with multiple long-duration small-amplitude pulsations (State B–2).

Figure 8 shows the wall temperature pulsation and the corresponding gas–liquid two-phase flow behavior for State B–1 for the flat two-phase thermosyphon, in which three wall temperature cycles (from 590–690 s) are enlarged. As shown in the figure, the early stage of a cycle (from point 1 to point 2, 590–604 s) can be characterized by large-amplitude fluctuations, and the energy continues to be absorbed and the temperature slowly increases without fluctuations. Bubble generation then begins on the evaporator surface, and this process lasts for 28 s, significantly longer than for State A, as the heat input increases. It is noteworthy that after the bubble generation has commenced, the wall temperature decreases rapidly in the evaporator section and increases in the condenser section (from point 2 to point 3). Subsequently, the temperature exhibits multiple small-magnitude fluctuations (from point 3 to point 5). This can be attributed to the random generation and detachment of bubbles on the evaporator surface and the scouring effect of the gas–liquid two-phase fluid on the condenser surface. In this case, the wall temperatures of both the evaporator and condenser sections pulsate with small amplitudes close to the steady state. However, as the heat input is still not sufficient, the proportion of small-magnitude temperature pulsations to the entire pulsation cycle is small, i.e., the number of large-magnitude temperature pulsations dominates a cycle.

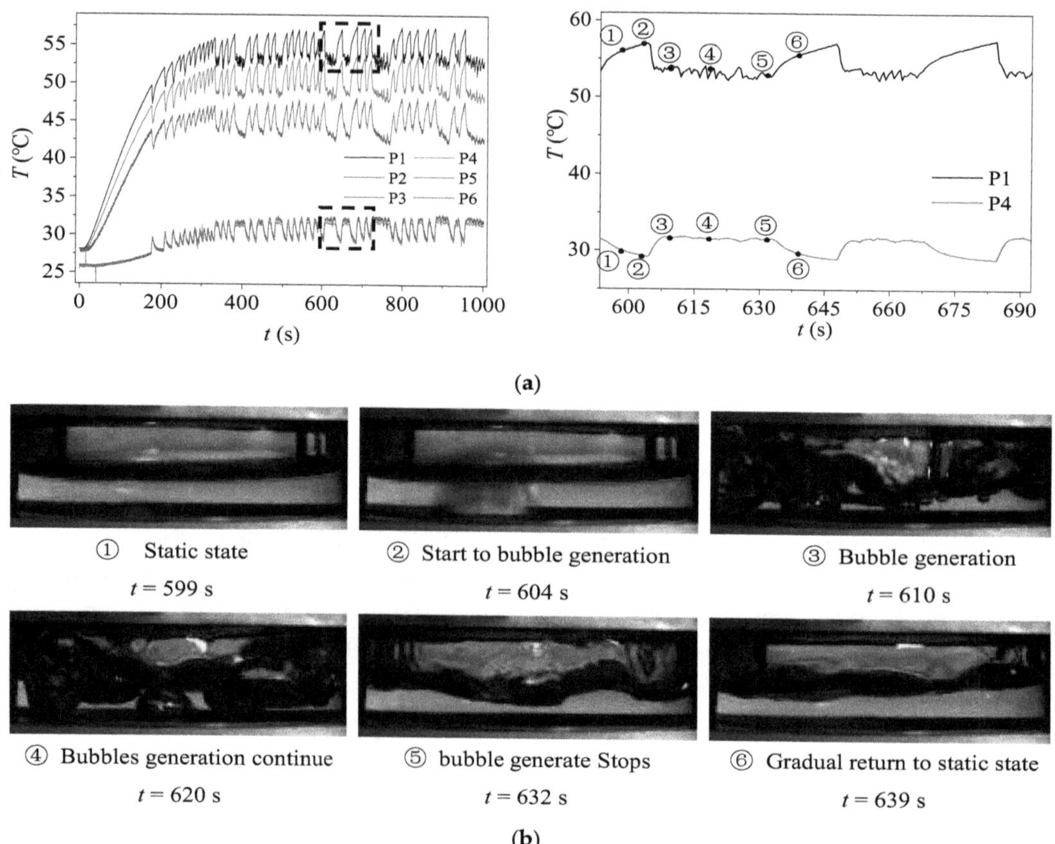

Figure 8. Alternate pulsation state (State B–1, φ = 47%, q = 10.9 W/cm^2): (**a**) temperature dynamic variation of evaporator wall; (**b**) gas–liquid two-phase behavior in one cycle.

With a further increase in the heat input, the operating state of State B–1 is transformed into State B–2. Figure 9 shows the wall temperature pulsations and the corresponding gas–liquid two-phase flow behavior in State B–2. Compared with State B–1, the duration of the small-magnitude pulsations is significantly greater in State B–2, and the duration of continuous bubble generation increases to approximately 100 s (from point 2 to point 6 in Figure 9a). The interval between bubbles decreased, and for the bulk of the time bubbles were generated continuously and rising from the evaporator surface, resulting in a significant fluid disturbance. According to the gas–liquid two-phase behavior images, it can also be seen that the disturbance of the working fluid amplifies with the heat input in the flat two-phase thermosyphon, which is favorable for the enhancement of heat transfer process. As a result, the heat input is efficiently transferred from the evaporator section to the condenser section through the intense gas–liquid phase change.

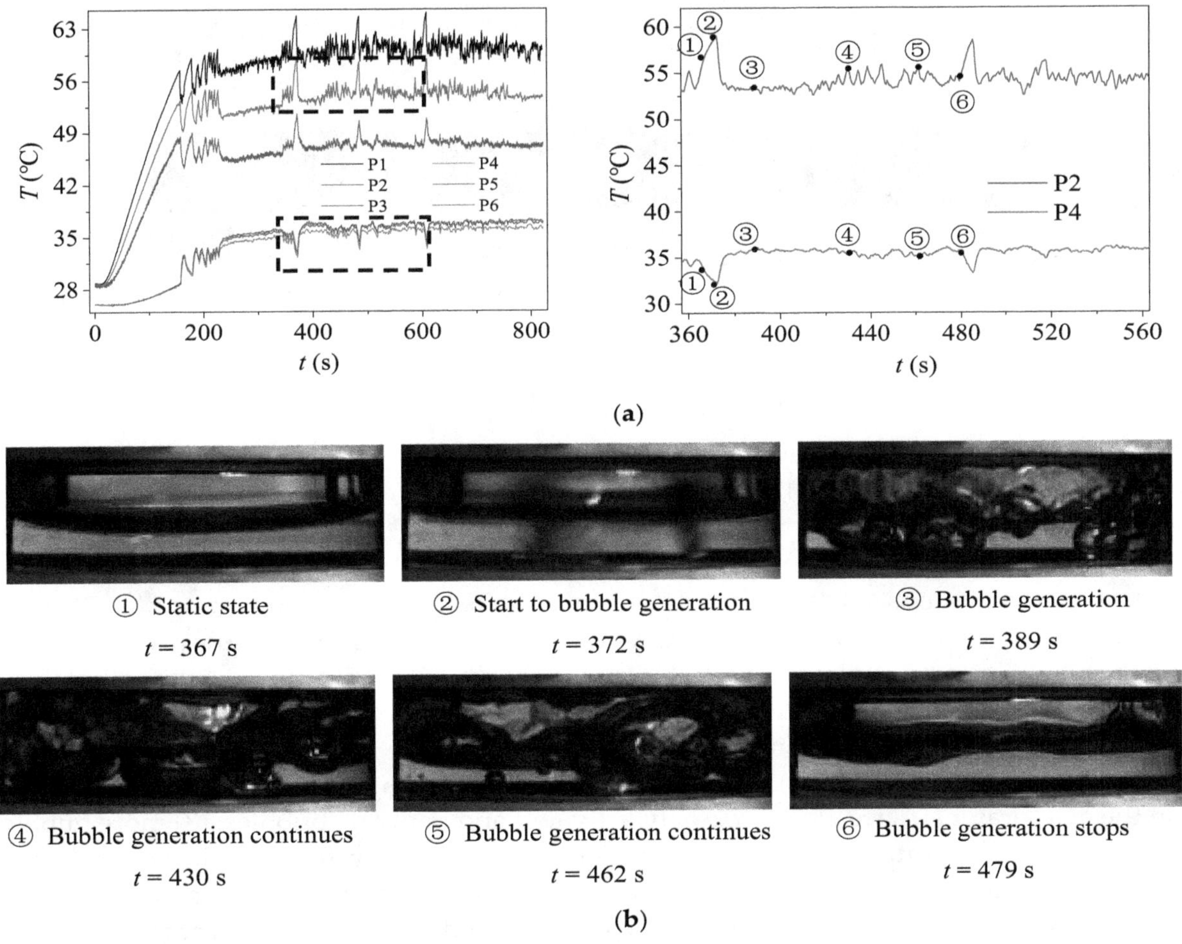

Figure 9. Alternate pulsation state (State B-2, φ = 47%, q = 16.8 W/cm^2): (**a**) characteristics of wall temperature pulsation; (**b**) gas–liquid two-phase behavior in one cycle.

3.2.3. Continuous Small-Amplitude Pulsation State (State C)

When the heat input is sufficiently great, the liquid is continuously boiling due to the intrinsically stochastic nature of liquid, resulting in continuous small-amplitude pulsation of wall temperature. In this case, the operating state enters a continuous small-amplitude pulsation state. Compared with the above operating states, there is no large-amplitude pulsation of wall temperature in State C as the boiling takes place continuously; there is no cyclical variation of wall temperature oscillation, and the wall temperature variation is random. Figure 10 shows the dynamic temperature variation of both the evaporator and condenser walls in State C. Once bubble generation begins in the cavity (point 2 in Figure 10), the bubbles are continuously generated on the evaporation wall. The wall temperature exhibits a small amplitude fluctuation, as opposed to a major change.

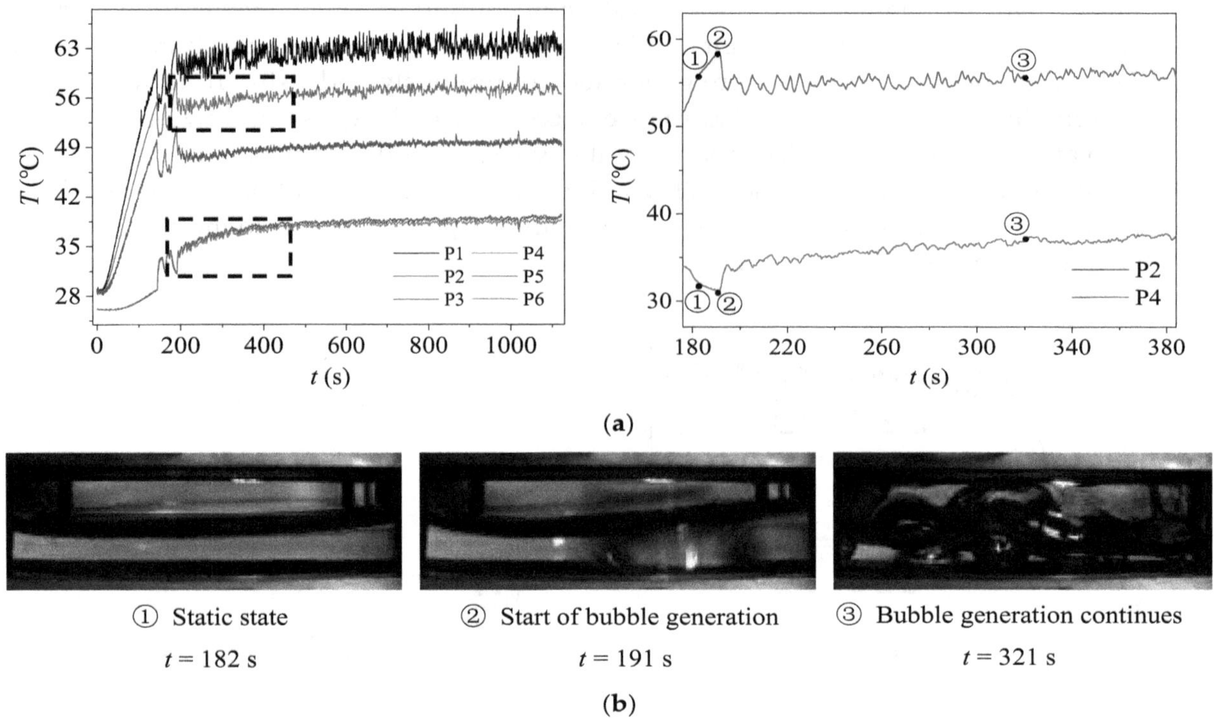

Figure 10. Continuous small-amplitude pulsation operation ($\varphi = 47\%$, $q = 19.9$ W/cm^2): (**a**) characteristics of wall temperature pulsation; (**b**) gas–liquid two-phase behavior from the static state to continuous bubble generation.

During the continuous small-amplitude pulsation state, there is no obvious cycle. Once bubble generation begins in the cavity, the bubbles continuously generate on the evaporation wall. This phenomenon can be explained as follows: (1) the vapor–liquid phase-change heat transfer between the evaporator and the condenser becomes stronger in State C, where the stable boiling and condensation in the cavity ensure efficient heat transfer; (2) the phase change processes (including evaporation and boiling) on the evaporator surface are strengthened, and the nucleation sites on the evaporator surface are more easily activated and generate bubbles; (3) more bubbles are generated on the evaporator surface (see point 3 in Figure 10) than for the other operating states, and the liquid is more disturbed by the bubble motion. In summary, greater energy transfer intensity between the evaporator and condenser sections is achieved in the flat two-phase thermosyphon through efficient vapor–liquid phase changes.

3.3. Thermal Resistance

According to the above analysis, the heat load directly determines the operation state of vapor–liquid two phase flows inside a flat two-phase thermosyphon. This inevitably affects the thermal performance of evaporation and condensation phase change. In order to analyze the relationship between the heat transfer performance and heat load, the total thermal resistance of the flat two-phase thermosyphon R is introduced, defined by

$$R = \Delta\overline{T}/Q \tag{1}$$

where $\Delta\overline{T} = \overline{T}_e - \overline{T}_c$ is the difference between the average temperature of the evaporator surface, $\overline{T}_e = (T_1 + T_2 + T_3)/3$, and the average temperature of the condenser surface, $\overline{T}_c = (T_4 + T_5 + T_6)/3$; and Q is the actual heat load of the thermosyphon.

Figure 11 describes the effect of heat load on total thermal resistance of a flat two-phase thermosyphon. During the continuous large-amplitude pulsation state (State A), the heat load is small, and the bubbles are generated intermittently on the evaporator surface, so the heat transfer regime is the alternation of natural convection and nucleate boiling. In addition, the natural convection occupies most of time. This leads to a large thermal resistance of two-phase thermosyphon. When the operation state transforms from State A to the alternate pulsation state (State B), more heat is transferred through the bubble generation on the evaporator surface, i.e., the duration of nucleate boiling is longer. In addition, the motion of the bubbles causes the disturbance of the liquid near the evaporator surface, which also enhances heat transfer of the thermosyphon. The nucleate boiling occupies the most of time for State B. Since the good heat transfer performance of nucleate boiling, the total thermal resistance of State B is smaller than that of State A.

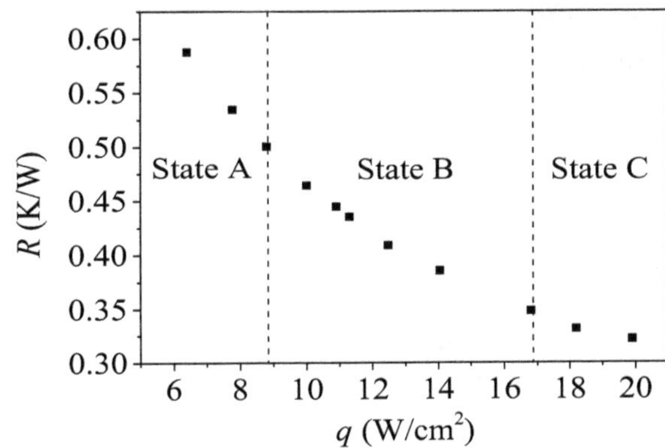

Figure 11. Effect of the heat load on total thermal resistance.

Unlike State A and B, during the continuous small-amplitude pulsation state (State C), the input heat is transferred wholly through nucleate boiling. Furthermore, the continuously generated bubbles cause a significant fluid disturbance, so the scouring effect of the gas–liquid two-phase fluid on the condenser surface make the condensate film thinner. This further reduces the thermal resistance as compared with State B. Therefore, as the thermal load increases, the thermal resistance of the flat two-phase thermosiphon decreases monotonously, thereby exhibiting better heat transfer performance. In other words, the thermal performance of alternate pulsation state in a flat two-phase thermosyphon is inferior to continuous small-amplitude pulsation state but superior to continuous large-amplitude pulsation state.

4. Conclusions

In this study, a flat two-phase thermosyphon with transparent sidewalls was manufactured for a visualization experiment. An experimental system was designed and conducted to investigate the phase-change heat transfer of the flat two-phase thermosyphon, with a particular focus on the startup mode and operating state. The dynamic temperature variations of the evaporator and condenser wall and the gas–liquid two-phase evolution in the flat two-phase thermosyphon were observed and analyzed. The primary conclusions were as follows:

1. Sudden startup and gradual startup were identified as the two types of startup modes in the flat two-phase thermosyphon, and the continuous large-amplitude pulsation state, alternate pulsation state, and continuous small-amplitude pulsation state were experienced in sequence with increasing heat load.

2. The continuous large-amplitude pulsation state occurred when a low heat load was applied to the evaporator section, in which the bubbles were generated intermittently.
3. The alternate pulsation state exhibited small fluctuations in addition to the large-amplitude fluctuations of the evaporator wall temperature because of the greater duration of boiling compared to the continuous large-amplitude pulsation state.
4. The continuous small-amplitude pulsation state occurred in the flat two-phase thermosyphon where the boiling occurred continuously and the wall temperature variation was random and exhibited no cyclical variations.

Author Contributions: W.Y. provided guidance and supervision. L.W. and Y.C. implemented the main research, discussed the results, and wrote the paper. S.W. and M.Z. collected the data. F.T. revised the manuscript. All authors read and approved the final manuscript.

Nomenclature

Variables

C_p	specific heat (J/kg·K)
P	measuring points
q	heat flux (W/cm^2)
Q	heat input (W)
R	thermal resistance
t	time (s)
T	temperature (°C)
$\overline{T}$	average temperature (°C)
x	horizontal coordinate (mm)
y	vertical coordinate (mm)
Subscripts	
c	condenser section
cool	cooling water
e	evaporator section
1~6	number of thermal couples
Greek symbols	
ρ	density (kg/m^3)
φ	filling ratio
Φ	diameter (mm)

References

1. Murshed, S.M.S.; Nieto de Castro, C.A. A critical review of traditional and emerging techniques and fluids for electronics cooling. *Renew. Sustain. Energy Rev.* **2017**, *78*, 821–833. [CrossRef]
2. Chen, Y.P.; Zhang, C.B.; Shi, M.H.; Wu, J.F. Three-dimensional numerical simulation of heat and fluid flow in noncircular microchannel heat sinks. *Int. Commun. Heat Mass Transf.* **2009**, *36*, 917–920. [CrossRef]
3. Muszynski, T.; Andrzejczyk, R. Heat transfer characteristics of hybrid microjet–microchannel cooling module. *Appl. Therm. Eng.* **2016**, *93*, 1360–1366. [CrossRef]
4. Zhang, C.; Chen, Y.; Shi, M. Effects of roughness elements on laminar flow and heat transfer in microchannels. *Chem. Eng. Process.* **2010**, *49*, 1188–1192. [CrossRef]
5. Chen, Y.P.; Deng, Z.L. Hydrodynamics of a droplet passsing through a microfluidic t-junction. *J. Fluid Mech.* **2017**, *819*, 401–434. [CrossRef]
6. Chen, Y.P.; Zhang, C.B.; Shi, M.H.; Wu, J.F.; Peterson, G.P. Study on flow and heat transfer characteristics of heat pipe with axial "Ω"–shaped microgrooves. *Int. J. Heat Mass Transf.* **2009**, *52*, 636–643. [CrossRef]
7. Zhang, C.; Chen, Y.; Wu, R.; Shi, M. Flow boiling in constructal tree-shaped minichannel network. *Int. J. Heat Mass Transf.* **2011**, *54*, 202–209. [CrossRef]
8. Deng, Z.; Liu, X.; Zhang, C.; Huang, Y.; Chen, Y. Melting behaviors of pcm in porous metal foam characterized by fractal geometry. *Int. J. Heat Mass Transf.* **2017**, *113*, 1031–1042. [CrossRef]

9. Zhang, C.B.; Deng, Z.L.; Chen, Y.P. Temperature jump at rough gas-solid interface in couette flow with a rough surface described by cantor fractal. *Int. J. Heat Mass Transf.* **2014**, *70*, 322–329. [CrossRef]
10. Zhang, C.; Chen, Y.; Deng, Z.; Shi, M. Role of rough surface topography on gas slip flow in microchannels. *Phys. Rev. E* **2012**, *86*. [CrossRef] [PubMed]
11. Sharma, C.S.; Tiwari, M.K.; Zimmermann, S.; Brunschwiler, T.; Schlottig, G.; Michel, B.; Poulikakos, D. Energy efficient hotspot-targeted embedded liquid cooling of electronics. *Appl. Energy* **2015**, *138*, 414–422. [CrossRef]
12. Chen, Y.; Zhang, C.; Shi, M.; Yang, Y. Thermal and hydrodynamic characteristics of constructal tree-shaped minichannel heat sink. *AIChE J.* **2010**, *56*, 2018–2029. [CrossRef]
13. Weibel, J.A.; Garimella, S.V. Chapter four–recent advances in vapor chamber transport characterization for high-heat-flux applications. In *Advances in Heat Transfer*, 1st ed.; Sparrow, E.M., Cho, Y.I., Abraham, J.P., Gorman, J.M., Eds.; Elsevier: New York, NY, USA, 2013; Volume 45, pp. 209–301.
14. Zhang, C.; Shen, C.; Chen, Y. Experimental study on flow condensation of mixture in a hydrophobic microchannel. *Int. J. Heat Mass Transf.* **2017**, *104*, 1135–1144. [CrossRef]
15. Chen, Y.; Gao, W.; Zhang, C.; Zhao, Y. Three-dimensional splitting microfluidics. *Lab Chip* **2016**, *16*, 1332–1339. [CrossRef] [PubMed]
16. Chen, Y.P.; Wu, R.; Shi, M.H.; Wu, J.F.; Peterson, G.P. Visualization study of steam condensation in triangular microchannels. *Int. J. Heat Mass Transf.* **2009**, *52*, 5122–5129. [CrossRef]
17. Zhang, M.; Liu, Z.; Ma, G. The experimental investigation on thermal performance of a flat two-phase thermosyphon. *Int. J. Therm. Sci.* **2008**, *47*, 1195–1203. [CrossRef]
18. Liu, Z.; Zheng, F.; Liu, N.; Li, Y. Enhancing boiling and condensation co-existing heat transfer in a small and closed space by heat-conduction bridges. *Int. J. Heat Mass Transf.* **2017**, *114*, 891–902. [CrossRef]
19. Chen, X.; Ye, H.; Fan, X.; Ren, T.; Zhang, G. A review of small heat pipes for electronics. *Appl. Therm. Eng.* **2016**, *96*, 1–17. [CrossRef]
20. Jouhara, H.; Chauhan, A.; Nannou, T.; Almahmoud, S.; Delpech, B.; Wrobel, L.C. Heat pipe based systems–advances and applications. *Energy* **2017**, *128*, 729–754. [CrossRef]
21. Zhang, G.; Liu, Z.; Wang, C. An experimental study of boiling and condensation co-existing phase change heat transfer in small confined space. *Int. J. Heat Mass Transf.* **2013**, *64*, 1082–1090. [CrossRef]
22. Zhang, G.; Liu, Z.; Wang, C. A visualization study of the influences of liquid levels on boiling and condensation co-existing phase change heat transfer phenomenon in small confined spaces. *Int. J. Heat Mass Transf.* **2014**, *73*, 415–423. [CrossRef]
23. Wu, J.; Shi, M.; Chen, Y.; Li, X. Visualization study of steam condensation in wide rectangular silicon microchannels. *Int. J. Therm. Sci.* **2010**, *49*, 922–930. [CrossRef]
24. Lu, L.; Liao, H.; Liu, X.; Tang, Y. Numerical analysis on thermal hydraulic performance of a flat plate heat pipe with wick column. *Heat Mass Transf.* **2015**, *51*, 1051–1059. [CrossRef]
25. Blet, N.; Lips, S.; Sartre, V. Heats pipes for temperature homogenization: A literature review. *Appl. Therm. Eng.* **2017**, *118*, 490–509. [CrossRef]
26. Do, K.H.; Kim, S.J.; Garimella, S.V. A mathematical model for analyzing the thermal characteristics of a flat micro heat pipe with a grooved wick. *Int. J. Heat Mass Transf.* **2008**, *51*, 4637–4650. [CrossRef]
27. Kim, H.J.; Lee, S.-H.; Kim, S.B.; Jang, S.P. The effect of nanoparticle shape on the thermal resistance of a flat-plate heat pipe using acetone-based Al_2O_3 nanofluids. *Int. J. Heat Mass Transf.* **2016**, *92*, 572–577. [CrossRef]
28. Deng, Z.; Zheng, Y.; Liu, X.; Zhu, B.; Chen, Y. Experimental study on thermal performance of an anti-gravity pulsating heat pipe and its application on heat recovery utilization. *Appl. Therm. Eng.* **2017**, *125*, 1368–1378. [CrossRef]
29. Xia, G.D.; Wang, W.; Cheng, L.X.; Ma, D.D. Visualization study on the instabilities of phase-change heat transfer in a flat two-phase closed thermosyphon. *Appl. Therm. Eng.* **2017**, *116*, 392–405. [CrossRef]
30. Chen, Y.; Yu, F.; Zhang, C.; Liu, X. Experimental study on thermo-hydrodynamic behaviors in miniaturized two-phase thermosyphons. *Int. J. Heat Mass Transf.* **2016**, *100*, 550–558. [CrossRef]
31. Liu, X.; Chen, Y.; Shi, M. Dynamic performance analysis on start-up of closed-loop pulsating heat pipes (clphps). *Int. J. Therm. Sci.* **2013**, *65*, 224–233. [CrossRef]

11

Effects of Welding Time and Electrical Power on Thermal Characteristics of Welding Spatter for Fire Risk Analysis

Yeon Je Shin and Woo Jun You *

Department of Architecture & Fire Safety, Dongyang University, Yeongju 36040, Korea; jeje5842@naver.com
* Correspondence: wjyou@dyu.ac.kr

Abstract: To predict the fire risk of spatter generated during shielded metal arc welding, the thermal characteristics of welding spatter were analyzed according to different welding times and electrical powers supplied to the electrode. An experimental apparatus for controlling the contact angle between the electrode and base metal as well as the feed rate was prepared. Moreover, the correlations among the volume, maximum diameter, scattering velocity, maximum number, and maximum temperature of the welding spatter were derived using welding power from 984–2067 W and welding times of 30 s, 50 s, and 70 s. It was found that the volume, maximum diameter, and maximum number of welding spatters increased proportionally as the welding time and electrical power increased, but the scattering velocity decreased as the particle diameter increased regardless of the welding time and electrical power. When the measured maximum temperature of the welding spatter was compared with an empirical formula, the accuracy of the results was confirmed to be within ±7% of the experimental constant $C = 112.414 \times P_e^{-0.5045}$. Results of this study indicate quantitatively predicting the thermal characteristics of welding spatter is possible for minimizing the risk of fire spread when the electrode type and welding power is known.

Keywords: shielded metal arc welding; welding spatter; electrode; electrical power; welding time

1. Introduction

Fire risks in construction sites may occur when flammable gases, liquids, or substances reach their ignition points owing to the scattered welding spatter [1–6]. Shielded metal arc welding (SMAW), a method of joining metals by generating an arc and heating the weld metal zone by applying electrical power between the base metal and electrode, has been widely used in industrial sites since the method of SMAW is applied to almost all repairing of cast iron in air or steel under water [7–10]. However, it involves the risk of fire spreading to nearby combustibles caused by high temperatures because the scattered welding spatters are larger comparable to those of gas metal arc welding (GMAW), which uses plasma [11,12]. Especially, the fire hazards from the SMAW at building construction sites can occur when welding spatters make contact with the inward of a pipe or other enclosed space filled with flammable vapor or liquid [12–19]. In addition, the polarity of electrode can cause changes in welding spatter diameter and number [7,19–23]. From the viewpoint of fire technology, analyzing the thermal characteristics of welding spatter is one of the widely used methods to predict fire spread, where related research has already been conducted.

Hagiwara et al. [24,25] conducted an experimental study on the particle size distribution of welding spatter according to the electrical power supplied to the electrode. They found that 90% of the particles had diameters of less than 1 mm and analyzed the fire spread phenomenon in combustibles, such as benzene, acetone, and urethane foam. This study, however, appears to have limitations in quantitatively analyzing the thermal characteristics of welding spatter crucial to fire risks, which are caused by the electrical power and depend on the particle size.

Hagimoto et al. [19] calculated the particle size according to the electrode diameter when the same electrical power was supplied to the electrode and found that approximately 80% of the particles were scattered to a distance of 0.5 m, 15% to 0.5–1.0 m, and 5% to more than 1 m. They reported that fire can spread to combustibles (urethane foam etc.) when large particles of diameters 0.9–3.0 mm are scattered to a distance of more than 3.5 m.

Brandi et al. [26] analyzed the correlation between the material properties of the electrode core and fire risks using standard mineral dressing techniques. They found that the porosity and density of the welding spatter varied according to the electrical power and stressed the importance of the electrode physical properties for satisfying the ignition requirements of combustibles.

Results from previous studies show that the conditions of fire spread to combustibles during welding vary due to the varying thermal characteristics of welding spatter depending on the electrical power [18,26,27]. Therefore, Shin and You [27] calculated the particle size distribution and mean particle size of welding spatter by assuming a steady-state maximum temperature of the welding spatter for igniting combustibles and proposed an equation for predicting the mean particle temperature based on the energy conservation relationship. According to them, predicting the maximum temperature of the welding spatter is possible when the electrical power, total volume, mean size, and scattering velocity of the particles are known. As these parameters (except the electrical power) vary depending on the electrical power, it is necessary to analyze the relationships among the main factors according to the experimental conditions. This is necessary for the quantitative analysis of the risk of fire spread due to scattered particles. Therefore, in this study, we propose a method for predicting fire risks by quantitatively deriving the thermal characteristics of welding spatter according to the welding time and electrical power.

2. Material and Methods

2.1. Theoretical Approach

Figure 1 shows the schematic of the total volume of the welding spatter during welding. The total mass of the welding spatter ($\Delta m_{p,total}$) can be calculated according to Equation (1) after measuring the mass melted on the base metal ($\Delta m_{b,p}$) and is dependent on the welding time (Δt) and electrical power (P_e). The core inside the electrode is made of steel (ρ_{iron} = 7860 kg/m^3) and the coating outside the electrode contains sodium silicate ($\rho_{\text{Sodium Silicate}}$ = 2400 kg/m^3). However, it is possible to calculate the volume of a single particle (ΔV_i) using the relationship $\Delta V_i = \Delta m_i/\rho_i$ only when the mixed ratios of materials are given for each welding spatter.

$$\Delta m_{p,total} = \Delta m_{el} - \Delta m_{b,p} \tag{1}$$

where Δm_{el}, $\Delta m_{b,p}$, and $\Delta m_{p,total}$ are the masses of the electrode and solidified weld metal attached to the base metal and total mass of scattered particles, respectively. In a previous study, the mean particle temperature was predicted by assuming the steady-state condition of the initial temperature of welding spatter as the maximum value for the ignition combustibles, as shown in Equation (2) [27].

$$T_{P,s} = T_\infty + \frac{P_e - \sigma\varepsilon A_{b,s}\left(T_{s,b}^4 - T_{sur}^4\right)}{NhA_{P,s}} \tag{2}$$

where $T_{p,s}$, T_∞, T_{sur}, P_e, σ, ε, $A_{b,s}$, $T_{s,b}$, N, h, and $A_{p,s}$ are the mean particle temperature, surrounding temperature, surface temperature, electrical power (P_e), Stefan–Boltzmann constant, emissivity, surface area and surface temperature of base metal, average number of particles, convective heat transfer coefficient, and surface area of particle, respectively. The mean particle size, $d_{p,m}$, and convective heat transfer coefficient, h, can be obtained using Equations (3) and (4), respectively [14,27,28].

$$N = \frac{6V_{p,total}}{\pi d_{p,m}^3} \tag{3}$$

$$Nu_D = \frac{hd_{p,m}}{k} = 2 + 0.6Re_D^{0.5}Pr^{1/3} \tag{4}$$

where $V_{p,total}$, $d_{p,m}$, Nu_D, k, Re, and Pr are the total volume of particles, mean diameter of particles, Nusselt number, thermal conductivity, Reynolds number ($Re_D = \rho u_{p,m} d_{p,m}/\mu$), and Prandtl number ($Pr = C_p\mu/k$)), respectively. Therefore, the temperature distribution prediction shown in Equation (2) is possible when $V_{p,total}$, $d_{p,m}$, and $u_{p,m}$ can be calculated using Equations (3) and (4). As the total volume, mean diameter, and scattering velocity of particles vary according to the welding time and electrical power, the functional relationship given by Equation (5) must be also determined [14,27].

$$N, d_m, h \sim f(P_e, \Delta t) \tag{5}$$

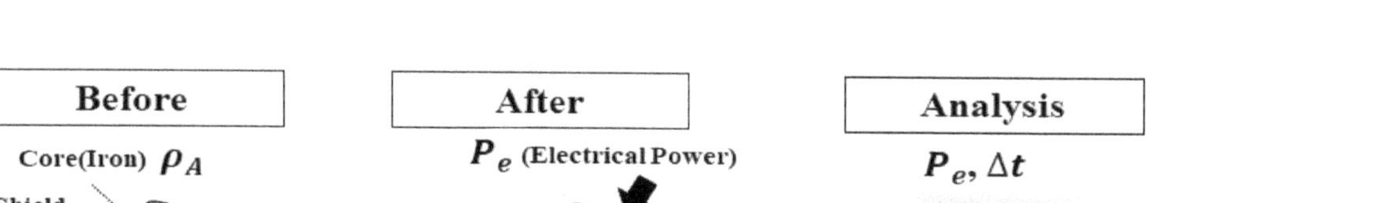
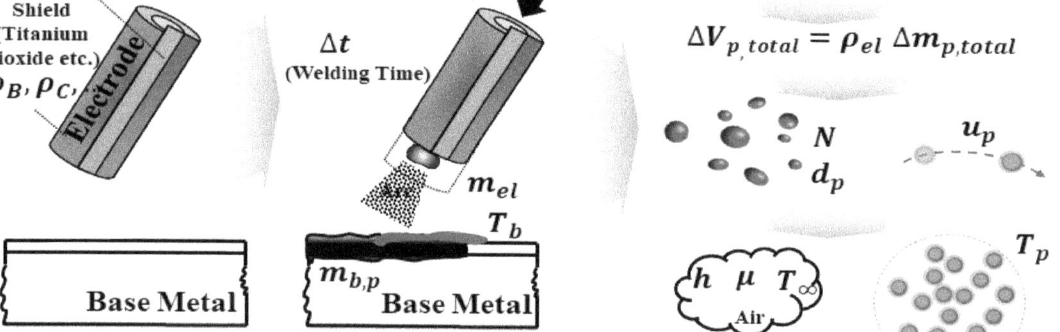

Figure 1. Schematic of welding spatter in shielded metal arc welding (SMAW) and assumptions for energy balance between particles and base metal.

2.2. Experimental Apparatus

Figure 2 shows the semi-automated SMAW experimental apparatus, which was constructed by maintaining perpendicularity between the electrode and base metal constant (welding angle θ = 90°) and specifying the maximum feed rate of the welding torch as 7 mm/s. This made it possible to analyze the size and scattering velocity of the particles. As shown in the figure, welding spatters were scattered under different welding times (Δt) and electrical power (P_e), and the scattering velocity and mean temperature of the welding spatters were measured using a high-speed camera (model: phantom Miro M/R/LC310, USA) and a thermal imaging camera (model: Fluke Tix501). The scattering velocity and mean temperature of the welding spatter were measured using a high-speed camera (model: phantom LC310) and a thermal imaging camera (model: Fluke Tix501). The particle size distribution was determined using Image J software after collecting all welding spatter in a 50 × 46 × 64 cm^3 acrylic box. Table 1 lists the specifications of the experimental apparatus and the experimental conditions for the average values of three times results are denoted in Table 2.

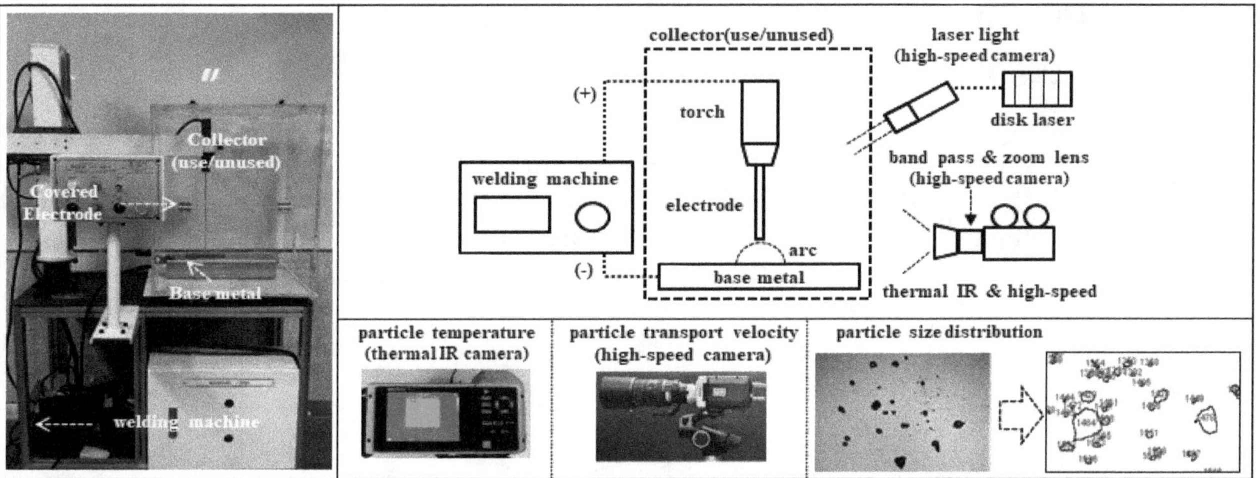

Figure 2. Schematic and images of experimental apparatus for welding spatter analysis.

Table 1. Specifications of experiment apparatus.

Equipment	Specification
Welding machine	Output current (20~220) A, Rated input voltage 220 V, Electric power (0~2.5) kW Rated duty cycle 60%, Model: Rolwal MMA-200E
Thermal imaging camera	Infrared resolution 640 × 480, Temp. measurement range −20 to 650 °C, Accuracy ±2 °C or 2%, Frame rate 60 Hz, Model: Fluke Tix 501
High-speed camera	Resolution 640 × 480, Sampling rate 10,000 fps, Model: phantom Miro M/R/LC310 Lens: Nikon 105 mm, 2× converter Band-pass filter: Φ 50 mm, 810 nm/12 nm, CN code 90022000
Electronic energy meter	230 AC, 60 Hz, 16 A/3680 W (Model: KEM2500)
Precision balance	Max. load weight 320 g, Accuracy 0.1 mg, Model: PX224KR
Electrode	High titanium oxide type electrode (AWS E-6013) Core: Iron (65–75%), Coating: Titanium dioxide (10–15%), Feldspar (5–10%), Mn (1–5%), Sodium silicate (1–5%), Limestone (1–5%), Mica (1–5%)

Table 2. Experimental conditions to study the effects of thermal characteristics of welding spatter on the welding time and electrical power.

Test Number	Welding Time, Δt (s)	Welding Current (A)	Welding Voltage (V)	Electrical Power, P_e (W)
Case #1	30			
Case #2	50	80	12	984
Case #3	70			
Case #4	30			
Case #5	50	100	13	1337.3
Case #6	70			
Case #7	30			
Case #8	50	130	14	1802.0
Case #9	70			
Case #10	30			
Case #11	50	150	17	2067.5
Case #12	70			

Welding polarity: DC-, Contact angle: 90°, Arc length: 5 mm, Base metal: Mild steel (SS400); Electrode diameter, d_{el}: 4.0 mm, Material properties of electrode given in Table 1.

3. Results and Discussion

3.1. Volume of Welding Spatter

Figure 3 shows the variation of the measured reduction rate (u_{rate}) of the electrode length according to the electrical power (P_e) in the case of welding times (Δt) of 30 s, 50 s, and 70 s. It is seen that as P_e increased, u_{rate} also increased proportionally as the mass of the electrode welded to the base metal ($\Delta m_{b,p}$) increased. When P_e was constant, a constant value of u_{rate} was calculated, which was consistent with that obtained using Equation (6) within ±4% for the average values u_{rate} regardless of Δt.

$$u_{rate} = a_1 + b_1 \times P_e \tag{6}$$

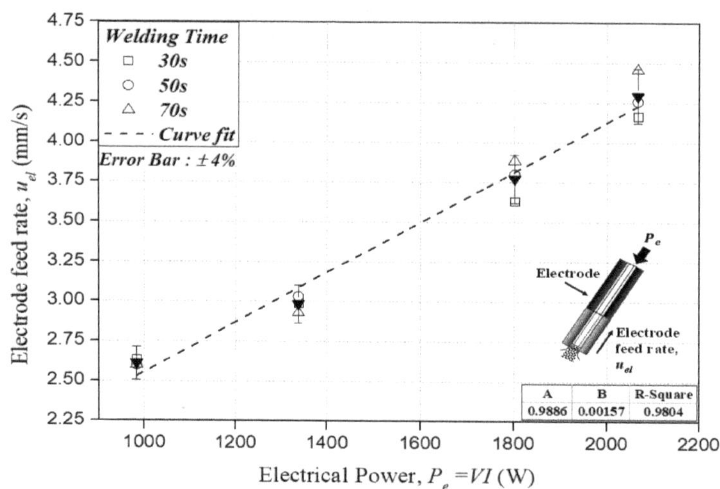

Figure 3. Electrode feed rate depending on the electrical power for welding time = 30 s, 50 s, and 70 s.

Here, a_1 and b_1 are experimental constants. It is estimated that a_1 = 0.989 mm/s and b_1 = 0.157×10^{-2} W-mm/s are obtained according to the base metal and electrode specifications listed in Table 1, and the electrical power ranges between 984 and 2067 W.

Figure 4 shows the variation in the measured mass loss of the electrode (Δm_{el}) and the mass welded to the base metal ($\Delta m_{b,p}$) according to the electrical power (P_e) for the welding times (Δt) of 30 s, 50 s, and 70 s. In this figure, the symbols enclosed in brackets represent Δm_{el}, which increased proportionally to u_{rate}. When Δt increased keeping P_e constant, Δm_{el} increased in proportion to u_{rate}. Therefore, when Equation (6) and the electrode density measured using a load cell (ρ = 4726 kg/m^3) are applied, Δm_{el} is given by Equation (7) expressed by the dotted line, which agrees with the measured value within an error range of approximately ±5%.

$$\Delta m_{el} = u_{rate} \times \Delta t \times \left(\frac{\pi}{4} d_{el}^2\right) \times \rho_{el} \tag{7}$$

where d_{el} is the electrode diameter is used as the reference value of 4.0 mm. Notably in the figure, the measured value of $\Delta m_{b,p}$ increased in proportion to the magnitudes of Δt and P_e, as shown by the closed symbol value. In addition, 88.6% Δm_{el} was found to be welded to the base metal on average. This result indicates that approximately 11.4% Δm_{el} was responsible for generating welding spatter when P_e was supplied to the electrode. The energy transmitted to the electrode can be simplified through the assumption shown in Equation (8).

$$\sigma \varepsilon A_{b,s}\left(T_{s,b}^4 - T_{sur}^4\right) \equiv 0.886 P_e \tag{8}$$

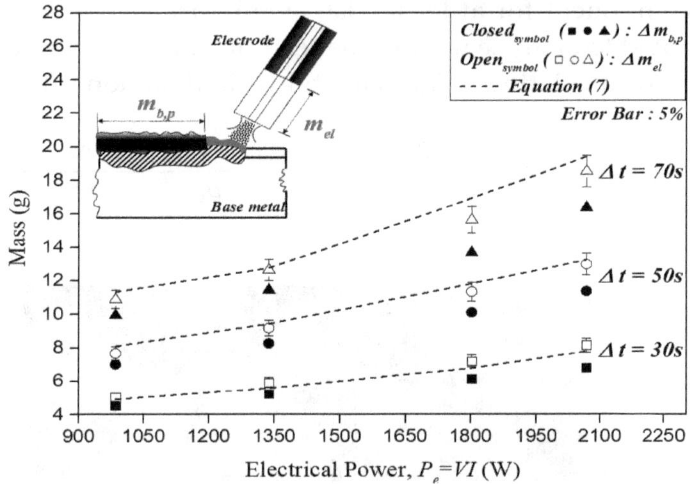

Figure 4. Experimental variation of Δm_{el} and $\Delta m_{b,p}$ according to P_e for Δt = 30 s, 50 s, and 70 s.

Figure 5a shows the variation of the total mass of the particles scattered from the electrode ($\Delta m_{p,total}$) using Equation (1), mass reduction of the electrode (Δm_{el}), and mass welded to the base metal ($\Delta m_{b,p}$) for the welding times (Δt) of 30 s, 50 s, and 70 s according to the electrical power. The result of curve-fitting the calculated values of $\Delta m_{p,total}$ with increasing electrical power (P_e) under the same Δt values is shown in Equation (9).

$$m_{p,total} = a_2 \cdot P_e^{b_2} \tag{9}$$

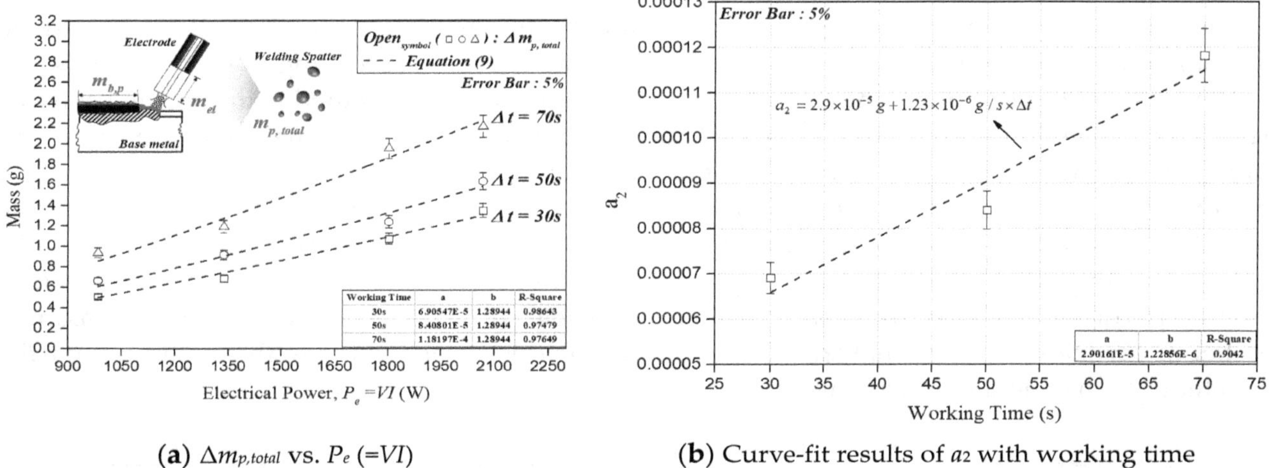

Working Time	a	b	R-Square
30s	6.90547E-5	1.28944	0.98643
50s	8.40801E-5	1.28944	0.97479
70s	1.18197E-4	1.28944	0.97649

a	b	R-Square
2.90161E-5	1.22856E-6	0.9042

(**a**) $\Delta m_{p,total}$ vs. P_e (=VI) (**b**) Curve-fit results of a_2 with working time

Figure 5. Effects of electrical power on the total mass of welding particles at welding times of 30 s, 50 s, and 70 s.

It was found that a_2 is related to Δt as shown in Figure 5b, and this tendency is shown in Equation (10) when b_2 is constant at 1.28944.

$$a_2 = 2.9 \times 10^{-5}\text{g} + 1.23 \times 10^{-6}\text{g/s} \times \Delta t \tag{10}$$

Figure 6 shows the density values (ρ_i) of a single scattered welding particle measured by calculating the mass ($m_{p,i}$) and volume ($\Delta V_{p,i}$) of the particle using diameters ($d_{p,i}$) of 1.736, 2.023, 2.294, and 2.352 mm. Because the electrode contains various metal components, as shown in Table 1, ρ_i may vary depending on the material composition inside the shield and core [26,29]. In particular, the mass proportions of metals that constitute each particle must be determined to obtain the total volume of scattered welding spatter ($\Delta V_{p,total}$), but limitations exist in analyzing the density when measuring each mixed component for at least 1000 small particles with a diameter of 0.1 mm or less. Therefore, we attempted to analyze the thermal characteristics of welding spatter by assuming $\rho_{p,total} \equiv \rho_{el}$ (4726 kg/m^3) and calculating $\Delta V_{p,total}$ as shown in Equation (11).

$$\Delta V_{p,total} = \frac{\Delta m_{p,total}}{\rho_{el}} = \frac{(2.9 \times 10^{-5} + 1.23 \times 10^{-6} \times \Delta t) \times P_e{}^{b_2}}{\rho_{el}} \tag{11}$$

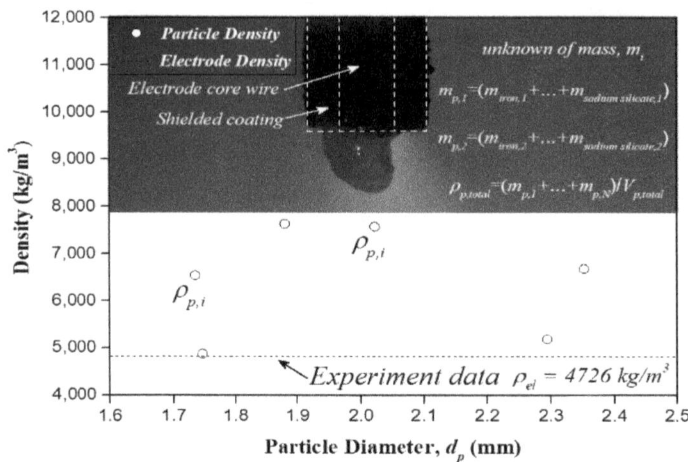

Figure 6. Measured value of one particle density using d_p = 1.73–2.35 mm.

3.2. Diameter and Number of Welding Spatters

Figure 7 shows the fraction (N_i/N_{total}) of the number of particles, which is the ratio of particles with diameter (N_i) to the total number of scattered particles (N_{total}), according to P_e at Δt = 30 s, 50 s, and 70 s. As mentioned before, the particle size distribution was determined using Image J software after collecting welding spatter in a 50 × 46 × 64 cm^3 acrylic space, and it was analyzed by excluding the diameters of 0.3 mm or less due to the resolution. It was found that the mean particle diameter ($d_{p,m}$) was approximately 0.3 mm regardless of Δt and P_e, which is similar to the results of previous studies ($d_{p,m}$ < 0.5 mm) [9,11]. Therefore, the mean number of particles (N) can be calculated using Equation (4). The main purpose of this study was to predict fire risks according to the thermal characteristics of scattered particles; however, the mean number of particles (N) was expressed using the maximum number of particles (N_{max}) to analyze the thermal characteristics according to the maximum particle diameter ($d_{p,max}$) generated during welding.

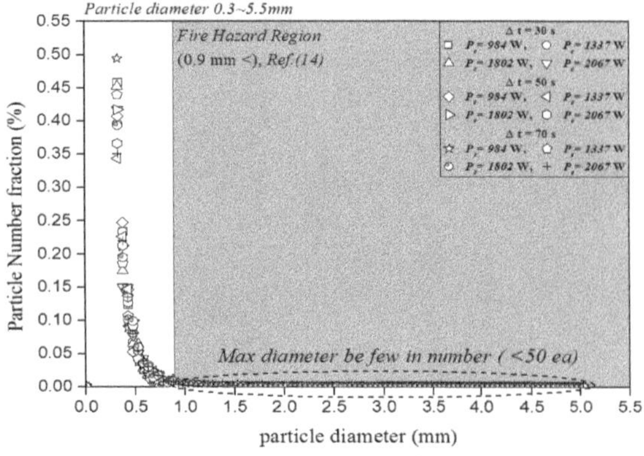

(**a**) Particle number fraction vs. particle diameter

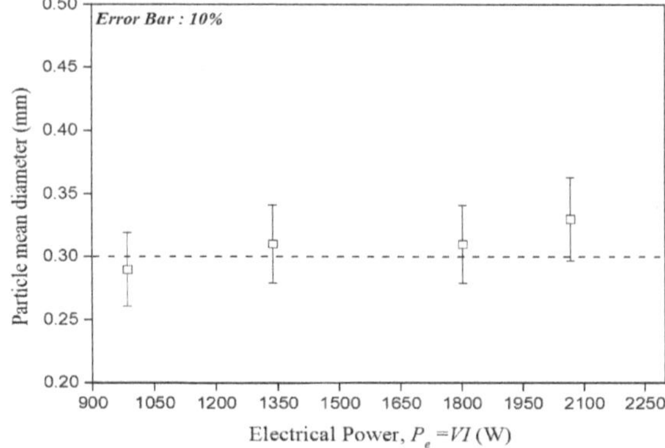

(**b**) Particle mean diameter vs. electrical power

Figure 7. Results of the (**a**) fraction of particle number, and (**b**) particle mean diameter when P_e = 984, 1337, 1802, and 2067 W and Δt = 30 s, 50 s, and 70 s.

As the maximum particle size, $d_{p,max}$, is still undetermined, it is necessary to analyze $d_{p,max}$ according to Δt and P_e to solve Equation (12).

$$N_{max} = \frac{6V_{total}}{\pi d_{p,max}^3},\ N = N_{max}\left(d_{p,max}/d_{p,m}\right)^3 \tag{12}$$

Figure 8 shows the results of analyzing the maximum diameter of scattered particles ($d_{p,max}$) according to the electrical power (P_e) at Δt = 30 s, 50 s, and 70 s. Each measured value represents the average of the maximum diameter obtained in three repeated experiments. Apparently, the size of the particles scattered from the electrode increased as Δt increased under a constant P_e because the temperature around the weld zone of the base metal increased. In addition, under the same Δt, the size of scattered particles increased in proportion to the melted mass of the electrode as P_e increased as shown in Equation (13).

$$d_{p,max} = a_3 \times P_e^{\,b_3} \tag{13}$$

where a_3 and b_3 are experimental constants. When $b_3 = 0.4825$, a_3 can be calculated by Equation (14) and plotted in Figure 8b.

$$a_3 = 2.194 \times 10^{-2} + 9.38 \times 10^{-4} \times \Delta t \tag{14}$$

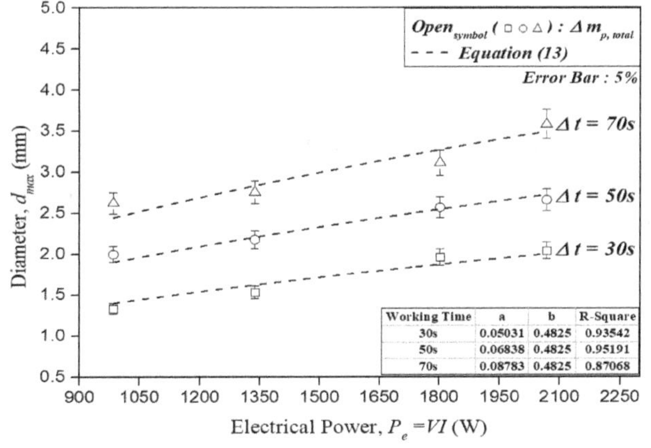

(**a**) Maximum particle diameter vs. electrical power

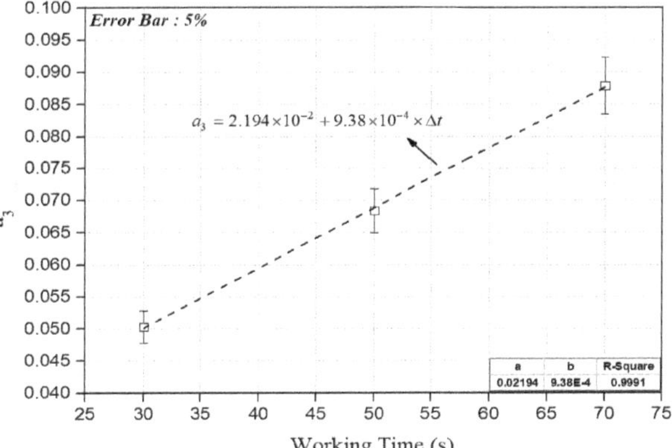

(**b**) The experiment coefficient of a3 vs. Welding time

Figure 8. Effects of electrical power on the distribution of the particles and the max diameter of the particles at welding time = 30 s, 50 s, and 70 s.

3.3. *Velocity of Welding Spatter*

Figure 9a shows the results of measuring the mean particle velocity according to the particle diameter under an electrical power of 984 W using a high-speed camera (model: phantom LC310) and particle tracking velocimetry (PTV). It is observed that the scattering velocity showed a tendency to decrease as the particle diameter increased under the experimental conditions of the welding time (Δt) and electrical power (P_e), as shown in Table 2. The correlation between the maximum diameter of scattered particles ($d_{p,max}$) and the scattering velocity ($u_{p,max}$) was analyzed, as shown in Figure 9b. $u_{p,max}$ decreased as $d_{p,max}$ increased with a difference of less than ±10% depending on the values of Δt and P_e, and the relationship shown in Equation (15) was found.

$$u_{p,max} = 1.3 \times d_{p,max}^{-1.004} \tag{15}$$

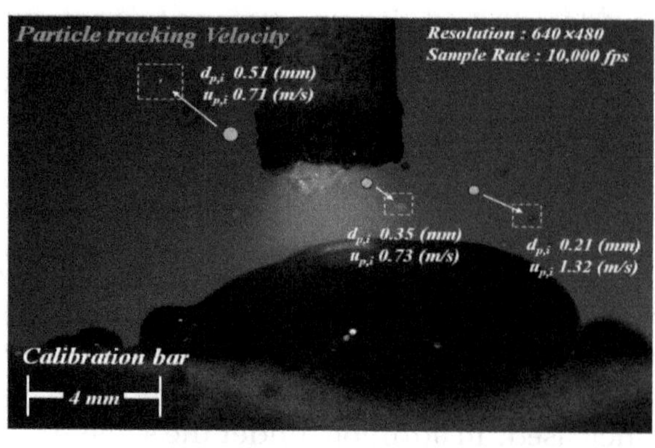

(**a**) Velocity contour of welding particle

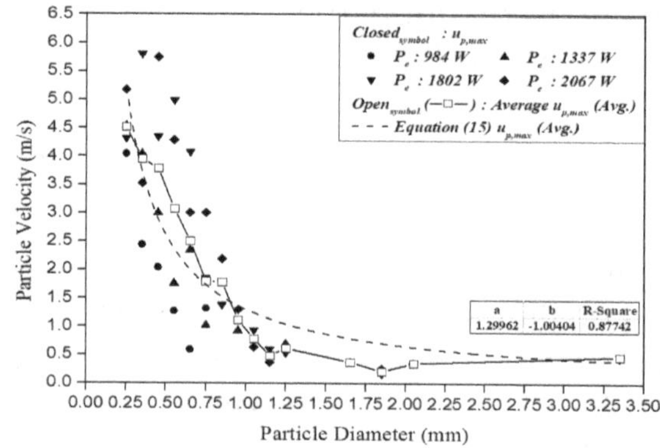

(**b**) Particle velocity vs. particle diameter

Figure 9. Results of the particle velocity according to the particle diameter.

3.4. Thermal Characteristics of Welding Spatter

Using the results of the total volume of particles ($V_{p,total}$), maximum number of particles (N_{max}), particle diameter ($d_{p,max}$), and scattering velocity ($u_{p,max}$) according to the Δt and P_e obtained in Section 3.1 to determine the convective heat transfer coefficient shown in Equation (4), Equation (16) can be formed.

$$h_{max} = \left(2 + 0.6Re_{d,max}^{0.5} Pr^{1/3}\right) k(T_{ref}) / d_{p,max}, \quad (16)$$

where $Re_{d,max}$ is the Reynolds number ($Re_{d,max} = \rho_p u_{p,max} d_{p,max}/\mu$) considering the maximum particle diameter ($d_{p,max}$). In particular, the thermal conductivity (k), specific heat (Cp), viscosity (μ), and density (ρ) of air vary depending on the reference temperature ($T_{ref} = (T_{p,s} + T_{\infty})/2$), as shown in Figure 10a. In this study, these can be calculated using Equations (17)–(20) based on the data given by this study [30].

$$k(T_{rep}) = -7.16 \times 10^{-3} + 1.72 \times 10^{-4} \times T - 2.45 \times 10^{-7} \times T^2 + 2.29 \times 10^{-10} \times T^3 - 9.81 \times 10^{-14} \times T^4 + 1.63 \times 10^{-17} \times T^5 \quad (17)$$

$$c_p(T_{rep}) = 1.05 - 2.89 \times 10^{-4} \times T + 6.83 \times 10^{-7} \times T^2 - 3.4 \times 10^{-10} \times T^3 + 2.25 \times 10^{-14} \times T^4 + 1.55 \times 10^{-17} \times T^5 \quad (18)$$

$$\mu(T_{rep}) = 4.26 \times 10^{-6} + 4.93 \times 10^{-8} \times T + -1.33 \times 10^{-11} \times T^2 + 2.36 \times 10^{-15} \times T^3 \quad (19)$$

$$\rho(T_{rep}) = 374.074 \times T^{-1.0114} \quad (20)$$

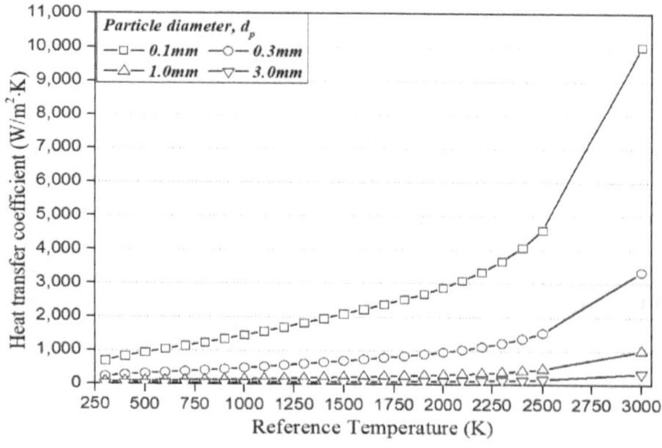

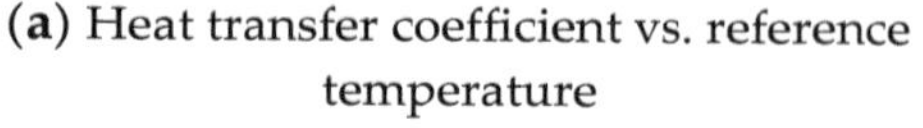
(**a**) Heat transfer coefficient vs. reference temperature

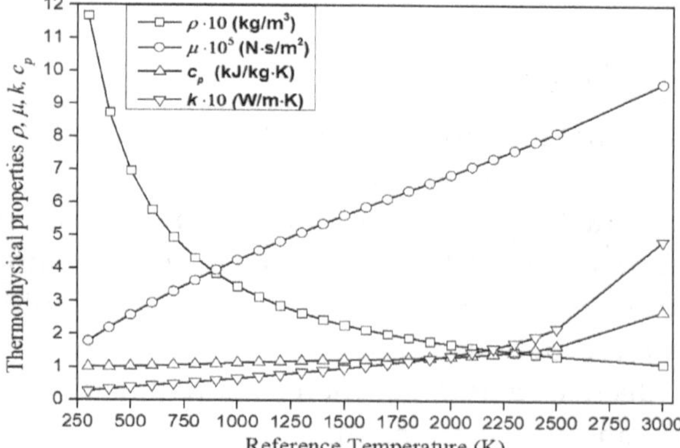

(**b**) Thermophysical properties (ρ, μ, k, c_P) vs. reference temperature

Figure 10. Effects of the heat transfer coefficient of the maximum particle dimeter and velocity according to thermophysical properties (ρ, μ, k, c_p).

Figure 10a shows the results of analyzing the convective heat transfer coefficient, h, according to T_{ref} when $d_{p,max}$ = 0.1, 0.3, 1.0, and 3.0 mm. h decreased as $d_{p,max}$ increased while the scattering velocity ($u_{p,max}$) decreased to 13.12, 4.35, 1.30, and 0.43 m/s at different $d_{p,max}$ values obtained by Equation (15). Therefore, Equation (2), for calculating the mean particle temperature considering the welding time and electrical power, can be expressed as in Equation (21).

$$T_{P,s} = T_\infty + \frac{0.13P_e}{N_{max} h_{max} A_{P,max} r_{ratio}}, \tag{21}$$

where N_{max}, h_{max}, $A_{p,max}$, and r_{ratio} are the total number of particles, heat transfer coefficient, surface area of a particle, and a constant calculated by replacing $d_{p,m}$ with $d_{p,max}$, respectively, when welding particles are at their maximum size. In particular, the mean number of particles (N) used in Equation (3) and the mean surface area ($A_{\mathrm{p,m}}$) are related as $N = N_{max}(d\_p,m/d\_p,max)^{-3}$ and $A_{p,m} = A_{p,max}(d\_p,m/d\_p,max)^2$, whereas the convective heat transfer coefficient (h) for determining the temperature of the welding spatter is given by $h = h_{max} \times (d\_p,m/d\_p,max)^{-0.5} \times (u_{p,m}/u_{p,max})^{0.5}$. Therefore, r_{ratio} is related as,

$$r_{ratio} \sim \left(\frac{u_{p,m}}{u_{p,max}}\right)^{0.5} \left(\frac{d_{p,m}}{d_{p,max}}\right)^{-1.5}, \tag{22}$$

where, $u_{p,m}/u_{p,max}$ represents the ratio of the mean velocity to the maximum velocity. As it varies depending on the diameter, it can be expressed as Equation (23).

$$\left(\frac{u_{p,m}}{u_{p,max}}\right)^{0.5} = C\left(\frac{d_{p,m}}{d_{p,max}}\right)^{0.5}, \tag{23}$$

where C is the experimental constant which may vary depending on the velocity difference. Because k, Cp, μ, and ρ used to obtain the convective heat transfer coefficient are functions of $T_{p,s}$ as shown in Equations (17)–(20), it is necessary to solve Equation (13) for the maximum particle size, Equation (15) for the maximum velocity, Equation (16) for the convective heat transfer coefficient, and Equation (23) to perform iterative calculations for predicting the maximum temperature of the welding spatter.

Figure 11a shows the results of maximum temperature ($T_{p,max}$) measured by capturing the welding spatter images accumulated on the acrylic collection plate at 60 fps for 70 s using a thermal imaging camera (Model: Fluck Ti520) at Δt = 70 s and P_e = 1337 W. As shown in the figure, the approximate maximum and mean particles were 432 °C and 347 °C, respectively.

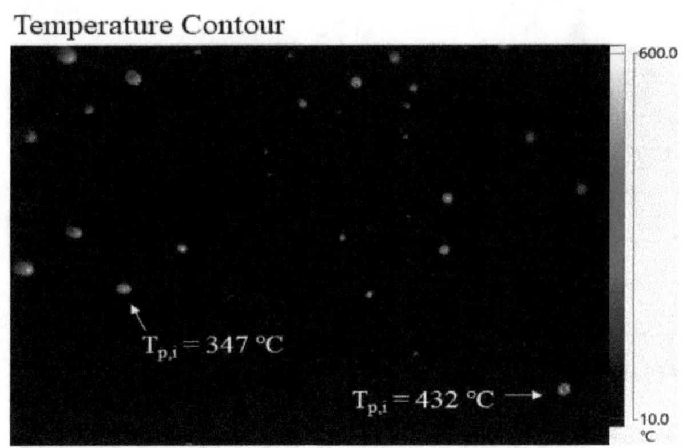

(a) Temperature contour of a welding particle

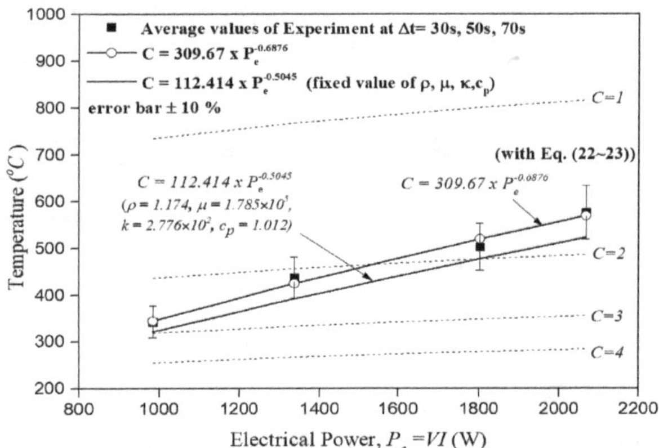

(b) Particle temperature vs. supply power

Figure 11. The results of temperature contour and the comparison of the prediction with experiment values of C.

Figure 11b shows the results calculated using the maximum particle temperature measurements and Equation (21) under the experimental conditions of Δt and P_e shown in Table 2. The experimental values agreed with the mean values with a difference of up to ±10% depending on Δt, and the temperature tended to increase as P_e increased. It should be noted that the maximum difference due to the time change was small (within ±2C) as shown in Equation (21) when C was constant, but the increasing tendency of the temperature in proportion to P_e was found to be consistent with the experimental values. However, when the values of C were 1, 2, 3, and 4, the slope at which the maximum particle temperature increased over the increase in P_e was smaller than the experimental value. This appears to be due to different values of C when the difference between the mean and maximum scattering velocities of the welding spatter increased along with P_e. At each P_e, the value of C can be obtained using Equations (24) and (25).

$$C_1 = 309.67 \times P_e^{-0.6876} \tag{24}$$

$$C_2 = 112.414 \times P_e^{-0.5045} \tag{25}$$

C_1 is a constant calculated by performing iterative calculations using Equations (17)–(20) for predicting the maximum particle temperature, whereas C_2 is calculated using the values of the density, thermal conductivity, viscosity coefficient, and specific heat at room temperature (298 K). Therefore, the maximum particle temperature can be predicted within an error range of approximately 5% using the equation to solve C_2.

Figure 12 shows the results of calculating the maximum temperature and diameter of the welding particles when the welding time (Δt) ranged from 30–70 s and P_e from 984–2067.5 W. "Fire hazard region" means the possibility of fire spreading to combustible materials such as polyurethan foam as mentioned in Ref [14,15,27], and "No ignition region" means the minimized conditions of fire spread. Based on previous studies, it can be confirmed that the maximum welding time and electrical power are 10 s and 1150 W, respectively, when the minimum particle size of welding spatter for the risk of fire spread is 0.9 mm, and the minimum temperature is 350 °C [19]. Therefore, the results of this study indicate that it is possible to calculate the electrical power for minimizing the risk of fire due to welding spatter when the electrode type and welding time are known.

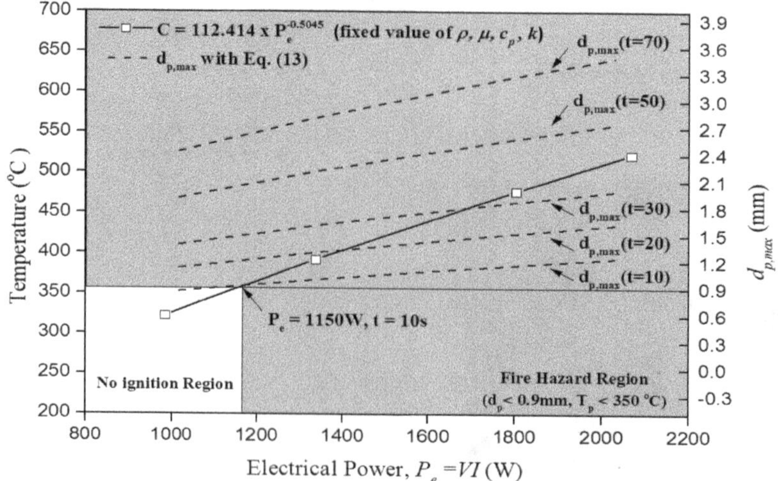

Figure 12. Predicted results of the maximum diameter and temperature of the welding particle depending on the electrical power and welding time.

4. Summary

In this study, the volume, maximum diameter, scattering velocity, and maximum number of welding spatter for shielded metal arc welding (SMAW) were analyzed according to the electrical power and welding time. When the electrical power was varied for welding times of 30 s, 50 s, and 70 s, the following results were derived.

First, when the mass of the electrode and scattered particles was calculated, an empirical formula was derived, which showed an increase in the mass of scattered particles when the electrical power increased at a constant welding time. In particular, the mass of scattered welding spatter represented approximately 11.45% of the total mass of the consumed electrode on average. The densities of the scattered particles were found to vary between 4876–7572 kg/m^3 depending on the volume fraction of the core and coating composition of the electrode as referred by manufacturer.

Second, it was found that the mean diameter of welding spatter was approximately 0.3 mm, which was constant regardless of the welding time and electrical power. The maximum particle size, which has an important impact on fire risks, however, showed a tendency to increase in proportion to the welding time and electrical power. An empirical formula considering the maximum particle size was also derived to predict the temperature of the scattered welding spatter.

Third, the scattering velocity differed with differences of up to ±91% according to the welding time and electrical power. This appears to be due to the fact that materials with significantly different densities were mixed, which affected the momentum of the welding spatter while they were generated from the electrode. However, the scattering velocity decreased as the particle diameter increased.

Fourth, empirical formulas for the volume, maximum diameter, and scattering velocity of the welding spatter according to the welding time and electrical power were derived and compared with the maximum temperature measurements during the welding process. Results showed a good agreement between the compared values within an error range of approximately 10%. After verifying this accuracy, the case in which the minimum temperature of welding spatter was 350 °C or higher and the particle size was 0.9 mm was analyzed. It was found that fire risks can be minimized when a maximum welding time of 10 s and maximum electrical power of 1150 W are used. It should be noted that the maximum temperature of the welding spatter increased in proportion to the electrical power regardless of the welding time. The results of this study are expected to be used as important data for quantitatively presenting measures to minimize the fire risk of welding spatter.

Author Contributions: Conceptualization, Y.J.S. and W.J.Y.; Data curation, W.J.Y.; Formal analysis, W.J.Y.; Funding acquisition, W.J.Y.; Investigation, Y.J.S.; Methodology, W.J.Y.; Project administration, W.J.Y.; Software, Y.J.S.; Supervision, W.J.Y.; Validation, Y.J.S. and W.J.Y.; Visualization, Y.J.S.; Writing—original draft, Y.J.S.; Writing—review & editing, W.J.Y. All authors have read and agreed to the published version of the manuscript.

References

1. Messler, R.W., Jr. *Principles of Welding: Processes, Physics, Chemistry, and Metallurgy*; Wiley: London, UK, 2008.
2. DuPont, J.N.; Lippold, J.C.; Kiser, S.D. *Welding Metallurgy and Weldability of Nickel-Base Alloys*; Wiley: London, UK, 2009.
3. Weman, K. *Welding Processes Handbook*, 2nd ed.; Cambridge Woodhead Publishing: Cambridge, UK, 2003.
4. Babrauskas, V. *Ignition Handbook: Principles and Applications to Fire Safety Engineering, Fire Investigation, Risk Management, and Forensic Science*; Fire Science Publishers: Issaquah, WA, USA, 2003.
5. National Fire Protection Association (NFPA). *NFPA Standard 51B: Standard for Fire Prevention During Welding, Cutting, and Other Hot Work, Technical Report*; National Fire Protection Association: Quincy, MA, USA, 2009.
6. Schonherr, W. Fire risks with welding torches and manual arc welding-globules, spatter and how far they can be thrown. *Schweiss. Schneid.* **1982**, *34*, E74–E77.
7. Tomków, J.; Fydrych, D.; Wilk, K. Effect of Electrode Waterproof Coating on Quality of Underwater Wet Welded Joints. *Materials* **2020**, *13*, 2947. [CrossRef] [PubMed]
8. Srisuwan, N.; Kumsri, N.; Yingsamphancharoen, T.; Kaewvilai, A. Hardfacing welded ASTM A572-based, high-strength low-alloy steel: Welding, characterization, and surface properties related to the wear resistance. *Metals* **2019**, *9*, 244. [CrossRef]
9. Omajene, J.E.; Martikainen, J.; Kah, P.; Pirinen, M. Fundamental difficulties associated with underwater wet welding. *Int. J. Eng. Res. Appl.* **2014**, *4*, 26–31.

10. Łabanowski, J.; Fydrych, D.; Rogalski, G.; Samson, K. Underwater welding of duplex stainless steel. *Solid State Phenom.* **2012**, *183*, 101–106. [CrossRef]
11. Janusas, G.; Jutas, A.; Palevicius, A.; Zizys, D.; Barila, A. Static and vibrational analysis of the GMAW and SMAW joints quality. *J. Vibroeng.* **2012**, *14*, 1220–1226.
12. Urban, J.L. Spot Ignition of Natural Fuels by Hot Metal Particles. Ph.D. Theses, University of California Berkeley, Berkeley, CA, USA, 2017.
13. Mikkelsen, K. An Experimental Investigation of Ignition Propensity of Hot Work Processes in the Nuclear Industry. Master's Thesis, University of Waterloo, Waterloo, ON, Canada, 2014.
14. Song, J.; Wang, S.; Chen, H. Safety distance for preventing hot particle ignition of building insulation materials. *Theor. Appl. Mech. Lett.* **2014**, *4*, 034005. [CrossRef]
15. Urban, J.L. Spot Fire Ignition of Natural Fuels by Hot Aluminum Particles. *Fire Technol.* **2017**, *54*, 797–808. [CrossRef]
16. Liu, Y.; Urban, J.L.; Xu, C.; Fernandez-Pello, C. Temperature and Motion Tracking of Metal Spark Sprays. *Fire Technol.* **2019**, *55*, 2143–2169. [CrossRef]
17. Kim, Y.S.; Eagar, T.W. Analysis of metal transfer in gas metal arc welding. *Weld. J.* **1993**, *72*, 269-s.
18. Tanaka, T. On the inflammability of combustible materials by welding spatter. *Rep. Natl. Res. Inst. Police Sci.* **1977**, *30*, 51–58.
19. Hagimoto, N.; Kinoshita, K. Scattering and Igniting Properties of Sparks Generated in an Arc Welding. In *6th Indo Pacific Congress on Legal Medicine and Forensic Sciences*; Indo Pacific Association of Law, Medeicine and Science: Kobe, Japan, 1998; pp. 863–866.
20. Shigeta, M.; Peansukmanee, S.; Kunawong, N. Qualitative and quantitative analyses of arc characteristics in SMAW. *Weld. World* **2016**, *60*, 355–361. [CrossRef]
21. Pistorius, P.G.; Liu, S. Changes in Metal Transfer Behavior during Shielded Metal Arc Welding. *Weld. J.* **1997**, *76*, 305–315.
22. Narasimhan, P.N.; Mehrotra, S.; Raja, A.R.; Vashista, M.; Yusufzai, M.Z.K. Development of hybrid welding processes incorporating GMAW and SMAW. *Mater. Today Proc.* **2019**, *18*, 2924–2932. [CrossRef]
23. Xu, X.; Liu, S.; Bang, K.S. Comparison of metal transfer behavior in electrodes for shielded metal arc welding. *Int. J. Korean Weld. Soc.* **2004**, *4*, 15–22.
24. Hagiwara, T.; Yamano, K.; Nishida, Y. Ignition risk to combustibles by welding spatter. *J. Jpn. Assn. Fire Sci. Eng.* **1982**, *32*, 8–12.
25. Kinoshita, K.; Hagimoto, Y. Temperature measurement of falling spatters of arc welding. In Proceedings of the 22nd Annual Meeting Japan Society for Safety Engineering, Japan Society for Safety Engineering, Tokyo, Japan; 1989; pp. 145–148.
26. Brandi, B.; Taniguchi, C.; Liu, S. Analysis of metal transfer in shielded metal arc welding. *Suppl. Weld. J.* **1991**, *70*, 261s–270s.
27. Shin, Y.J.; You, W.J. Analysis of the thermal characteristics of welding spatters in SMAW using simplified model in fire technology. *Energies* **2020**, *13*, 2266. [CrossRef]
28. Ranz, W.E.; Marshall, W.R. Evaporation from drop, Part 1. *Chem. Eng. Prog.* **1952**, *48*, 141–146.
29. Marques, E.S.V.; Silva, F.J.G.; Paiva, O.C.; Pereira, A.B. Improving the mechanical strength of ductile cast iron welded joints using different heat treatments. *Materials* **2019**, *226*, 2263. [CrossRef] [PubMed]
30. Incropera, F.P.; Lavine, A.S.; Bergman, T.L.; DeWitt, D.P. *Principles of Heat and Mass Transfer*; Wiley: Hoboken, NJ, USA, 2013.

12

Sensitivity of Axial Velocity at the Air Gap Entrance to Flow Rate Distribution at Stator Radial Ventilation Ducts of Air-Cooled Turbo-Generator with Single-Channel Ventilation

Yong Li [1,2], Weili Li [1] and Ying Su [1,*]

1 School of Electrical Engineering, Beijing Jiaotong University, Beijing 100044, China
2 China North Vehicle Research Institute, Beijing 100072, China
* Correspondence: 15117377@bjtu.edu.cn

Abstract: In the design and calculation of a 330 MW water-water-air cooling turbo-generator, it was found that the flow direction of the fluid in the local stator radial ventilation duct is opposite to the design direction. In order to study what physical quantities are associated with the formation of this unusual fluid flow phenomenon, in this paper, a 100 MW air-cooled turbo-generator with the same ventilation structure as the abovementioned models is selected as the research object. The distribution law and pressure of the fluid in the stator radial ventilation duct and axial flow velocity at the air gap entrance are obtained by the test method. After the calculation method is proved correct by experimental results, this calculation method is used to calculate the flow velocity distribution of the outlets of multiple radial ventilation ducts at various flow velocities at air gap inlets. The relationship between the flow distribution law of the stator ventilation ducts and the inlet velocity of the air gap is studied. The phenomenon of backflow of fluid in the radial ventilation duct of the stator is found, and then the influence of backflow on the temperature distribution of stator core and winding is studied. It is found that the flow phenomenon can cause local overheating of the stator core.

Keywords: air-cooled steam turbine generator; single-channel ventilation; backflow; radial ventilation duct; fluid field

---

1. Introduction

Regarding the research on the temperature field of large turbo-generators, most scholars have mainly focused on the research on the fluid field and the temperature field [1–3]. Some research mainly focused on the heat transfer relationship between the temperature rise of structural parts such as the stator winding and fluid flow. There were also some studies on the internal fluid distribution law of generators, and the main research direction was the rational design of the ventilation system of the generator [4–6]. Unlike the above research contents, the main content of this paper is about which physical quantities of the fluid are related to the fluid flow distribution rules in the ventilation duct of the generator and which significant changes will occur with the changes of these physical quantities. The accurate measurement results of the fluid distribution in the ventilation duct of the stator core are the prerequisites for the accurate calculation of the above research contents.

In this paper, a 100 MW single-channel ventilation turbo-generator set is taken as the research object, and the experimental measurement method [7–9] of flow in stator radial duct distribution and air gap flow velocity of ventilation duct is proposed. The fluid pressure and flow velocity measuring sensors are buried at the generator stator ventilation duct and the air gap entrance. After measuring the pressure and flow speed at these positions, it is found that the fluid velocity at the local stator radial ventilation duct is significantly lower. Then the calculation results of fluid flow velocity in different ventilation ducts of the generator stator during the test conditions are obtained by the three-dimensional fluid field calculation method. The calculated results are in good agreement with the experimental results after comparison.

Though our analysis, it is found that the air gap inlet velocity is the main factor affecting the distribution of the fluid flow rate in the stator ventilation duct. Based on this calculation method, the results of velocity distribution in different radial ventilation ducts of generator stator are calculated at three different flow velocities at the air gap entrance. It is also found that with the increase of the velocity at the air gap entrance, the phenomenon of backflow in the stator iron core ventilation duct will occur locally.

2. Experimental Measurement and Result Analysis

2.1. Test and Measurement of the Velocities at Air Gap Inlet and Radial Ventilation Duct Outlets in the Generator

The schematic diagram of single ventilation is shown in Figure 1. There are 76 radial ventilation ducts in the stator of the test generator. Due to the large number of them, the test basically follows the principle of embedding a wind speed measuring component in every other wind duct. Due to the existence of the step structure on both sides of the stator core end, the fluid is blocked by the core step in this area, resulting in a large change in the fluid flow velocity in this area. Therefore, the measuring elements are embedded densely at the air gap entrance of the generator. The measuring component adopts the pitot tube, and the pitot tube is connected with two pressure measuring tubes at the tail. The measured full pressure and static pressure are led to the outside of the generator through the pressure tube, which is convenient for connecting the pressure collector and obtaining the test result. The pitot tubes are buried in the ventilation duct yokes of the stator core and care to ensure the same embedment depth and axial position. In the axial direction, the buried pitot tube is located in the axial center of the ventilation duct, and it is not affected by the boundary layer with low flow velocity at the wall surface. Each pitot tube in the radial direction is oriented toward the center of the circle to ensure the consistency between the measurement sensor and the direction angle of the flow velocity. In this way, the measuring point placed in the ventilation duct can clearly and effectively measure the wind speed and pressure in the entire wind channel, ensuring the consistency of data. The specific installation positions are shown in Figures 2 and 3.

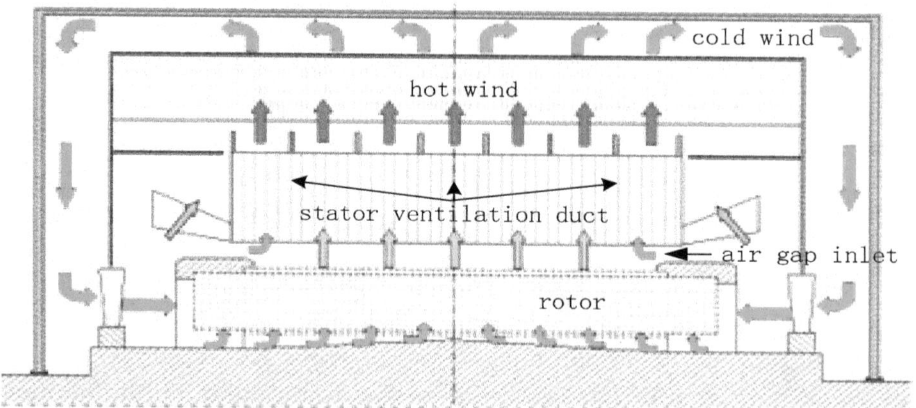

Figure 1. Single-channel ventilation air-cooled turbo-generator ventilation system.

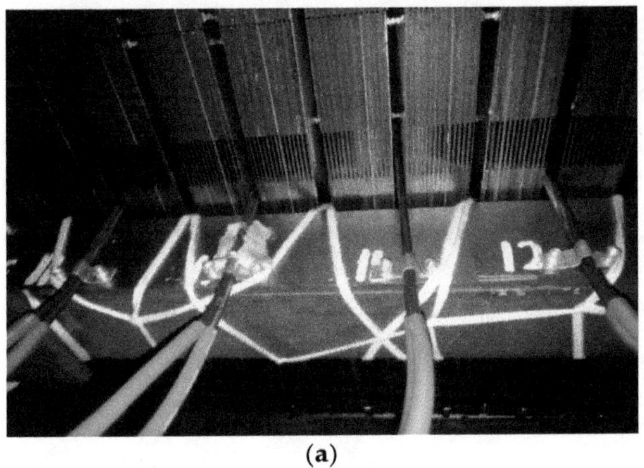

(a)

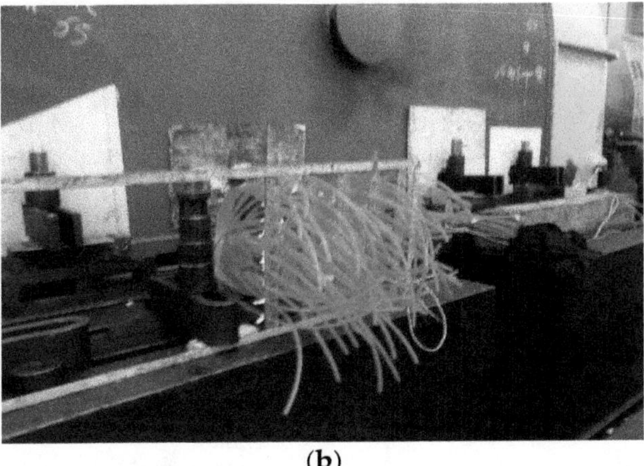

(b)

Figure 2. Placement and measuring position of pitot tube in the stator ventilation ducts. (**a**) Placement position of pitot tube; (**b**) measuring position of ventilation duct.

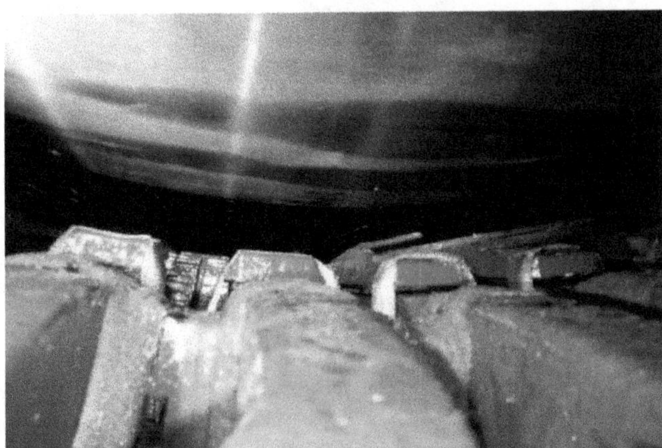

Figure 3. The sensor placement at the air gap entrance.

Figure 3 shows the embedding position of the wind speed sensor at the air gap entrance. The sensor is located at the entrance of the generator air gap. Considering the influence of rotating air flow at the air gap entrance, the installation angle between the wind sensor direction and the axial direction is 40 degrees. Since the air volume passing through this position accounts for about 70% of the total air volume of the generator, it is very important to accurately measure the wind speed at this inlet position. At the same time, in single-channel ventilation of turbo-generator, the flow resistance at the air gap entrance is the biggest in the whole ventilation system. It is necessary to bury the pressure measuring point here, which is also of great guiding significance to verify the calculated value of the whole ventilation system design. This measurement will provide important inlet boundary conditions for the following finite element calculations.

2.2. Stator Radial Ventilation Duct Outlet Flow Rate Measurement Results

Table 1 shows that the static pressure at the measuring position of the stator core yoke ranges from 1000 to 1500 Pa. From the distribution of pressure difference, when the fluid is located at the end of both sides of the stator core, the pressure difference is small, and it is bigger close to the stator core center. According to the Bernoulli equation of fluid flow, it can be known that: This pressure difference is actually the dynamic pressure of the fluid, and this value can reflect the fluid velocity at the measurement position.

$$p_{total} - p_{static} = p_{Dynamic} = \frac{1}{2}\rho v^2 \tag{1}$$

Table 1. Pressure and flow velocity measurement results in stator radial ventilation ducts.

No. of Ventilation Ducts	Total Pressure Pa	Static Pressure Pa	Pressure Difference Pa	Calculated Wind Speed m/s
2	1514	1464	50	8.8
3	1324	1256	68	10.3
4	1188	1178	10	3.9
5	1122	1117	5	2.8
6	1153	1143	10	3.9
8	1205	1145	60	9.6
9	1178	1125	53	9.1
11	1173	1136	37	7.6
13	1193	1144	49	8.7
15	1238	1153	85	11.5
18	1231	1118	113	13.2
20	1191	1088	103	12.6
22	1201	1085	116	13.4
24	1231	1092	139	14.7
26	1211	1097	114	13.3
28	1270	1103	167	16.1
30	1257	1089	168	16.1
33	1317	1117	200	17.6
34	1367	1097	270	20.5
36	1286	1094	192	17.3
38	1259	1112	147	15.1
40	1359	1103	256	19.9
42	1304	1093	211	18.1
44	1263	1094	169	16.2
46	1353	1108	245	19.5
49	1142	1281	−139	14.7
51	1359	1152	207	17.9
53	1289	1144	145	15.0
55	1281	1134	147	15.1
57	1259	1140	119	13.6
59	1260	1124	136	14.5
61	1311	1162	149	15.2
64	1298	1206	92	11.9
66	1284	1209	75	10.8
68	1280	1216	64	10.0
70	1256	1203	53	9.1
71	1256	1212	44	8.3
73	1205	1191	14	4.7
74	1019	1172	−153	15.4
75	1259	1220	39	7.8
76	1576	1515	61	9.7

The calculated static pressure at the generator fan inlet is about 4500 Pa. It can be inferred from the static pressure measured by the sensor in Table 1 that the sum of fluid flow resistance at the generator end, air gap, core tooth, and other positions is 3000 Pa. The static pressure in the ventilation duct on both sides of the generator end is large, but the difference between the total pressure and the static pressure is small. The static pressure at the ventilation duct of the generator center is small, but the difference between the total pressure and the static pressure is large, indicating that the flow velocity in most ventilation ducts in the central area of the generator stator is larger than that at the end. Detailed analysis of the data in the table: It was found that the pressure difference in radial ventilation duct no. 49 and 74 was negative. After rechecking the measuring element, it was found that an installation error caused the negative measuring result.

In order to facilitate intuitive analysis, the measurement results are shown in Figure 4. The wind speed measurement results in the stator core ventilation duct have a total of 76 ventilation ducts throughout the core. The wind speed measurement components are densely packed at the ends and relatively less in the middle. It can be seen that the fluid radial velocities in the entire core ventilation ducts show a higher speed in the middle and a lower speed at both sides of the end. This indicates that the flow velocity from the two ends to the center gradually increases, and the maximum flow velocity is about 20 m/s.

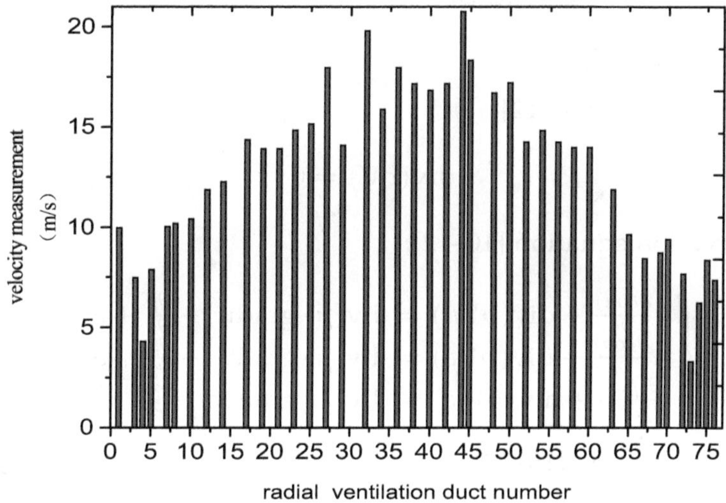

Figure 4. The distribution of wind speed values in the stator radial ventilation duct at 3000 r/min.

It can also be found that the flow velocity in the first three ventilation ducts is high on both sides of the stator, while the flow velocity in the fourth, fifth, and sixth ventilation ducts decreased rapidly, down to less than 4 m/s. It is not difficult to infer that the stator cores and winding near the fourth to the seventh ventilation ducts will have a higher temperature rise due to the lower wind speed.

The reason for the decrease of the fourth, fifth, and sixth ventilation duct wind speeds on both sides of the stator is that this position is the beginning of the normal stator ventilation duct. These ventilation duct structures are different from the first, second, and third which have the characteristics of the step structure, shown in Figure 5. The wind in the first three ventilation ducts, due to the iron core step structure block, has a higher wind speed in radial ventilation ducts. After passing through the first three ventilation ducts, the flow duct structure area is smooth in the axial direction and the flow becomes unobstructed. The wind has a high axial velocity when it flows in the air gap. This will cause the static pressure around the gap to drop. Thus, it results in a very low wind speed in radial ventilation ducts at this location. As the capacity of the generator increases, the fluid velocity at the entrance of the air gap increases, and this problem becomes more serious.

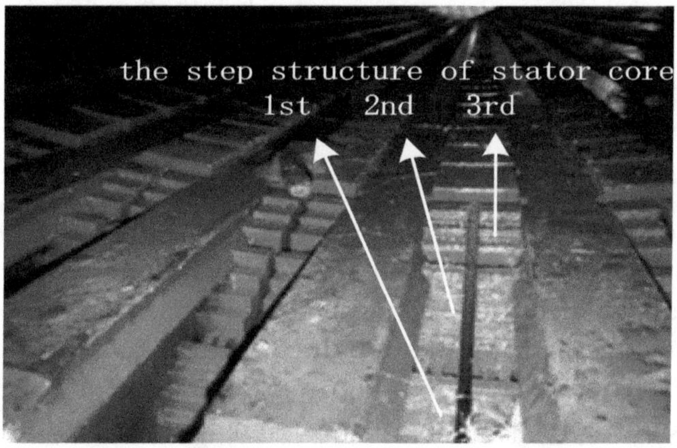

Figure 5. The step structure of stator end.

In order to study the influence of different axial velocities at the air gap entrance on the radial velocity of fluid in the adjacent ventilation duct, the finite element method is used to calculate the test condition.

3. Study on Sensitivity of Air Gap Inlet Flow Velocity to Stator Radial Ventilation Duct Flow Rate Distribution

3.1. Mathematical Model and Physical Model Description

3.1.1. Mathematical Model Description

Mass conservation equation [10–12]:

$$\mathrm{div}(\rho V) = 0 \tag{2}$$

Momentum conservation equation [10–12]:

$$\begin{cases} \frac{\partial(\rho u)}{\partial t} + \mathrm{div}(\rho V u) = \mathrm{div}(\mu \cdot \mathrm{grad u}) - \frac{\partial p}{\partial x} + S_u \\ \frac{\partial(\rho v)}{\partial t} + \mathrm{div}(\rho V v) = \mathrm{div}(\mu \cdot \mathrm{grad v}) - \frac{\partial p}{\partial x} + S_v \\ \frac{\partial(\rho w)}{\partial t} + \mathrm{div}(\rho V w) = \mathrm{div}(\mu \cdot \mathrm{grad w}) - \frac{\partial p}{\partial x} + S_w \end{cases} \tag{3}$$

Energy conservation equation [10–12]:

$$\frac{\partial(\rho T)}{\partial t} + \mathrm{div}(\rho V T) = \mathrm{div}\left(\frac{\lambda}{c}\mathrm{grad}T\right) + S_T \tag{4}$$

where ρ is fluid density; t is the time; V is the relative fluid velocity vector; u, v, and w are the components of V in x, y, and z axes; μ is viscosity coefficient; p is static pressure acting on a micro cell in air; S_u, S_v, and S_w are the source items of the momentum equation; λ is thermal conductivity; c is specific heat in constant pressure; S_T is the ratio of the heat generated in the unit volume and specific heat c.

Since the fluid in the stator calculation domain has a turbulent flow and the air in the radial ventilating duct of the generator can be regarded as incompressible, a standard k-ε model in the commercial solver "FLUENT" is used to solve the turbulence movement [10–12].

$$\begin{cases} \frac{\partial(\rho k)}{\partial t} + div(\rho k V) = div\left[\left(\mu + \frac{\mu_t}{\sigma_k}\right)gradk\right] + G_k - \rho\varepsilon \\ \frac{\partial(\rho \varepsilon)}{\partial t} + div(\rho V \varepsilon) = div\left[\left(\mu + \frac{\mu_t}{\sigma_\varepsilon}\right)grad\varepsilon\right] + G_{1\varepsilon}\frac{\varepsilon}{k}G_k - G_{2\varepsilon}\rho\frac{\varepsilon^2}{k} \end{cases} \tag{5}$$

where k is the turbulent kinetic energy, ε is the diffusion rate, G_k is turbulent generation rate, μ_t is turbulent viscosity coefficient, $G_{1\varepsilon}$ and $G_{2\varepsilon}$ are constant; σ_k and σ_ε are, respectively, the equation k and equation ε of Planck's constant turbulence.

3.1.2. Physical Model Description, Mesh Analysis, and the Boundary Conditions

a. Physical Model Description

The model of Figure 6a includes 76 air gap inlets and radial ventilation duct outlets, the stator core is near the ventilation ducts, and the fluid flow in the ventilation ducts will take away the loss in the core. The model of Figure 6b includes: The upper- and lower-layer bar of the generator stator slots and their main insulation, wedges, and ventilation ducts fluid area. The first three ventilation ducts of the iron core have the step structure shown in Figure 5, and their lengths in the radial direction are sequentially increased, and the corresponding fluid area here is sequentially reduced.

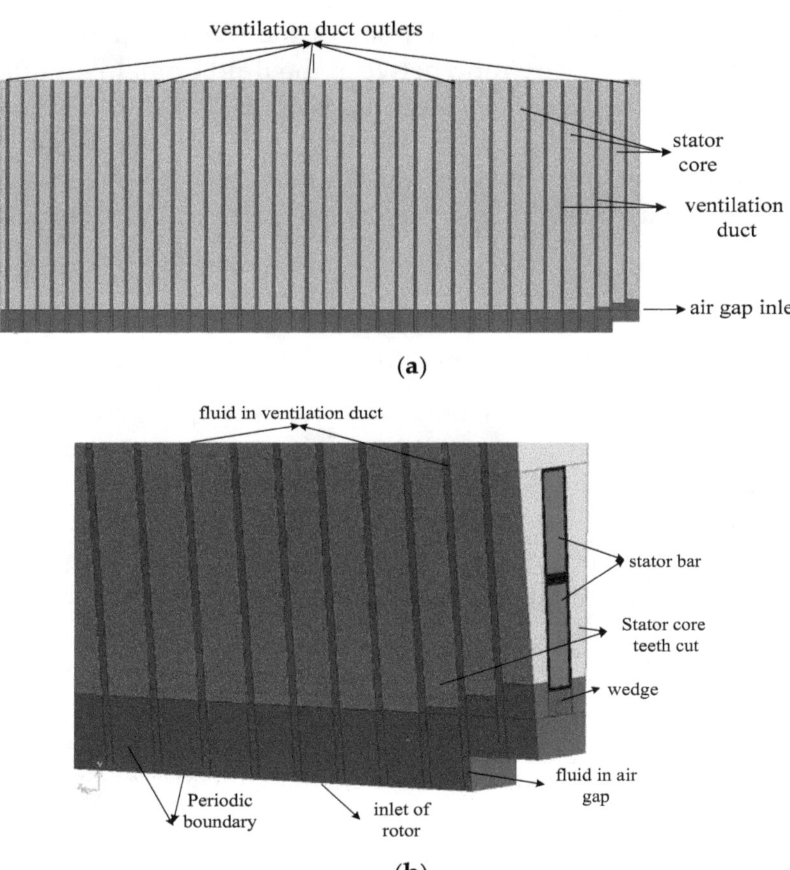

Figure 6. Physical model of stator ventilation system. (**a**) Calculation model of fluid field for 100 MW air-cooled turbo-generator; (**b**) partial display of calculation model.

b. Grid Mesh and Memory Usage

All the models adopted eight-node hexahedral mesh generation, as shown in Figure 7. The mesh generation was regular, and there were no over-sized or under-sized cells of the generation volume or area. The statistical results of the number of nodes and memory are shown in Table 2.

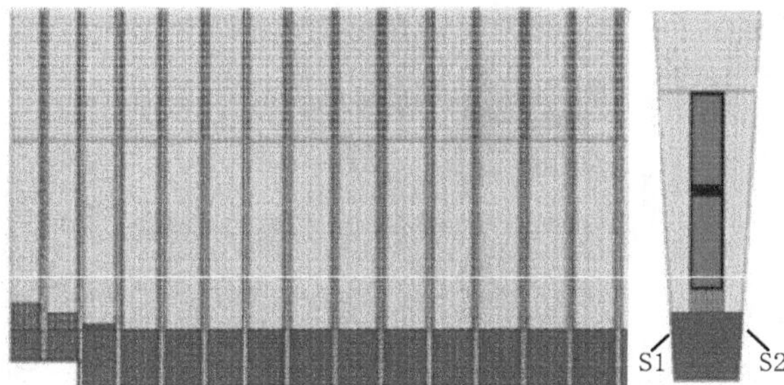

Figure 7. Model axial–radial and circumferential sections grid mesh.

Table 2. Statistical results of the number of nodes and memory.

	Cells	Faces	Nodes	Edges
Number Used	561,180	1,830,765	692,014	0
Mbytes Used	102	136	29	0
Number Allocated	561,180	1,830,765	692,014	0

In order to improve the computational accuracy of the fluid domain, the grid density in the fluid domain was dense, the grid density ratio between the fluid domain and the solid domain was about 10:1 in the axial direction. In the radial direction, the grid density was dense at the air gap and sparse at the yoke. While ensuring the accuracy of calculation, this effectively reduces the amount of computation and shortens the time. The grid mesh accuracy of this calculation method has been verified by the grid encryption method through previous work, which can better meet the calculation accuracy [13].

c. Boundary Conditions

The calculation boundary conditions are given as follows [14–16]:

(1) The air gap inlet is set as the velocity boundary condition, which is given according to the measurement results: The angle between the measuring element direction (Figure 3) and the axial direction is about 40 degrees, and the measured velocity is 74 m/s. Therefore, after triangle calculation, the inlet velocity of the air gap entrance is given as 56 m/s.
(2) The 38 radial ventilation outlets (half of generator) are basically balanced with the surrounding atmospheric pressure. Given the pressure boundary condition, the value is equivalent to one atmospheric pressure.
(3) The air gap exists in the whole circumferential direction, but the calculation model includes two additional surfaces, S1, S2 (shown in Figure 7), at the air gap that will increase the frictional resistance when fluid flows. In fact, this frictional resistance does not exist, so the periodic boundary conditions given for these two sides, eliminate the influence of frictional resistance on fluid flow.
(4) The velocity inlet of the rotor is given as the velocity boundary condition, and the velocity value is given according to the calculation results of the rotor ventilation calculation.

3.2. Comparison of Fluid Calculation Results and Measured Values at Test Conditions

The comparison of the test results and the calculation results of stator ventilation system are shown in Figure 8.

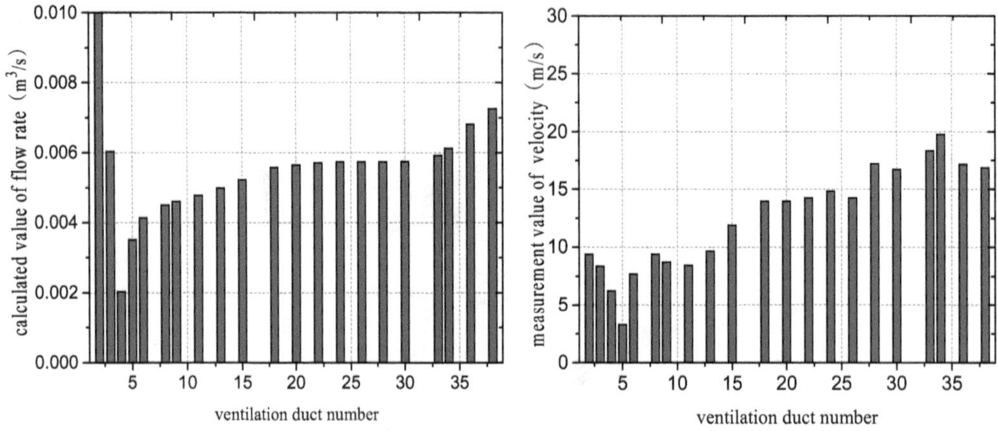

Figure 8. Comparison of calculated and measured values of fluid flow rate in the ventilation duct.

Because the fluid distribution is not uniform even in the same ventilation duct of the stator, the flow velocity value of one measuring point in the ventilation duct cannot completely reflect the distribution law of fluid velocity in this ventilation duct. Furthermore, the flow rate of the test result in this ventilation duct cannot be accurately obtained. However, since the fluid flow velocity measurement points are buried in the same position in each radial ventilation duct of the generator, the test velocity distribution law can indirectly reflect the flow rate distribution law in the stator radial ventilation ducts. The test results of flow velocities in different ventilation ducts and FEM calculation results of flow rates in different ventilation ducts are comparable.

It can be seen intuitively from Figure 8 that, in the fourth ventilation duct of the stator, the test results and calculated results both show a small flow rate at this position. In addition, the flow trend of fluid in different ventilation ducts is consistent. They all showed that the flow rate in the first three radial ventilation ducts is large, and then there is a trend of rapid increase after the flow rate decreases. The calculated flow distribution results are in agreement with the experimental measurements, which proves the accuracy of the calculation method.

3.3. Study on Sensitivity of Axial Velocity at the Air Gap Entrance to Flow Rate Distribution at Stator Radial Ventilation Ducts

In order to analyze the effect of different axial velocities at the air gap entrance on flow rate distribution at the stator radial ventilation ducts, this section gives four different wind speeds at the air gap entrance, which are 56, 70, 85, and 100 m/s respectively. Through the calculation of finite element theory, the distribution law of fluid flow in different radial ventilation ducts can be determined at the four different air gap inlet flow velocities.

When the air gap inlet velocity is 56 m/s, the flow velocity distribution of the fluid in the air gap and in each radial ventilation duct can be seen from Figure 9. After the fluid enters the air gap, the maximum flow velocity is 71 m/s, located below the second ventilation duct. According to the analysis of fluid continuity theory, the flow area in this place is the smallest position in the air gap, so the fluid velocity is the highest here. After fluid flowing through this position, the cross-sectional area in the air gap suddenly increases, but the fluid distribution does not rapidly expand as the air gap space increases, and the flow velocity in the air gap is still high. The velocity distribution trend shows the higher velocity in upper layer of the air gap, and the lower velocity in the lower layer. As the fluid in the air gap flows into the radial ventilation ducts sequentially, the flow velocity in the air gap becomes smaller and smaller. From the 38 radial flow velocity distributions, it can be seen directly that the velocity in the fourth and fifth ventilation ducts are lower, while the velocity of the first, second, 37th, and 38th are higher.

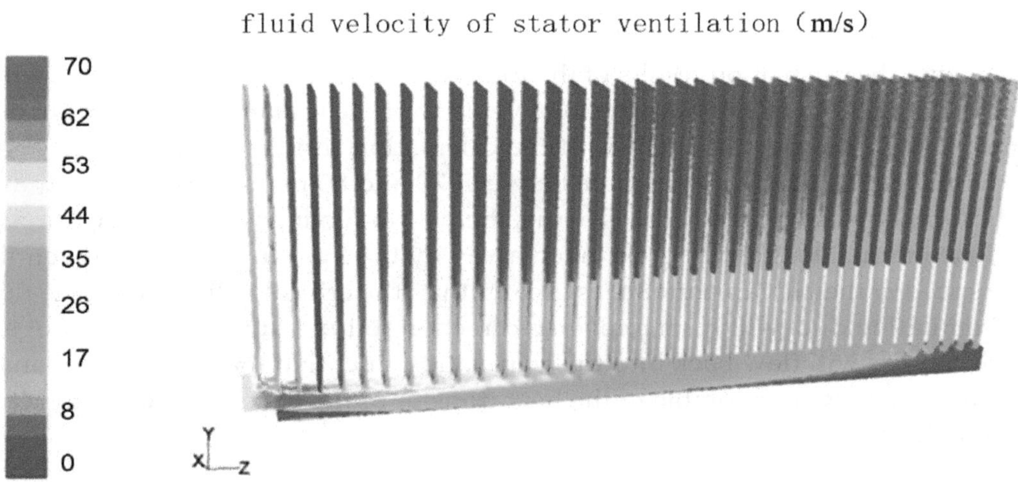

Figure 9. Fluid velocity distribution of stator air gap and the ventilation duct.

Similar to the above flow law, for the convenience of comparison, the fluid flow distribution in each radial ventilation duct is shown in Figure 10 under four different air gap entrance velocities.

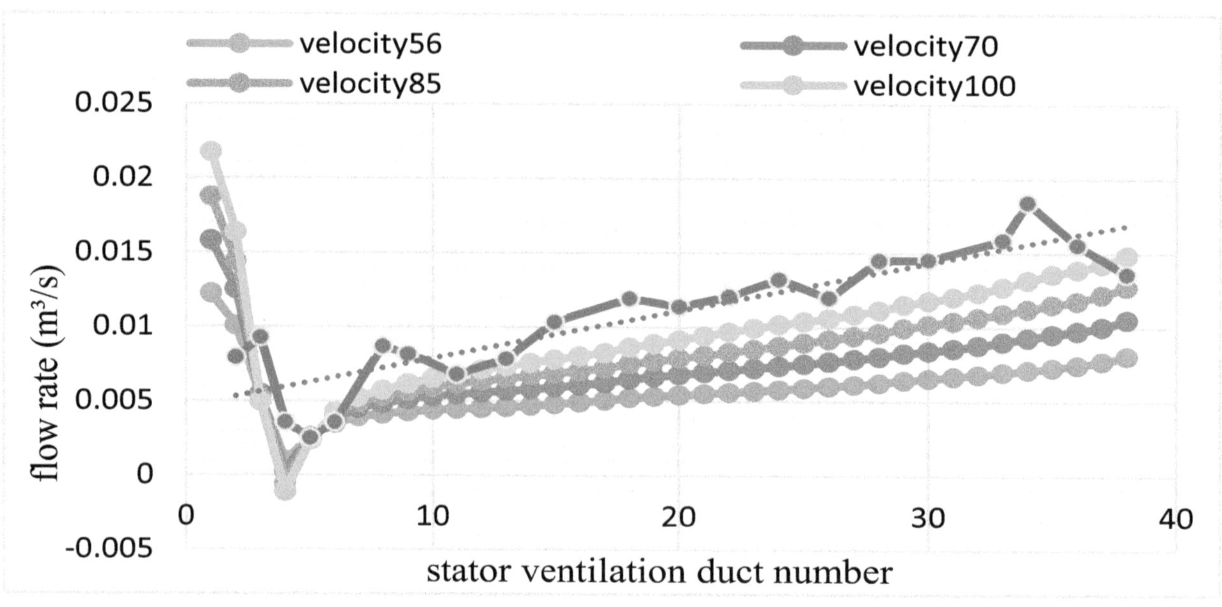

Figure 10. Flow rate distribution in the radial ventilation ducts at different air gap entrance velocities.

In Figure 10, the horizontal axis is the stator ventilation duct number. There is a total of six curves in the figure. Among them four curves are the four different inlet wind speed calculated values of the fluid velocity distribution in each ventilation duct of the stator; the other two curves are when the air gap inlet is 56 m/s, the flow rate test measured value and linear fit value at different radial ventilation ducts of the stator. From the calculated-value curves it can be seen that, with the increase of the flow velocities at the air gap entrance, the flow velocity in the other radial ventilation ducts except the fourth, fifth, and sixth ducts increases correspondingly; while the flow velocity in the fourth, fifth, and sixth ventilation ducts has the opposite distribution law, where the flow velocity does not increase but decreases.

After analyzing the calculation results of the fourth ventilation duct flow velocity, when the inlet speed is 56, 70, 85, and 100 m/s, the fluid average velocity of the radial ventilation duct outlets is 1.1, 0.24, −0.6, and −1.3 m/s. From the analysis of the calculation results, it can be seen that when the air gap inlet wind speed increases to a value between 70 and 85 m/s, the wind speed in the fourth ventilation duct will be close to zero. That is to say, there is no cooling wind blowing through the core and winding, the internal loss can only be taken away by the fluid in the adjacent ventilation duct; and when the speed at the air gap entrance is greater than the critical value above, the fluid velocity in the radial ventilation duct will form a reverse-direction reflux. At this time, the hot air from the other ventilation ducts will flow back into this ventilation duct. No matter which of any the above situations occurs, it is very unfavorable to the heat transfer of the generator stator structure and the safe operation of the generator.

3.4. The Influence of Backflow in the Radial Ventilation Duct on the Temperature of the Stator Winding and Core

Figure 11 shows the calculation results of the generator stator temperature field when the air gap inlet velocity is 56 m/s. It can be seen from the figure that the temperature of the stator core near the fourth radial ventilation duct is significantly higher than that of the other stator cores, but the temperature of the upper and lower windings of the stator is less affected by the low wind speed in this region.

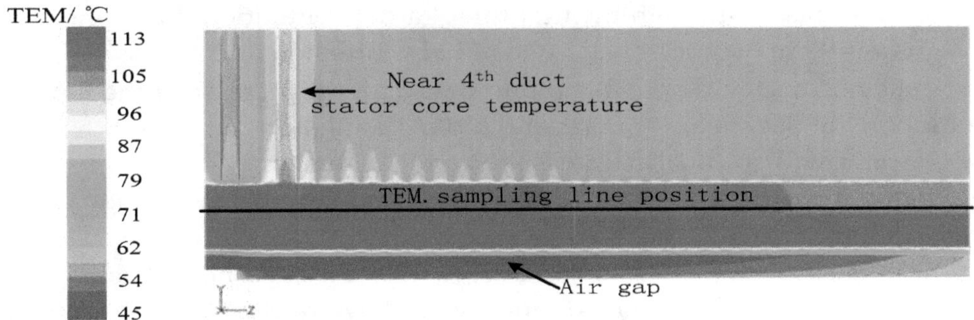

Figure 11. Temperature field of stator at air gap inlet velocity 56 m/s.

In order to analyze the axial distribution of the temperature of the stator core and winding in the case of backflow in the radial ventilation duct, Figure 12 shows the axial temperature distribution curve of the sampling line with the same radial height. The sampling line position is shown in Figure 11. When backflow occurs in the stator radial ventilation duct, the calculated temperatures of the stator core, winding, and cooling air along the axial direction are collected from the sampling line. At the same time, the corresponding air gap inlet velocity was 85 m/s.

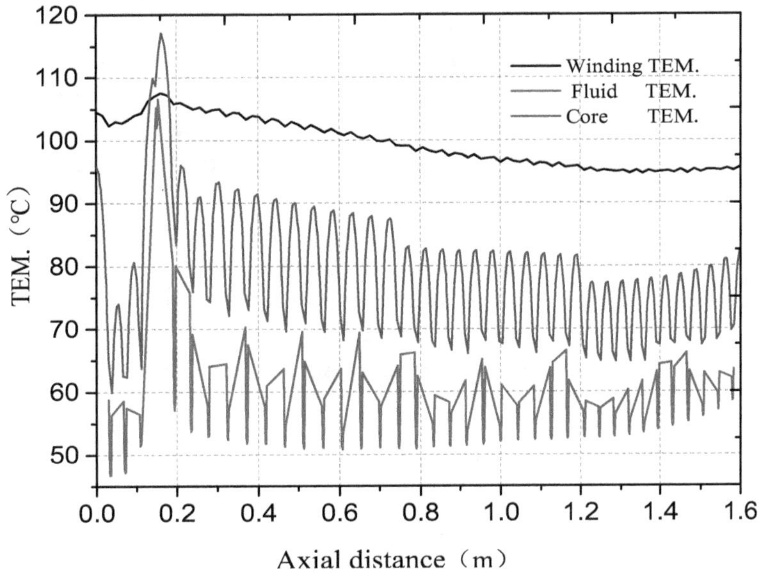

Figure 12. Winding and core temperature distribution at air gap inlet velocity 85 m/s.

It can be seen from Figure 12 that when backflow occurs in the local radial ventilation duct of the stator core, the negative pressure brings the hot wind at the back of the stator yoke into one or several radial ventilation ducts of the stator yoke to cool the stator core, resulting in a significant rise in the local stator core temperature, which reaches 115 °C. The core temperature is even higher than the average temperature of the stator windings. However, the average temperature of the stator winding is less affected by backflow. Therefore, the backflow of the fluid in the ventilation duct will have a direct impact on the temperature rise of the stator core, which is likely to produce local overheating of the core and affect the safe operation of the unit.

4. Conclusions

In this paper, a 100 MW air-cooled turbo-generator is taken as the research object. Through experimental measurement and finite element analysis of the model, the following conclusions are obtained:

1. Through the measurement of the fluid flow law in the stator radial ventilation duct, it is found

that after the fluid passes through the first three radial ventilation ducts, the wind speed will decrease significantly in the back one or two radial ventilation ducts.

2. Through calculation and research, it is found that with the increase of the flow velocity at the air gap entrance, the flow velocity in several radial ventilation ducts after the step structure of the stator core end will gradually decrease to zero. At this time, if flow velocity at the air gap inlet continues to increase, the velocity direction of fluid in radial ventilation ducts will reverse, forming a backflow.
3. The backflow of the fluid in the ventilation duct will have a direct impact on the temperature rise of the local stator core, which is likely to produce local overheating of the core and affect the safe operation of the unit.

Author Contributions: Y.L. analyzed the data, contributed analysis tools and wrote the paper; W.L. conceived and designed the experiments; Y.S. performed the experiments.

References

1. Pickering, S.J.; Lampard, D.; Shanel, M. Modelling Ventilation and Cooling of The Rotors of Salient Pole Machines. In Proceedings of the IEEE International Electric Machines and Drives Conference (IEMDC), Cambridge, MA, USA, 17–20 June 2001; pp. 806–808. [CrossRef]
2. Shanel, M.; Pickering, S.J.; Lampard, D. Conjugate Heat Transfer Analysis of a Salient Pole Rotor in an Air-Cooled Synchronous Generator. In Proceedings of the International Electric Machines and Drives Conference, Madison, WI, USA, 1–4 June 2003; Volume 2, pp. 737–741. [CrossRef]
3. Shoulu, G.; Yichao, Y.; Yuzheng, L. Review on ventilation system of large turbine generator. *Energy Res. Inf.* **2004**, *4*, 195–200.
4. Shuang, L.; Zhihua, A.; Guangyu, Q. Research and performance verification of 150 MW air-cooled turbo-generator ventilation system. *Large Mot. Technol.* **2007**, *1*, 8–10.
5. Dongping, Z.; Rongshan, C. Calculation of ventilation and temperature rise of 100–200 MW air-cooled turbo-generator. *Power Gener. Equip.* **2006**, *3*, 193–195.
6. Guangde, L.; Weihong, Z. Ventilation system design of air-cooled turbo-generator. *Large Mot. Technol.* **1998**, *4*, 13–16.
7. Yamamoto, M.; Kimura, M. Ventilation and cooling technology of a new series of double-pole air-cooled steam turbine generator (Translated by Cai Qianhua). *Foreign Large Electr. Mot.* **2000**, *3*, 16–18.
8. Chauveau, E.; Zaim, E.H.; Trichet, D.; Fouladgar, J. A statistical of temperature calculation in electrical machines. *IEEE Trans. Mag.* **2000**, *36*, 1826–1829. [CrossRef]
9. Fujita, M.; Kabata, Y.; Tokumasu, T.; Kakiuchi, M.; Shiomi, H.; Nagano, S. Air-Cooled Large Turbine Generator with Multiple-Pitched Ventilation Ducts. In Proceedings of the IEEE International Conference on Electric Machines and Drives, San Antonio, TX, USA, 15 May 2005; pp. 910–917. [CrossRef]
10. Zixiong, Z.; Zengnan, D. *Viscous Fluid Mechanics*; Tsinghua University Press: Beijing, China, 1998; pp. 255–279.
11. Borishenko, A.N. *Aerodynamics and Heat Transfer in Motor*; Machinery Industry Press: Beijing, China, 1985.
12. Changming, Y. *Heat Conduction and Its Numerical Analysis*; Tsinghua University Press: Beijing, China, 1981; pp. 1–7.
13. Huo, F.; Li, Y.; Li, W. Calculation and analysis of stator ventilation structure optimization scheme of large air-cooled turbo-generator. *Chin. J. Electr. Eng.* **2010**, *6*, 95–103.
14. Li, H.J.W. Calculation and analysis of fluid velocity and fluid temperature in large air-cooled turbo-generator stator. *Proc. CSEE* **2006**, *26*, 168–173. [CrossRef]
15. Li, W.; Ding, S.; Zhou, F. Diagnostic numerical simulation of large hydro-generator with insulation aging. *Heat Transf. Eng.* **2008**, *29*, 902–909. [CrossRef]
16. Li, W.; Hou, Y. Heating analysis of stator strands of large hydro-generator based on numerical method. *Proc. CSEE* **2001**, *21*, 115–1186. [CrossRef]

13

Thermal Analysis Strategy for Axial Permanent Magnet Coupling Combining FEM with Lumped-Parameter Thermal Network

Xikang Cheng, Wei Liu *, Ziliang Tan, Zhilong Zhou, Binchao Yu, Wenqi Wang, Yang Zhang and Sitong Liu

Key Laboratory for Precision and Non-Traditional Machining Technology of the Ministry of Education, Dalian University of Technology, Dalian 116024, China; xikangc@mail.dlut.edu.cn (X.C.); tanziliang2020@mail.dlut.edu.cn (Z.T.); zzl666@mail.dlut.edu.cn (Z.Z.); yubinchao@mail.dlut.edu.cn (B.Y.); wqwang19@mail.dlut.edu.cn (W.W.); zy2018@dlut.edu.cn (Y.Z.); liusitong@mail.dlut.edu.cn (S.L.)

* Correspondence: lw2007@dlut.edu.cn

Abstract: Thermal analysis is exceptionally important for operation safety of axial permanent magnet couplings (APMCs). Combining a finite element method (FEM) with a lumped-parameter thermal network (LPTN) is an effective yet simple thermal analysis strategy for an APMC that is developed in this paper. Also, some assumptions and key considerations are firstly given before analysis. The loss, as well as the magnetic field distribution of the conductor sheet (CS) can be accurately calculated through FEM. Then, the loss treated as source node loss is introduced into the LPTN model to obtain the temperature results of APMCs, where adjusting conductivity of the CS is a necessary and significant link to complete an iterative calculation process. Compared with experiment results, this thermal analysis strategy has good consistency. In addition, a limiting and safe slip speed can be determined based on the demagnetization temperature permanent magnet (PM).

Keywords: thermal analysis; axial permanent magnet coupling (APMC); eddy current; finite element method (FEM); lumped-parameter thermal network (LPTN)

1. Introduction

Permanent magnet couplings (PMCs) as a new transmission topology can be employed in several industrial applications, such as conveyors, fans, pumps and braking devices [1–3]. Also, PMCs offer many advantages, such as no physical contact, soft starting, shock isolation, and misalignment tolerance, which all taken together provide better protection for mechanical systems [4,5]. Typically, there are two configurations of PMCs: axial [6–8] and radial [9–11] type, both with the above-mentioned advantages. In this work, we pay attention to axial permanent magnet coupling (APMC) as depicted in Figure 1. It is divided into two parts: one is the permanent magnet (PM) module including a PM holder and several PMs (generally Nd-Fe-B type), and the other is the conductor sheet (CS) module, generally manufactured with copper. Additionally, there are two other iron yokes to make the magnetic fluxes close corresponding the PM holder and the CS, respectively.

In [3,6,12,13], the operating principle of APMCs was introduced in detail. On account of having the slip ($s = n_{in} - n_{out}$) between the CS module and the PM module, the eddy currents, which are generated on the CS, interact with the original magnetic field of PMs and induce an effective torque. Meanwhile, the temperature rise, especially at the low-slip condition, is inevitable due to the generated-currents. Surpassing the thermal limit of PMs involves the risk of irreversible demagnetization, while the excessive temperature can also damage other components. Beside the precaution of each component destruction, intensive thermal stress can shorten equipment lifetime.

Therefore, thermal analysis is absolutely necessary both for the design stage and the monitoring stage of APMCs. Given the high-speed rotation characteristics of APMCs, analyzing the components' temperature imposes a greater challenge.

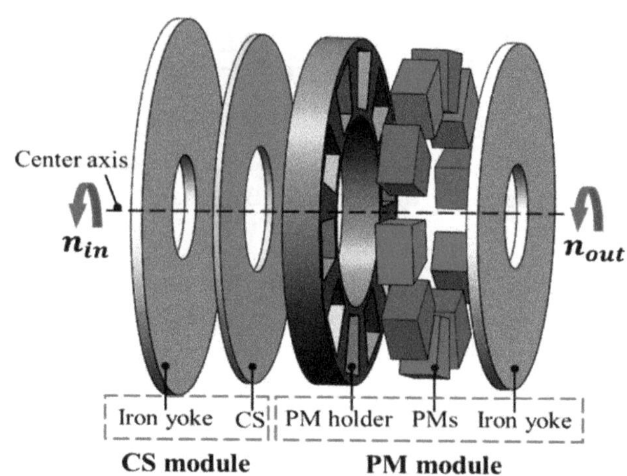

Figure 1. Configuration of an axial permanent magnet coupling.

Today, APMCs have been investigated for a long time, and a number of papers can be discovered in [14–22]. Research in [16] and [17], based on a two-dimensional (2-D) approximation of the magnetic field distribution, a practical and effective analytical calculation approach for the torque performance of APMCs, was proposed. Given the three-dimensional (3-D) edge effects and curvature effects, a 3-D analytical model was developed in [19] and [22] to compute the torque and the axial force. Obviously, all of studies mentioned above are focused on the torque analysis ignoring the thermal influence since APMCs belong to a kind of transmission device ultimately.

Unfortunately, thermal analysis to APMCs is not easy work due to the property variation of each component with temperature that result in mathematical difficulties. Hence, it is then acceptable to find very little literature about this. Here, good news is that there are some methods from PM motors to be borrowed [23–26]. In general, thermal analysis can be segmented into two primary means: numerical methods and thermal network. The former, such as finite-element method (FEM), not only can obtain the temperature distribution accurately, but also can take the actual 3D geometry and material properties into account. However, for the analysis objects, which have complex topologies or contain components equipped with highly diverse thermal characteristics, FEM is difficult to simplify [27]. Consequently, the mesh is refined and enormous, resulting in having a heavy expense in computer resources and being time-consuming, which then hampers the application of FEM in rapid optimization [28]. The latter is a lumped–parameter approach offering the advantages of economy, flexibility and simplicity. Nevertheless, it cannot acquire the temperature distribution in detail and heavily depends on the precise thermal parameters of the loss and heat coefficient (conduction, radiation and convection) [29,30]. For the aforementioned studies, however, the study methods are not fully applicable to the thermal analysis of APMCs.

From the perspective of engineering application and operation safety for APMCs, it is important to have a simplified and accurate thermal analysis strategy in order to quickly obtain the temperature distribution of APMCs. In this paper, FEM and lumped-parameter thermal network (LPTN) are combined to effectively calculate the temperature results of APMCs. The remainder of the paper is organized as follows: Section 2 presents the geometry of the studied APMC. In Section 3, the proposed strategy and assumptions are offered. The magnetic field model, including the loss results is established in Section 4. Then, Section 5 gives the LPTN model to obtain the temperature results. Finally, Section 6 concludes the work of this paper.

2. Geometry of the Studied APMC

Figure 2 provides the geometry of the studied APMC as well as its exploded view and geometrical parameters, where the pole-pairs number of the PMs is supposed to be 6. Here, l_g is adjustable and generally ranges from 3 mm to 8 mm. In the PM holder, the fan-shaped PMs are arranged in accordance with a N/S alternating sequence. The major parameters of the studied APMC are given in Table 1. These parameters are derived from some reliable engineering design experience and its detailed design thought can be referred to [3,4,6]. Given the adjustable air-gap and these parameters, the manufactured prototype of the studied APMC was built and put on the cast-iron platform, as shown in Figure 3.

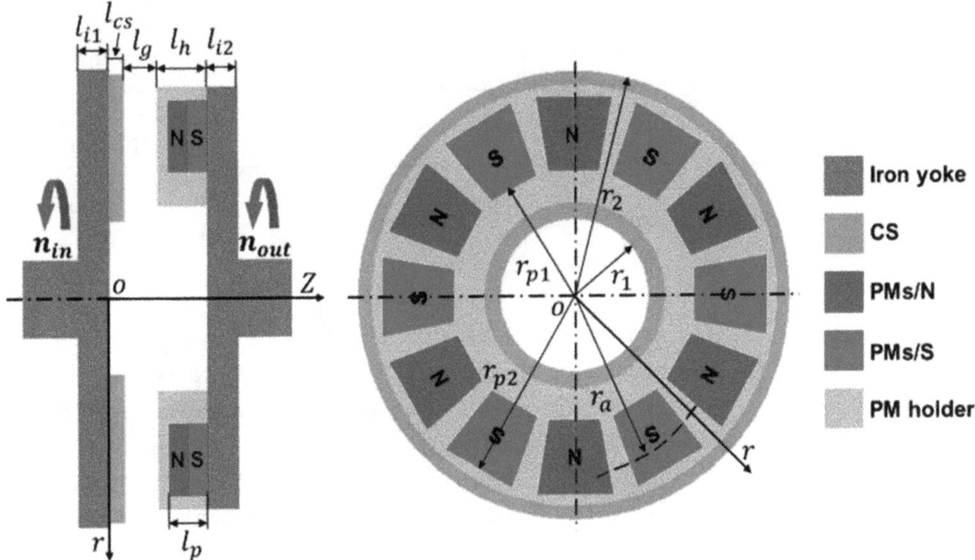

Figure 2. The geometry of the studied APMC ($p = 6$) with its exploded view and geometrical parameters.

Figure 3. The manufactured prototype of the studied APMC ($l_g = 30$ mm).

Table 1. Parameters of the studied APMC.

Symbol	Meaning	Value
l_{i1}	Thickness of the iron yoke (CS side)	10 mm
l_{i2}	Thickness of the iron yoke (PM side)	10 mm
l_{cs}	Thickness of the CS	6 mm
l_g	Thickness of the air-gap	3–8 mm
l_h	Thickness of the PM holder	26 mm
l_p	Thickness of the PM	25 mm
r_1	Inside radius of the CS	87.5 mm
r_2	Outside radius of the CS	187.5 mm
r_{p1}	Inside radius of the PM	115 mm
r_{p2}	Outside radius of the PM	165 mm
r_a	Average radius of the PM	140 mm
H_p	Coercive force of the PM	−900 KA/m
σ_{cs}	Conductivity of the CS	5.8×10^7 S/m (20 °C)

3. Proposed Strategy and Assumptions

3.1. Proposed Strategy

Figure 4 describes the procedure of the proposed thermal analysis strategy. Since the conductivity of the CS is closely related to temperature, this procedure is an iterative updating calculation. After initialization and referring to the parameters in Table 1, the magnetic field result is obtained as well as the loss employing FEM. Then, regarding the loss as the heat source, the LPTN model is employed to calculate the temperature distribution. Thanks to the temperature obtained each time being different, so in each iteration, the conductivity of the CS was updated according to its temperature characteristic. Based on the updated conductivity, the thermal analysis re-executed until the temperature convergence threshold is satisfied. Herein, ΔT is the percentage error of each component temperature relative to the previous calculation, and T_s is the set convergence threshold, such as 1%.

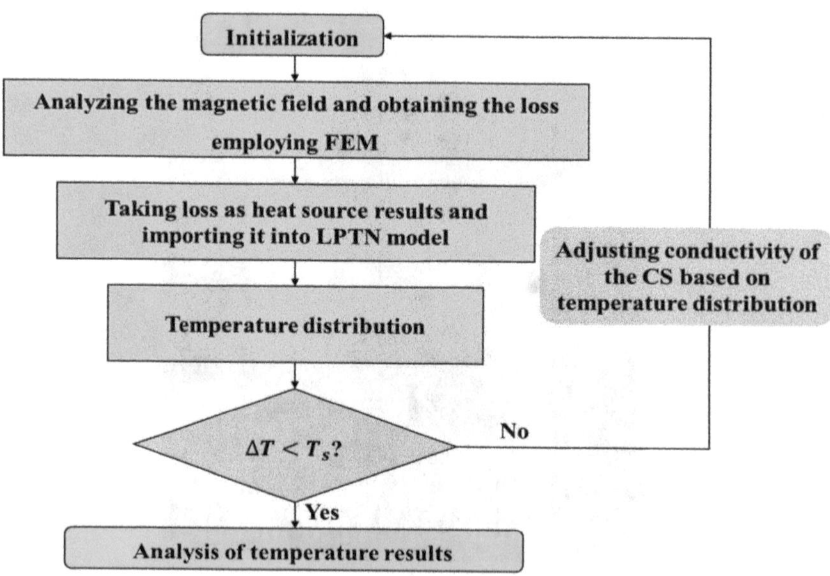

Figure 4. The procedure of the thermal analysis strategy.

3.2. Assumptions

1. From Figure 2, the geometry of the APMC is approximatively centrosymmetric. Also, the parameters of every component along the circumferential direction (r) were uniform. Therefore, the loss and temperature distributions were centrosymmetric as well.
2. The APMC operated in the steady-state with a certain slip speed (s), so it is reasonable to think the air in the air-gap as stable, whereby the temperature distribution in the air-gap was also the same.
3. Considering the skin effect, the loss is mainly concentrated upon the CS, and other losses, were ignored.

4. Magnetic Field Model

4.1. Key Considerations

In this section, we employ the FEM software Ansoft Maxwell 3D (v16, ANSYS Inc, Canonsburg, Pennsylvania, USA.) to model the magnetic field of the APMC. Assigning H_p and σ_{cs} in Table 1 to the PMs and the CS respectively is the first important consideration, where σ_{cs} is a parameter to be adjusted from Figure 4 and valued at room temperature (20 °C) initially. As important as the PM and the CS, the magnetic characteristic of the iron yokes (B-H curve) should also be considered in the FEM, as exposed in Figure 5. Additionally, Figure 6 shows the mesh size for all the components, following a certain proportion. Here, from the perspective of simulation precision, the CS and the PMs are meshed detailedly. Figure 7 shows the mesh in 3D of the FEM model, where this model includes 570,979 nodes and 119,316 elements.

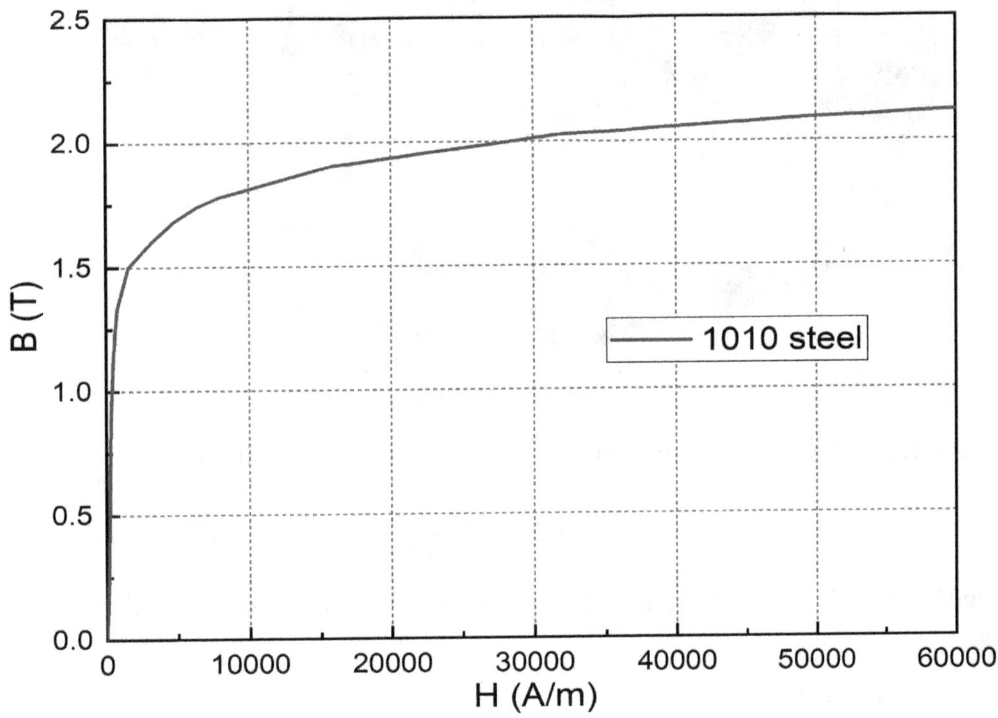

Figure 5. Nonlinear B–H curve (1010 steel) for the iron yokes.

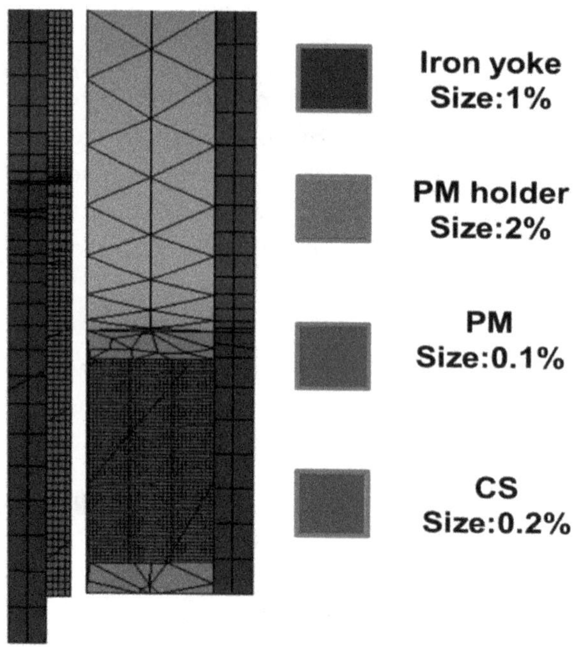

Figure 6. The mesh size for all the components used in the FEM.

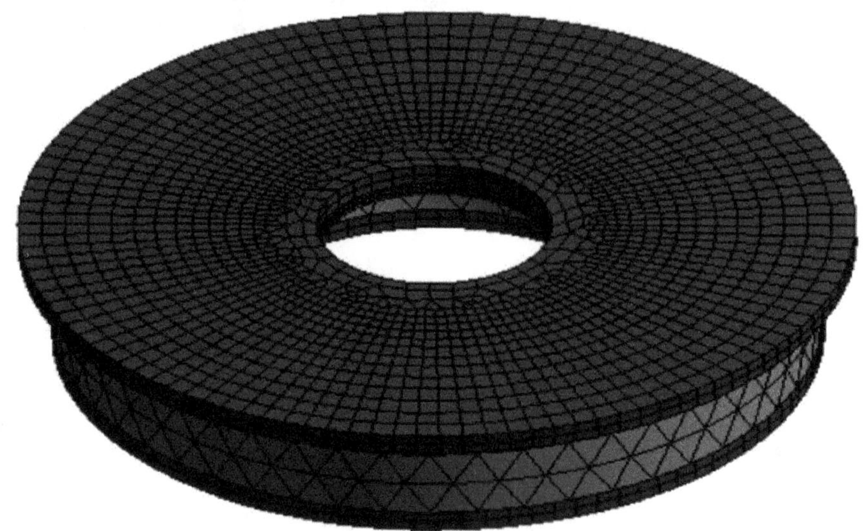

Figure 7. Mesh in 3D of the FEM model.

Besides, considering the convergence of this FEM results, the simulation step can be given by

$$t_{step} = (1.2 \sim 1.5)\frac{60}{1000s} \tag{1}$$

wherein, the coefficient (1.2~1.5) indicates that the CS module has rotated 1.2 to 1.5 turns relative to the PM module.

4.2. Analysis of Magnetic Field on the CS

The magnetic flux density and the eddy current are significant for the loss analysis of the APMC. As presented in Figure 8a, the magnetic flux density distributions on the surface of the CS seen from z-axis correspond to the alternating distributions of the PMs (N/S), in which the slip speed is set to 50 r/min. Obviously, the peak value of the magnetic density on the surface of the CS is 0.62T, while the location of the minimum value is between N-pole and S-pole of the PMs. Also, we can obtain the eddy current distributions on the surface of the CS again this FEM model, as shown in Figure 8b. Different from the magnetic flux density, the center of the eddy currents is situated between N-pole and S-pole of the PMs. Corresponding to the number of the PMs, the number of the eddy currents is *2p*. In addition, the adjacent eddy currents are in opposite direction owing to the alternating magnetic poles.

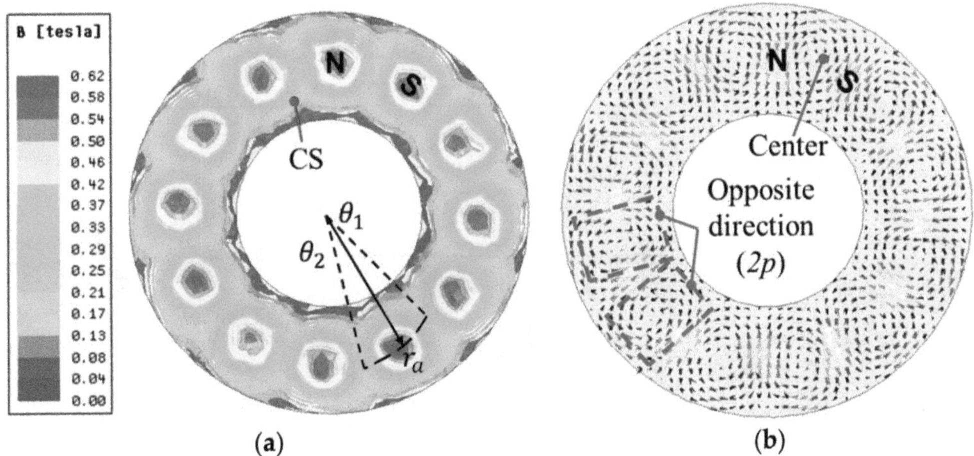

Figure 8. Magnetic field analysis (z-axis, p = 6 and s = 50 r/min): (**a**) the magnetic flux density distributions, (**b**) the eddy current distributions.

Clearly, corresponding to the distributions in Figures 8 and 9 depicts the relationship for the magnetic flux density (B_{cs}) and the eddy current density (J_{cs}) of the CS varying with different positions (θ) wherein $\theta_1 = 0°$ and $\theta_2 = 30°$ (360°/2p). Thanks to Faraday's law of electromagnetic induction, the region with the strongest magnetic flux density is the weakest eddy current density.

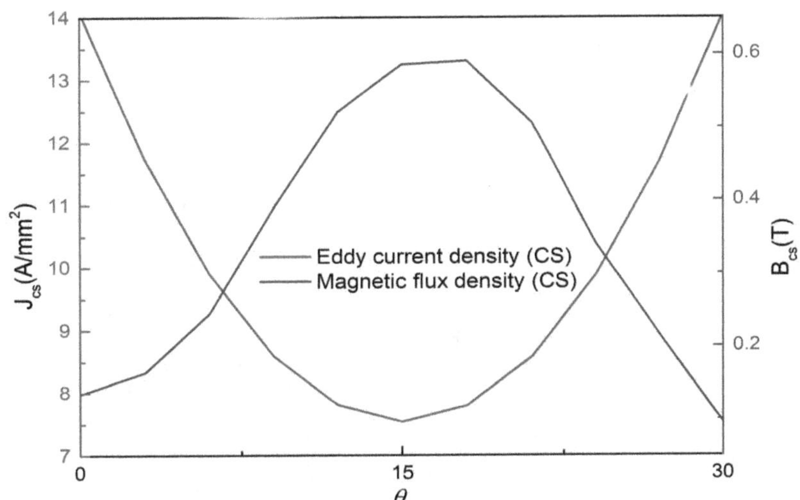

Figure 9. The curves for depicting the distributions on the CS.

4.3. Loss Calculation

Accurately calculating the loss is essential to study the temperature rise for ensuring reliable operation. On account of skin effect, the eddy currents mainly concentrated on the surface of the CS. Therefore, we only consider the eddy current loss generated on the CS and ignore other losses, like mechanical loss. Based on the magnetic field analysis and Equation (2) belonging to an embedded algorithm in the FEM, the loss results under different air-gap lengths and slips are displayed in Figure 10. Obviously, with the increase of the slip speed (s), the loss increases continuously, while the loss decreases gradually with the increase of the air-gap length.

$$P_{cs} = 1/\sigma_{cs} \int_V |\mathbf{J}| dV \quad (2)$$

where V is the volume of an integral region, and $\mathbf{J}$ is the vector of eddy current density.

$$\begin{cases} P_{loss} = k_c P_{cs} \\ k_c = 1 - \frac{\tanh(\pi w_{pm}/2\tau_p)}{(\pi w_{pm}/2\tau_p)[1+\tanh(\pi w_{pm}/2\tau_p)\tanh(\pi w_{cs}/\tau_p)]} \end{cases} \quad (3)$$

where $w_{pm} = r_{p2} - r_{p1}$ and $w_{cs} = (r_2 - r_1 - w_{pm})/2$.

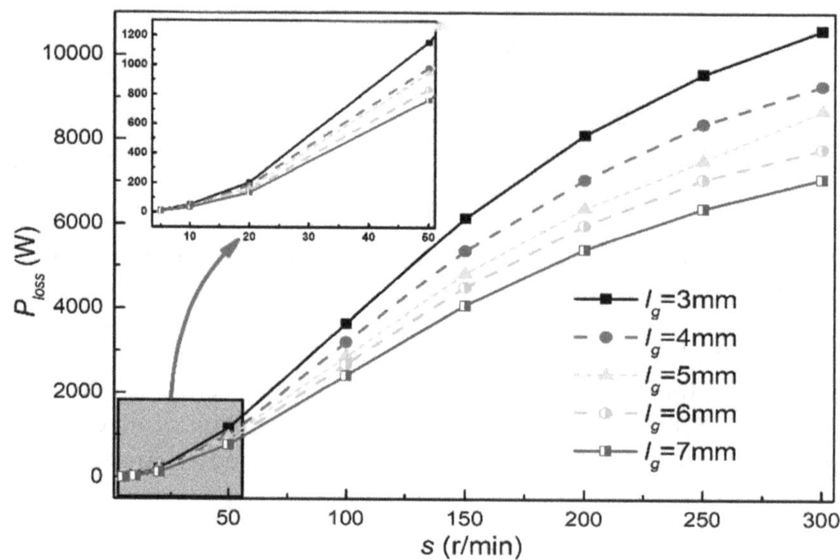

Figure 10. The loss results under different air-gap lengths and slips using FEM.

However, the loss results in Figure 9 do not take into account 3-D edge effect. In view of this, Figure 11 shows the real eddy current paths on the CS where only the central region plays a role. In order to solve this problem, the correction factor (k_c), also called Russell–Norsworthy correction factor [6,11], is adopted as Equation (3).

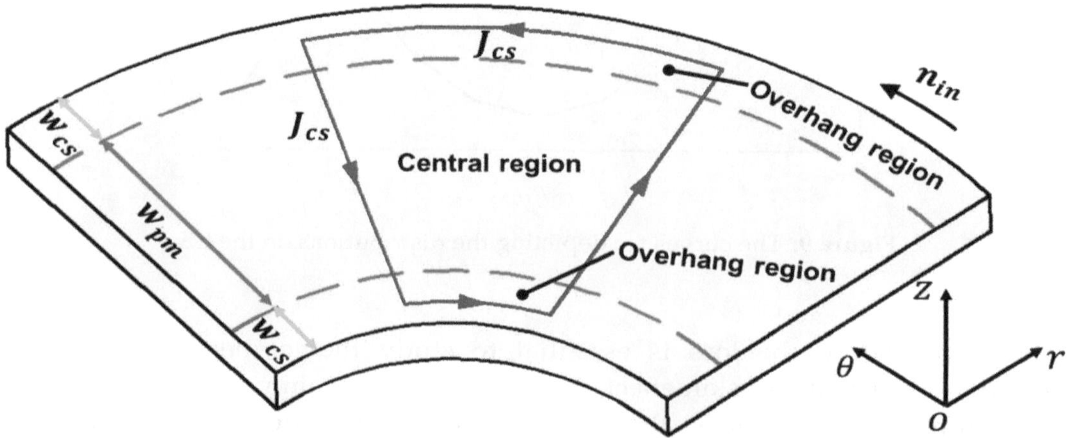

Figure 11. The real eddy current paths on the CS.

5. LPTN Model

5.1. LPTN Model

The proposed LPTN model, including thermal nodes distribution and equivalent LPTN are exposed in Figure 12. Because of the rotational axisymmetry, only three PMs are presented in Figure 12a, where the view is obtained from the average radius of the PM ($r = r_a$). In this model, the thermal modes are located at different components involving ambient, iron yoke, CS, air-gap, PM holder, PM an iron yoke. Totally, there are 22 nodes (0~22) in this thermal network. In addition, the thermal nodes in Figure 12b are divided into red and black types, in which the power loss computed in the previous section are injected into the red nodes. It is worth noting that only conduction heat transfer and convection heat transfer are considered, not existing radiation heat transfer nearly.

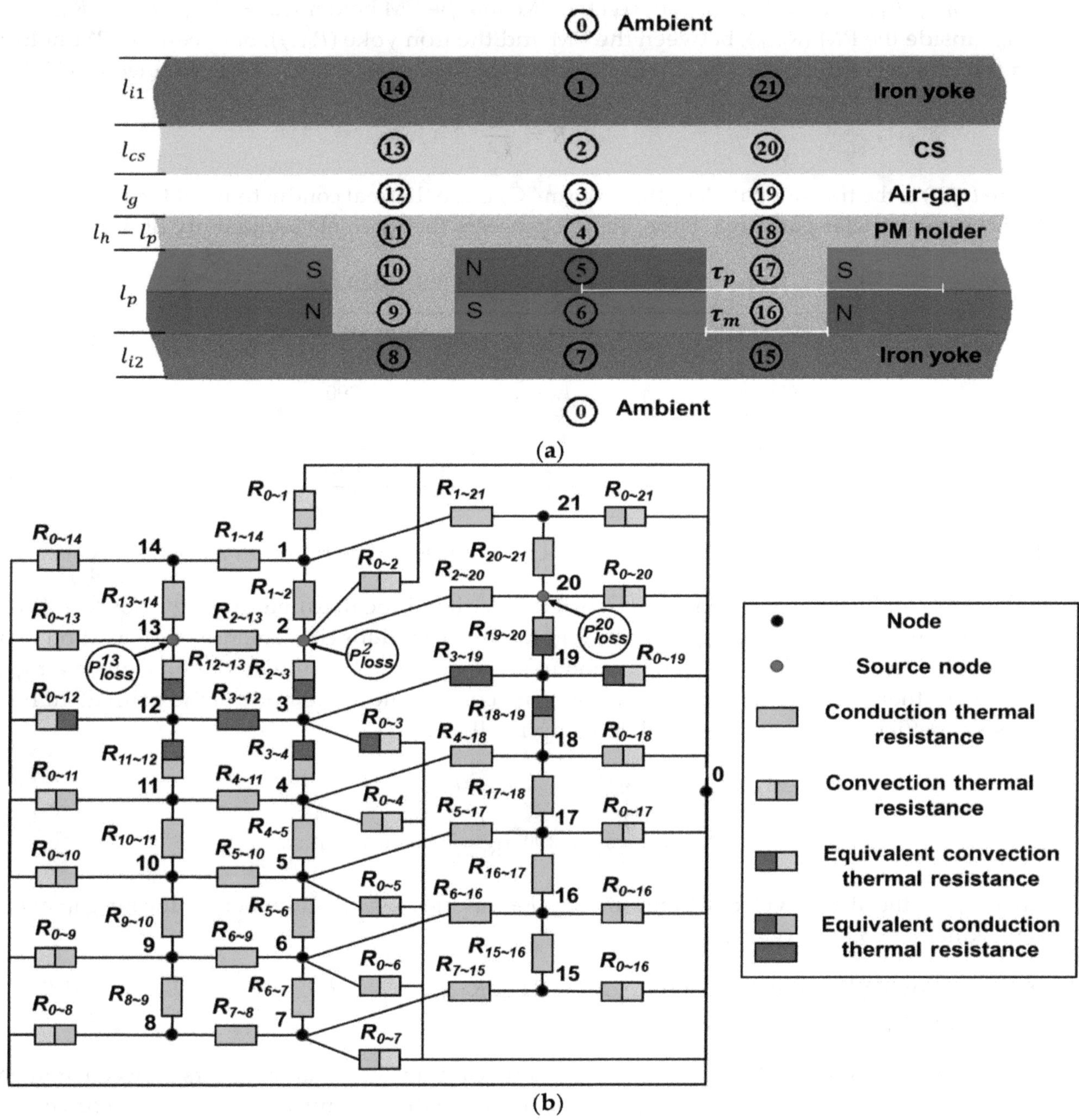

Figure 12. Proposed LPTN model: (**a**) thermal nodes distribution, (**b**) equivalent LPTN.

In addition, providing that the APMC is divided into $2p$ parts, the nodes 2, 13, and 20 in Figure 12b are given different proportional losses respectively, as follows:

$$\begin{cases} P_{loss}^{13} = P_{loss}^{20} = \alpha P_{loss}/2p \\ P_{loss}^{2} = (1-\alpha)P_{loss}/2p \end{cases} \tag{4}$$

where $\alpha = \tau_m/\tau_p$, representing the non-PM proportion after being divided into $2p$ parts.

5.2. Calculation for Thermal Resistances

1. Conduction heat transfer: The region not in contact with air belongs to conduction heat transfer, as follows: inside the iron yoke ($R_{1\sim14}$, $R_{1\sim21}$, $R_{7\sim8}$ and $R_{7\sim15}$); inside the CS ($R_{2\sim13}$ and $R_{2\sim20}$); between the iron yoke and the CS ($R_{1\sim2}$, $R_{13\sim14}$ and $R_{20\sim21}$); inside the PM holder ($R_{4\sim11}$, $R_{4\sim18}$, $R_{9\sim10}$, $R_{10\sim11}$, $R_{16\sim17}$ and $R_{17\sim18}$); between the PM and the PM holder ($R_{4\sim5}$, $R_{5\sim10}$, $R_{5\sim17}$, $R_{6\sim9}$ and $R_{6\sim16}$); inside the PM ($R_{5\sim6}$); between the PM and the iron yoke ($R_{6\sim7}$); between the PM holder and the iron yoke ($R_{8\sim9}$ and $R_{15\sim16}$). Conduction thermal resistances can be obtained as [23,29]

$$R = \frac{L}{kA} \tag{5}$$

where L (m) is the transfer path length, k (W/(m·°C) is the thermal conductivity of the material and A (m^2) is the transfer path area. Here, Table 2 presents the thermal conductivity of the materials.

Table 2. Thermal Conductivity of the Materials.

Material	Symbol	Value (W/(m·°C))
Steel	k_{st}	36
Copper	k_{cop}	390
Aluminum	k_{al}	237
Nd-Fe-B	k_{nd}	9
Air	k_a	0.026

Using Equation (5), all the conduction thermal resistances can be found in Appendix A.

2. Equivalent conduction/convection heat transfer: Since it is difficult to determine the flow condition in the air-gap, the air in the air gap can be regarded as a solid for convenience of calculation, that is, the convection heat transfer phenomenon in the air-gap can be simulated with a given air gap equivalent thermal conductivity. For the air-gap, corresponding to node 3, the equivalent heat transfer coefficient (h_{air}) can be calculated empirically as

$$\begin{cases} v_{air} = \pi(n_{in} - n_{out})(r_2 + r_{p2})/60 \\ \mathrm{Re}_{air} = v_{air} l_g/\gamma \\ h_{air} = 0.0019(l_g/r_2)^{-2.9084} \mathrm{Re}_{air}{}^{0.4614 \ln[3.33361(l_g/r_2)]} \end{cases} \tag{6}$$

where v_{air} is the airflow velocity in the air gap, Re_{air} is the Reynolds number for the air-gap and γ is the kinematic viscosity of air.

Then, employing Equations (5) and (6), $R_{2\sim3}$, $R_{3\sim4}$, $R_{3\sim12}$, $R_{3\sim19}$, $R_{11\sim12}$, $R_{12\sim13}$, $R_{18\sim19}$ and $R_{19\sim20}$ can be found in Appendix B.

3. Convection heat transfer: In Figure 12, convection heat transfer certainly occurs between ambient node 0 and other nodes. Generally, as fluid flows, natural convection or forced convection would occur. The former is caused by the nonuniformity of the temperature or concentration; the latter relies on an external force, such as a pump or fan. For the studied APMC, although there is

no pump or fan, the heat transfer belongs to forced convection because of the rotational motion of the CS and the PM holder. Convection thermal resistances can be obtained as [29]

$$R = \frac{1}{hA} \tag{7}$$

where h (W/(m·°C)) is the convection heat transfer coefficient, and A (m2) is the transfer path area.

In the ambient, corresponding to node 0, the convection heat transfer coefficient (h_{am}) can be calculated by

$$\begin{cases} h_{am} = 13.3(1 + {v_{am}}^{0.5}) \\ v_{am} = \pi(n_{in} + n_{out})(r_2 + r_{p2})/120 \end{cases} \tag{8}$$

wherein v_{am} is the airflow velocity in the ambient.

Based on Equations (6) to (8), the convection thermal resistances can be found in Appendix C.

5.3. Calculation for Temperature Rise

For the LPTN model, the temperature rise of each node can be solved by

$$[G][T] = [P] \tag{9}$$

where $[P]$ is the column matrix of nodal power loss, $[G]$ is the thermal conductance matrix and $[T]$ is the temperature rise vector. The thermal conductance matrix can be defined by

$$[G] = \begin{bmatrix} \sum\limits_{i=1}^{n} \frac{1}{R_{1\sim i}} & -\frac{1}{R_{1\sim 2}} & \cdots & -\frac{1}{R_{1\sim n}} \\ -\frac{1}{R_{2\sim 1}} & \sum\limits_{i=1}^{n} \frac{1}{R_{2\sim i}} & \cdots & -\frac{1}{R_{2\sim n}} \\ \vdots & \vdots & \ddots & \vdots \\ -\frac{1}{R_{n\sim 1}} & -\frac{1}{R_{n\sim 2}} & \cdots & \sum\limits_{i=1}^{n} \frac{1}{R_{n\sim i}} \end{bmatrix} \tag{10}$$

where n is the number of all the nodes for the LPTN model, and $R_{i\sim j}$ represents the thermal resistance between adjacent nodes i and node j. Besides, $1/R_{i\sim j}$ is normally ignored because there is no heat exchange between itself and $R_{i\sim j} = R_{j\sim i}$.

Accounting for the effect of ambient temperature, Equation (9) is not perfect to solve the temperature rise of the APMC. Thus, a modified equation for Equation (9) is offered as

$$[T] = \{[P] - T_0[G_0]\}[G]^{-1} \tag{11}$$

where T_0 is the local ambient temperature in accordance with the room temperature of the experiment, and $[G_0] = [-1/R_{1\sim 0}, -1/R_{2\sim 0}, \ldots, -1/R_{n\sim 0}]$.

Besides, the node temperature in the same region is averaged as the calculation result. For example, the average value of nodes 2, 13 and 20 is taken as the temperature result of the CS.

5.4. Adjusting the Conductivity of the CS

Additionally, the column matrix of nodal power loss is temperature-dependent and require temperature update in the thermal analysis strategy. As a result, an iterative coupled electromagnetic and thermal modeling, corresponding to Figure 4, must be employed in this section. Here, the conductivity of the CS is the most crucial factor for updating the nodal power loss and temperature. Figure 13 gives the relationship between the conductivity of the CS (σ_{cs}) and different temperature (T_{cs}), which provides an effective reference for adjusting the conductivity of the CS. This iteration process is generally over until a set convergence threshold (T_s) is met, such as 1%.

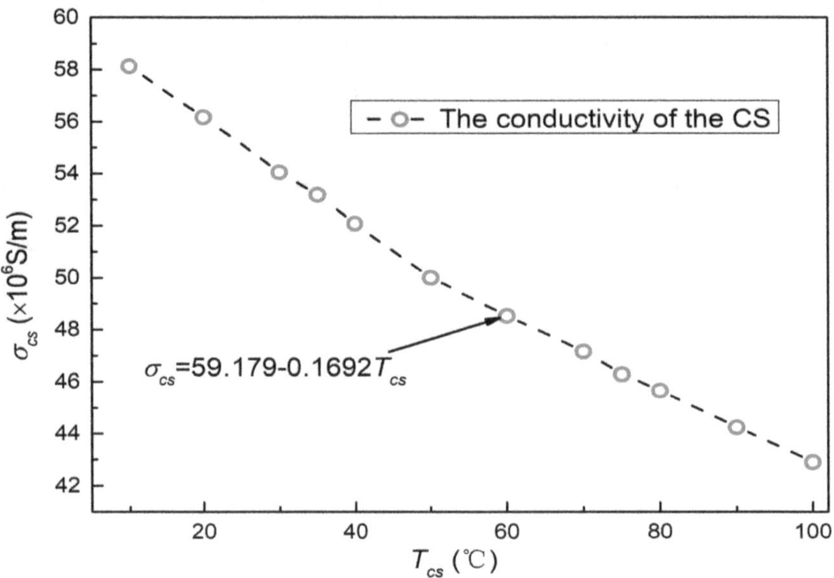

Figure 13. The relationship between the conductivity of the CS and different temperature.

6. Experiment Verification

In order to verify the thermal analysis strategy proposed in this paper, the experiment platform for the studied APMC is built in Figure 14. Since the APMC is in the state of high-speed rotation, the traditional static temperature measurement method cannot be carried out. Thus, a high-precision thermal camera, labeled as Telops FAST V100k, (Telops Inc, Montreal, Quebec, Canada.) is used to measure the temperature. The thermal camera was set up half a meter away from the APMC. After the APMC was in stable operation, the PM module and CS module were photographed. Since the interior of the components could not be photographed, surface temperature data was used as experimental results. However, the experimental results are generally obtained when the APMC operates for one hour to reach the thermal balance. Also, the cross-section of the LPTN model inside the components is about 10 mm away from the surface. Therefore, it is considered that the error between the internal temperature of this cross-section and surface temperature is not large.

Figure 14. The experiment platform for the studied APMC.

Besides, if the uniformity of measured temperature is well, and the fluctuation is less than 1%, the measured results can be used, as shown in Figure 15. Also, Figure 16 presents a measurement result by thermal camera in the case of 'l_g = 5 mm and s = 20 r/min'. The measured results can be obtained by taking the points. Specifically, in the post-processing of the thermal camera software, some points in the photographed area are selected as the temperature results of each component.

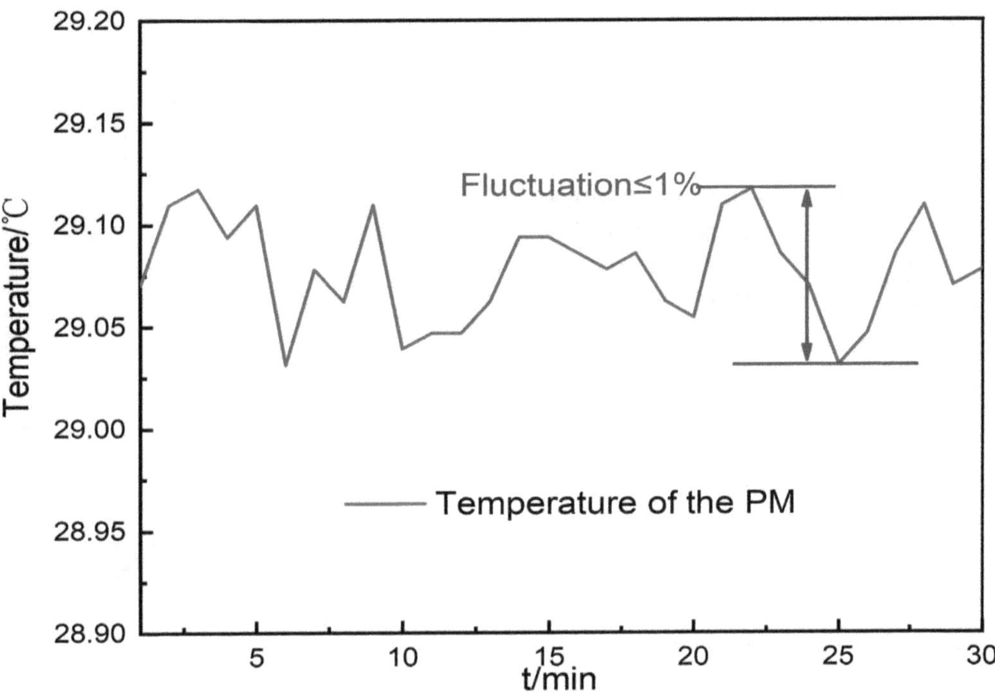

Figure 15. Temperature variation of the PM within 30 min (s = 25 r/min).

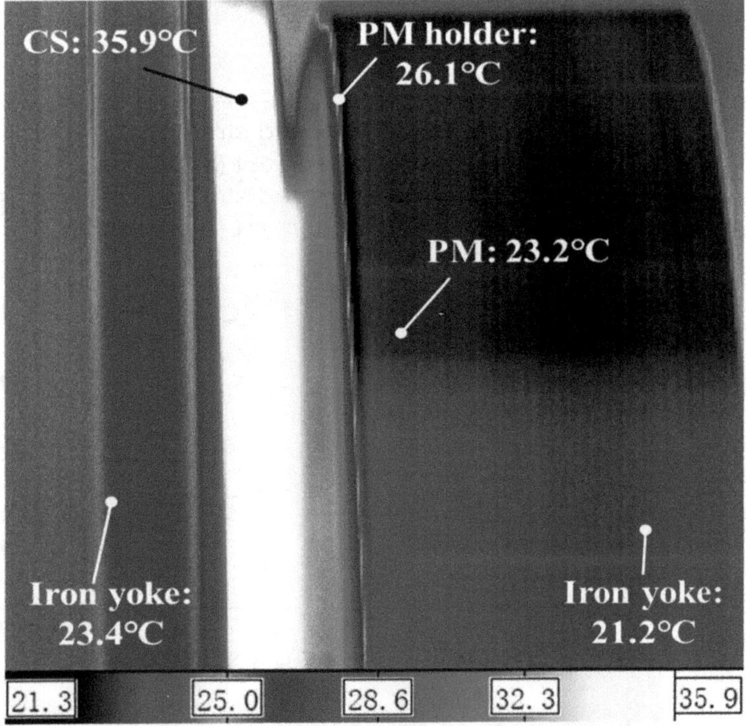

Figure 16. Temperature measurement results obtained by thermal camera (l_g = 5 mm and s = 20 r/min).

Here, the temperature of the CS and the PM is of greatest concern. As shown in Figure 17, compared with the experiment results at different slip speed, the proposed thermal analysis strategy for APMCs are in good agreement, and the relative error is within 6.7%. Moreover, the demagnetization temperature of the PM used in this paper is 180 °C. Therefore, the slip speed of 120r/min is the limit from Figure 17. If selected PM has a higher demagnetization temperature, a greater slip speed can be achieved.

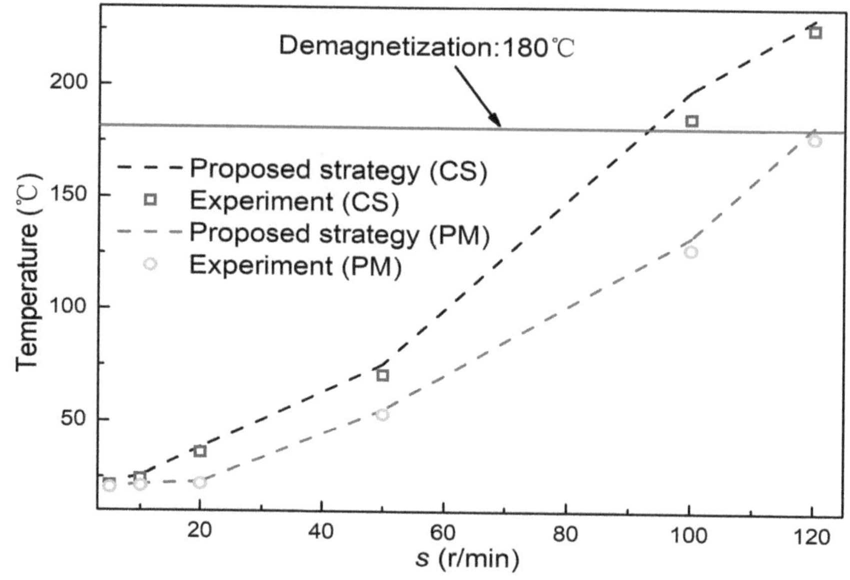

Figure 17. Comparison between the proposed strategy and experiment (l_g = 5 mm).

In order to further verify the accuracy of the strategy, Figure 18 presents the comparison between the proposed strategy and experiment under different air-gaps and slip speeds. With the increase of l_g, the temperature of PM decreases gradually. As can be seen from Figure 18, the strategy has a good coincidence with the experimental results, with the error less than 6.7%.

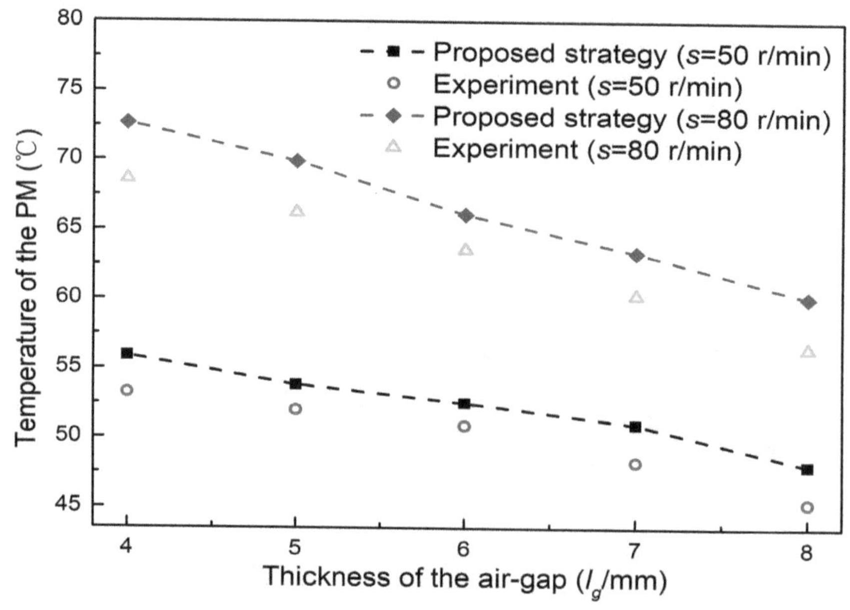

Figure 18. Comparison between proposed strategy and experiment under different air-gaps and slip speeds.

In the practical design and optimization of APMCs, the pole-pairs number of the PMs and the thickness of the CS are two important parameters. Besides, the temperature of the PM is key for the safe operation of APMCs due to demagnetization effect. Through the proposed strategy, Figure 19 presents the relationship between the temperature of the PM and the pole-pairs number of the PMs (p) at s = 120 r/min. From Figure 19, with the increase of p, the temperature of the PM increases gradually. Taking the demagnetization effect into account, selecting p as 6 is appropriate. Moreover, by the proposed strategy, Figure 20 presents the relationship between the temperature of the PM and the thickness of the CS (l_{cs}) at s = 120 r/min. From Figure 20, with the increase of l_{cs}, the temperature of the PM decreases gradually. Taking the demagnetization effect into account, selecting l_{cs} ≥6 is appropriate. However, in order to avoid excessive mass, selecting l_{cs} as 6 is optimal.

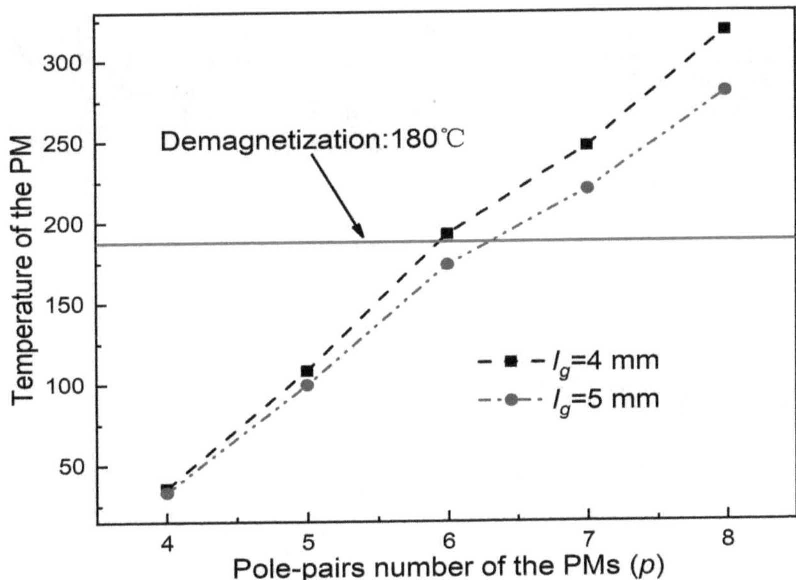

Figure 19. The relationship between the temperature of the PM and the pole-pairs number of the PMs. (s = 120 r/min).

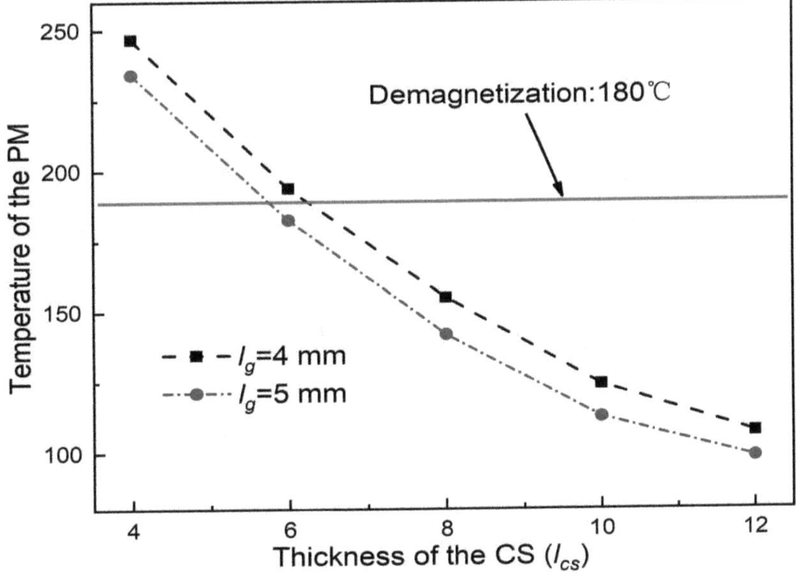

Figure 20. The relationship between the temperature of the PM and the thickness of the CS (s = 120 r/min).

7. Conclusions

A thermal analysis strategy is proposed to obtain the temperature results for APMCs, which combines FEM with LPTN. Firstly, the manufactured prototype of the studied APMC is built as well as giving its parameters. This proposed strategy is an iterative process considering some assumptions. Secondly, the magnetic field employing the FEM is offered to obtain the loss generated on the CS, where the magnetic field of the CS is analyzed. Then, the source nodes are assigned losses to calculate the matrix of the LPTN model and get the temperature results by adjusting the conductivity of the CS. Finally, the strategy is verified by experiment results where its relative error is less than 6.7%. In summary, the strategy developed in this paper can provide constructive references for operation safety of APMCs.

Author Contributions: Conceptualization, X.C. and W.L.; formal analysis, X.C. and Z.T.; funding acquisition, W.L.; investigation, X.C., Z.T., Z.Z., and B.Y.; methodology, X.C.; resources, B.Y., W.W., Y.Z., and S.L.; validation, X.C. and W.L.; writing—original draft, X.C. and W.L.; writing—review and editing, Z.Z., W.W., Y.Z., and S.L. All authors have read and agreed to the published version of the manuscript.

Nomenclature

n_{in}	rotational speed of the CS module
n_{out}	rotational speed of the PM module
s	Slip speed ($n_{in} - n_{out}$)
p	pole-pairs number of the PMs
l_{i1}	thickness of the iron yoke (CS side)
l_{i2}	thickness of the iron yoke (PM side)
l_{cs}	thickness of the CS
l_g	thickness of the air-gap
l_h	thickness of the PM holder
l_p	thickness of the PM
r_1	inside radius of the CS
r_2	outside radius of the CS
r_{p1}	inside radius of the PM
r_{p2}	outside radius of the PM
r_a	average radius of the PM
H_p	coercive force of the PM
σ_{cs}	conductivity of the CS
R, $R_{i\sim j}$	thermal resistance
P_{cs}	loss generated on the CS
T, T_0	temperature, ambient temperature
τ_p	length between centers of adjacent PMs
τ_m	length between adjacent PMs

Appendix A

The conduction thermal resistances can be calculated in detail as:

$$\begin{cases} R_{1\sim2} = \frac{l_{i1}/2}{k_{st}(\tau_p-\tau_m)(r_{p2}-r_{p1})} + \frac{l_{cs}/2}{k_{cop}(\tau_p-\tau_m)(r_{p2}-r_{p1})} \\ R_{13\sim14} = R_{20\sim21} = \frac{l_{i1}/2}{k_{st}\tau_m(r_{p2}-r_{p1})} + \frac{l_{cs}/2}{k_{cop}\tau_m(r_{p2}-r_{p1})} \\ R_{1\sim14} = R_{1\sim21} = \frac{(\tau_p-\tau_m)/2+\tau_m/2}{k_{st}l_{i1}(r_{p2}-r_{p1})} \\ R_{7\sim8} = R_{7\sim15} = \frac{(\tau_p-\tau_m)/2+\tau_m/2}{k_{st}l_{i2}(r_{p2}-r_{p1})} \\ R_{2\sim13} = R_{2\sim20} = \frac{(\tau_p-\tau_m)/2+\tau_m/2}{k_{cop}l_{cs}(r_{p2}-r_{p1})} \\ R_{4\sim11} = R_{4\sim18} = \frac{(\tau_p-\tau_m)/2+\tau_m/2}{k_{al}(l_h-l_p)(r_{p2}-r_{p1})} \\ R_{9\sim10} = R_{16\sim17} = \frac{l_p/2}{k_{al}\tau_m(r_{p2}-r_{p1})} \\ R_{10\sim11} = R_{17\sim18} = \frac{l_p/4+(l_h-l_p)/2}{k_{al}\tau_m(r_{p2}-r_{p1})} \\ R_{4\sim5} = \frac{(l_h-l_p)/2}{k_{al}(\tau_p-\tau_m)(r_{p2}-r_{p1})} + \frac{l_p/4}{k_{nd}(\tau_p-\tau_m)(r_{p2}-r_{p1})} \\ R_{5\sim10} = R_{5\sim17} = R_{6\sim9} = R_{6\sim16} = \frac{(\tau_p-\tau_m)}{k_{nd}l_p(r_{p2}-r_{p1})} + \frac{\tau_m}{k_{al}l_p(r_{p2}-r_{p1})} \\ R_{5\sim6} = \frac{l_p/2}{k_{nd}(\tau_p-\tau_m)(r_{p2}-r_{p1})} \\ R_{6\sim7} = \frac{l_{i2}/2}{k_{st}(\tau_p-\tau_m)(r_{p2}-r_{p1})} + \frac{l_p/4}{k_{nd}(\tau_p-\tau_m)(r_{p2}-r_{p1})} \\ R_{8\sim9} = R_{15\sim16} = \frac{l_{i2}/2}{k_{st}\tau_m(r_{p2}-r_{p1})} + \frac{l_p/4}{k_{al}\tau_m(r_{p2}-r_{p1})} \end{cases} \tag{A1}$$

Appendix B

$R_{2\sim3}$, $R_{3\sim4}$, $R_{3\sim12}$, $R_{3\sim19}$, $R_{11\sim12}$, $R_{12\sim13}$, $R_{18\sim19}$ and $R_{19\sim20}$ can be calculated in detail as:

$$\begin{cases} R_{2\sim3} = \frac{l_{cs}/2}{k_{cop}(\tau_p-\tau_m)(r_{p2}-r_{p1})} + \frac{l_g/2}{h_{air}(\tau_p-\tau_m)(r_{p2}-r_{p1})} \\ R_{3\sim4} = \frac{l_g/2}{h_{air}(\tau_p-\tau_m)(r_{p2}-r_{p1})} + \frac{(l_h-l_p)/2}{k_{al}(\tau_p-\tau_m)(r_{p2}-r_{p1})} \\ R_{3\sim12} = R_{3\sim19} = \frac{(\tau_p-\tau_m)/2+\tau_m/2}{h_{air}l_g(r_{p2}-r_{p1})} \\ R_{11\sim12} = R_{18\sim19} = \frac{l_g/2}{h_{air}\tau_m(r_{p2}-r_{p1})} + \frac{(l_h-l_p)/2}{k_{al}\tau_m(r_{p2}-r_{p1})} \\ R_{12\sim13} = R_{19\sim20} = \frac{l_{cs}/2}{k_{cop}\tau_m(r_{p2}-r_{p1})} + \frac{l_g/2}{h_{air}\tau_m(r_{p2}-r_{p1})} \end{cases} \tag{A2}$$

Appendix C

The convection thermal resistances can be calculated in detail as:

$$\begin{cases} R_{0\sim1} = \frac{l_{i1}/2}{k_{st}(\tau_p-\tau_m)(r_{p2}-r_{p1})} + \frac{1}{h_{am}(\tau_p-\tau_m)(r_{p2}-r_{p1})} \\ R_{0\sim2} = \frac{1}{(1-\alpha)h_{am}\pi r_2 l_{cs}/p} \\ R_{0\sim3} = \frac{1}{(1-\alpha)h_{am}\pi(r_2+r_{p2})l_g/2p} \\ R_{0\sim4} = \frac{1}{(1-\alpha)h_{am}\pi r_{p2}(l_h-l_p)/p} \\ R_{0\sim5} = R_{0\sim6} = \frac{1}{(1-\alpha)h_{am}\pi r_{p2}l_p/2p} \\ R_{0\sim7} = \frac{l_{i2}/2}{k_{st}(\tau_p-\tau_m)(r_{p2}-r_{p1})} + \frac{1}{h_{am}(\tau_p-\tau_m)(r_{p2}-r_{p1})} \\ R_{0\sim8} = R_{0\sim15} = \frac{l_{i2}/2}{k_{st}\tau_m(r_{p2}-r_{p1})} + \frac{1}{h_{am}\tau_m(r_{p2}-r_{p1})} \\ R_{0\sim9} = R_{0\sim10} = R_{0\sim16} = R_{0\sim17} = \frac{1}{\alpha h_{am}\pi r_{p2}l_p/2p} \\ R_{0\sim11} = R_{0\sim18} = \frac{1}{\alpha h_{am}\pi r_{p2}(l_h-l_p)/p} \\ R_{0\sim12} = R_{0\sim19} = \frac{1}{\alpha h_{am}\pi(r_2+r_{p2})l_g/2p} \\ R_{0\sim13} = R_{0\sim20} = \frac{1}{\alpha h_{am}\pi r_2 l_{cs}/p} \\ R_{0\sim14} = R_{0\sim21} = \frac{l_{i1}/2}{k_{st}\tau_m(r_{p2}-r_{p1})} + \frac{1}{h_{am}\tau_m(r_{p2}-r_{p1})} \end{cases} \tag{A3}$$

References

1. Ye, L.Z.; Li, D.S.; Ma, Y.J.; Jiao, B.F. Design and Performance of a Water-cooled Permanent Magnet Retarder for Heavy Vehicles. *IEEE Trans. Energy Convers.* **2011**, *26*, 953–958. [CrossRef]
2. Li, Y.B.; Lin, H.Y.; Huang, H.; Yang, H.; Tao, Q.C.; Fang, S.H. Analytical Analysis of a Novel Brushless Hybrid Excited Adjustable Speed Eddy Current Coupling. *Energies* **2019**, *12*, 308. [CrossRef]
3. Mohammadi, S.; Mirsalim, M.; Vaez-Zadeh, S.; Talebi, H.A. Analytical Modeling and Analysis of Axial-Flux Interior Permanent-Magnet Couplers. *IEEE Trans. Ind. Electron.* **2014**, *61*, 5940–5947. [CrossRef]
4. Lubin, T.; Mezani, S.; Rezzoug, A. Simple Analytical Expressions for the Force and Torque of Axial Magnetic Couplings. *IEEE Trans. Energy Convers.* **2012**, *27*, 536–546. [CrossRef]
5. Wang, S.; Guo, Y.C.; Cheng, G.; Li, D.Y. Performance Study of Hybrid Magnetic Coupler Based on Magneto Thermal Coupled Analysis. *Energies* **2017**, *10*, 1148. [CrossRef]
6. Wang, J.; Zhu, J.G. A Simple Method for Performance Prediction of Permanent Magnet Eddy Current Couplings Using a New Magnetic Equivalent Circuit Model. *IEEE Trans. Ind. Electron.* **2018**, *65*, 2487–2495. [CrossRef]
7. Dai, X.; Liang, Q.H.; Cao, J.Y.; Long, Y.J.; Mo, J.Q.; Wang, S.G. Analytical Modeling of Axial-Flux Permanent Magnet Eddy Current Couplings With a Slotted Conductor Topology. *IEEE Trans. Magn.* **2016**, *52*, 15. [CrossRef]
8. Lubin, T.; Mezani, S.; Rezzoug, A. Experimental and Theoretical Analyses of Axial Magnetic Coupling Under Steady-State and Transient Operations. *IEEE Trans. Ind. Electron.* **2014**, *61*, 4356–4365. [CrossRef]
9. Mohammadi, S.; Mirsalim, M. Double-sided permanent-magnet radial-flux eddy-current couplers: Three-dimensional analytical modelling, static and transient study, and sensitivity analysis. *IET Electr. Power Appl.* **2013**, *7*, 665–679. [CrossRef]
10. Ravaud, R.; Lemarquand, V.; Lemarquand, G. Analytical Design of Permanent Magnet Radial Couplings. *IEEE Trans. Magn.* **2010**, *46*, 3860–3865. [CrossRef]
11. Mohammadi, S.; Mirsalim, M.; Vaez-Zadeh, S. Nonlinear Modeling of Eddy-Current Couplers. *IEEE Trans. Energy Convers.* **2014**, *29*, 224–231. [CrossRef]
12. Fei, W.Z.; Luk, P.C.K. Torque Ripple Reduction of a Direct-Drive Permanent-Magnet Synchronous Machine by Material-Efficient Axial Pole Pairing. *IEEE Trans. Ind. Electron.* **2012**, *59*, 2601–2611. [CrossRef]
13. Min, K.C.; Choi, J.Y.; Kim, J.M.; Cho, H.W.; Jang, S.M. Eddy-Current Loss Analysis of Noncontact Magnetic Device with Permanent Magnets Based on Analytical Field Calculations. *IEEE Trans. Magn.* **2015**, *51*, 4. [CrossRef]
14. Choi, J.Y.; Jang, S.M. Analytical magnetic torque calculations and experimental testing of radial flux permanent magnet-type eddy current brakes. *J. Appl. Phys.* **2012**, *111*, 3. [CrossRef]
15. Dai, X.; Cao, J.Y.; Long, Y.J.; Liang, Q.H.; Mo, J.Q.; Wang, S.G. Analytical Modeling of an Eddy-current Adjustable-speed Coupling System with a Three-segment Halbach Magnet Array. *Electr. Power Compon. Syst.* **2015**, *43*, 1891–1901. [CrossRef]
16. Wang, J.; Lin, H.Y.; Fang, S.H.; Huang, Y.K. A General Analytical Model of Permanent Magnet Eddy Current Couplings. *IEEE Trans. Magn.* **2014**, *50*, 9. [CrossRef]
17. Lubin, T.; Rezzoug, A. Steady-State and Transient Performance of Axial-Field Eddy-Current Coupling. *IEEE Trans. Ind. Electron.* **2015**, *62*, 2287–2296. [CrossRef]
18. Dolisy, B.; Mezani, S.; Lubin, T.; Leveque, J. A New Analytical Torque Formula for Axial Field Permanent Magnets Coupling. *IEEE Trans. Energy Convers.* **2015**, *30*, 892–899. [CrossRef]
19. Lubin, T.; Rezzoug, A. 3-D Analytical Model for Axial-Flux Eddy-Current Couplings and Brakes Under Steady-State Conditions. *IEEE Trans. Magn.* **2015**, *51*, 12. [CrossRef]
20. Wang, J.; Lin, H.Y.; Fang, S.H. Analytical Prediction of Torque Characteristics of Eddy Current Couplings Having a Quasi-Halbach Magnet Structure. *IEEE Trans. Magn.* **2016**, *52*, 9. [CrossRef]
21. Li, Z.; Wang, D.Z.; Zheng, D.; Yu, L.X. Analytical modeling and analysis of magnetic field and torque for novel axial flux eddy current couplers with PM excitation. *AIP Adv.* **2017**, *7*, 13. [CrossRef]
22. Lubin, T.; Rezzoug, A. Improved 3-D Analytical Model for Axial-Flux Eddy-Current Couplings with Curvature Effects. *IEEE Trans. Magn.* **2017**, *53*, 9. [CrossRef]

23. Huang, X.Z.; Li, L.Y.; Zhou, B.; Zhang, C.M.; Zhang, Z.R. Temperature Calculation for Tubular Linear Motor by the Combination of Thermal Circuit and Temperature Field Method Considering the Linear Motion of Air Gap. *IEEE Trans. Ind. Electron.* **2014**, *61*, 3923–3931. [CrossRef]
24. Lu, Y.P.; Liu, L.; Zhang, D.X. Simulation and Analysis of Thermal Fields of Rotor Multislots for Nonsalient-Pole Motor. *IEEE Trans. Ind. Electron.* **2015**, *62*, 7678–7686. [CrossRef]
25. Vese, I.C.; Marignetti, F.; Radulescu, M.M. Multiphysics Approach to Numerical Modeling of a Permanent-Magnet Tubular Linear Motor. *IEEE Trans. Ind. Electron.* **2010**, *57*, 320–326. [CrossRef]
26. Zhu, X.Y.; Wu, W.Y.; Yang, S.; Xiang, Z.X.; Quan, L. Comparative Design and Analysis of New Type of Flux-Intensifying Interior Permanent Magnet Motors with Different Q-Axis Rotor Flux Barriers. *IEEE Trans. Energy Convers.* **2018**, *33*, 2260–2269. [CrossRef]
27. Le Besnerais, J.; Fasquelle, A.; Hecquet, M.; Pelle, J.; Lanfranchi, V.; Harmand, S.; Brochet, P.; Randria, A. Multiphysics Modeling: Electro-Vibro-Acoustics and Heat Transfer of PWM-Fed Induction Machines. *IEEE Trans. Ind. Electron.* **2010**, *57*, 1279–1287. [CrossRef]
28. Sun, X.K.; Cheng, M. Thermal Analysis and Cooling System Design of Dual Mechanical Port Machine for Wind Power Application. *IEEE Trans. Ind. Electron.* **2013**, *60*, 1724–1733. [CrossRef]
29. Mo, L.H.; Zhang, T.; Lu, Q. Thermal Analysis of a Flux-Switching Permanent-Magnet Double-Rotor Machine with a 3-D Thermal Network Model. *IEEE Trans. Appl. Supercond.* **2019**, *29*, 5. [CrossRef]
30. Valenzuela, M.A.; Ramirez, G. Thermal Models for Online Detection of Pulp Obstructing the Cooling System of TEFC Induction Motors in Pulp Area. *IEEE Trans. Ind. Appl.* **2011**, *47*, 719–729. [CrossRef]

14

Entropy Generation in a Dissipative Nanofluid Flow under the Influence of Magnetic Dissipation and Transpiration

Dianchen Lu [1], Muhammad Idrees Afridi [2], Usman Allauddin [3], Umer Farooq [1] and Muhammad Qasim [1,*]

1. Department of Mathematics, Faculty of Science, Jiangsu University, Zhenjiang 212013, China; dclu@ujs.edu.cn (D.L.); umer_farooq@comsats.edu.pk (U.F.)
2. Department of Computing, Abasyn University, Islamabad 45710, Pakistan; Muhammad.idrees@abasynisb.edu.pk
3. Department of Mechanical Engineering, NED University of Engineering & Technology, Karachi 75270, Pakistan; usman.allauddin@neduet.edu.pk

* Correspondence: mqasim@ujs.edu.cn

Abstract: The present study explores the entropy generation, flow, and heat transfer characteristics of a dissipative nanofluid in the presence of transpiration effects at the boundary. The non-isothermal boundary conditions are taken into consideration to guarantee self-similar solutions. The electrically conducting nanofluid flow is influenced by a magnetic field of constant strength. The ultrafine particles (nanoparticles of Fe_3O_4/CuO) are dispersed in the technological fluid water (H_2O). Both the base fluid and the nanofluid have the same bulk velocity and are assumed to be in thermal equilibrium. Tiwari and Dass's idea is used for the mathematical modeling of the problem. Furthermore, the ultrafine particles are supposed to be spherical, and Maxwell Garnett's model is used for the effective thermal conductivity of the nanofluid. Closed-form solutions are derived for boundary layer momentum and energy equations. These solutions are then utilized to access the entropy generation and the irreversibility parameter. The relative importance of different sources of entropy generation in the boundary layer is discussed through various graphs. The effects of space free physical parameters such as mass suction parameter (S), viscous dissipation parameter (Ec), magnetic heating parameter (M), and solid volume fraction (ϕ) of the ultrafine particles on the velocity, Bejan number, temperature, and entropy generation are elaborated through various graphs. It is found that the parabolic wall temperature facilitates similarity transformations so that self-similar equations can be achieved in the presence of viscous dissipation. It is observed that the entropy generation number is an increasing function of the Eckert number and solid volume fraction. The entropy production rate in the $Fe_3O_4-H_2O$ nanofluid is higher than that in the $CuO-H_2O$ nanofluid under the same circumstances.

Keywords: nanofluid; heat transfer; entropy generation; viscous dissipation; magnetic heating

---

1. Introduction

The Navier-Stokes equations, which are second-order nonlinear partial differential equations, govern the viscous fluid–fluid flow. The exact solution of the complete Navier–Stokes equations has not yet been computed. However, closed-form solutions can be established in certain physical circumstances under reasonable suppositions [1–5]. Exact solutions are important since such solutions can be utilized to validate asymptotic analytical and numerical solutions. Crane [6] found the closed-form solution of the simplified Navier-Stokes equations under the boundary layer approximations to analyze the flow over a stretched surface. Some researchers determined the closed-form solutions of boundary layer flow after the pioneering work of Crane with various physical conditions [7–11].

It is essential to examine heat transfer issues in industrial engineering. Recently, heat transfer analysis has been limited to the first law of thermodynamics, which only concerns energy conservation during the interactions of the systems and surroundings. It deals solely with the amount of energy regardless of its quality. Moreover, the first law does not distinguish between heat and work. It assumes that work and heat are fully interchangeable, but work is high-quality energy and can be fully converted into heat, while heat is low-quality energy and cannot be fully converted into work. Heat is an unorganized form of energy. The law of entropy shows that the entropy increase in the cold object is higher than the decrease of entropy in the hot object. This means that the final state is more random in the thermodynamic system. This analysis suggests that the heat transfer phenomenon decreases energy quality or increases the system entropy. To investigate this energy quality reduction, Bejan [12,13] proposed a method called entropy minimization that is based on the law of entropy. The law of entropy (second law of thermodynamics) is used to maintain energy quality [14–20]. In addition to heat transfer, frictional heating and magnetic dissipation also generate entropy in fluid flow problems [21–25].

Conventional working fluids such as kerosene, gasoline, water, engine oil, and fluid mixtures have exceptionally poor thermal conductivity, as demonstrated by the vast number of industries dealing with these conventional working fluids. However, due to their inefficiency in thermal conductivity, they face several problems. The use of nanoscale elements in base fluids is one of the most important techniques used to resolve this deficiency. Such a mixture of nanometer-sized particles and a working fluid is called a nanofluid. In comparison to base liquids, nanofluids possess high thermal conductivity [26–32]. Many researchers firmly agree on the remarkable characteristics of nanofluids. Over the past two decades, this new type of fluid has attracted the attention of many researchers. Nanofluid studies have a variety of important applications, such as product provision for cancer, cooling systems, nuclear power plant cooling, and computer equipment cooling. Hsiao [33] conducted stagnation nanofluid energy conversion analysis for the conjugate problem of conduction–convection and heat source/sink. Ma et al. [34] explored the gravitational convection term of heat management in a shell and tube heat exchanger filled with a $Fe_3O_4 - H_2O$ nanoliquid by utilizing a lattice Boltzmann scheme. Wakif et al. [35] reported the impacts of thermal radiation and surface roughness on the complex dynamics of water transporting alumina and copper oxide nanoparticles. Hsiao [36] reported nanofluid flow for conjugating mixed convection and radiation with interactive physical characteristics. In a channel with active heaters and coolers, a numerical simulation was introduced by Ma et al. [37] to examine the impacts of magnetic field on heat transfer in a $MgO - Ag - H_2O$ nanoliquid. Prasad et al. [38] examined the upper-convected Maxwell three-dimensional rotational flow with a convective boundary condition and zero mass flux for the concentration of nanoparticles. Frictional heating is the conversion of fluid kinetic energy to heat due to the frictional forces between all the neighboring fluid layers. Frictional heating is the main factor in the study of heat transfer in boundary layer flows. Since large velocity gradients exist within the boundary layer, the viscous dissipation effects cannot be neglected. When there is a viscous dissipation, a term for viscous dissipation is incorporated into the energy equation [39–46].

In this research, the exact solutions of transformed nonlinear dimensionless momentum and energy equations that occur in the magnetohydrodynamic (MHD) boundary layer flow of nanofluid are obtained. The goal of the work, apart from providing a benchmark solution for numerical simulation, is the parametric analysis of entropy generation. The work also describes how boundary conditions facilitate similarity transformations to get self-similar equations. The literature review reveals that nonsimilar problems are treated as self-similar problems. Furthermore, the entropy generation analysis exists in literature, but the analysis is limited to the low temperature difference between the boundary and bulk fluid. The present work is free from such a constraint and is valid for both low and high temperature differences. In addition, the terms for frictional heating and magnetic dissipation are added to the energy equation and the expression for entropy generation. To the best of our knowledge, no one has reported the exact solutions for nanofluid flow induced by a linearly stretching surface with a parabolic temperature profile at the boundary. Obtained exact solutions are used for calculating entropy generation and the Bejan number. Visual representations are used to investigate the effects of physical parameters on the nanofluid flow, thermal field, entropy generation profile, and Bejan number.

2. Statement of the Problem and Governing Equations

Consider the electrically conducting and dissipative nanofluid flow over a stretching surface as shown in Figure 1. The nanofluid is supposed to be a mixture of base fluid (water) and nanoparticles Fe_3O_4/CuO. The Cartesian coordinate system (X, Y) is chosen in such a way that the $X-$ axis is taken along the solid boundary and the $Y-$ axis is normal to it. Let $U_w(X) = U_oX$ be the velocity of the stretching boundary and $T_w(X) = T_b + C_oX^2$ be the temperature variation at the surface of the stretching boundary; here, T_b and the subscript w represent the bulk fluid temperature and the condition at the solid boundary, while U_o and C_o represent the dimensional constants. The imposed magnetic field is constant and of strength B_o. The generalized Ohm's law in the absence of an electrical field is $\vec{j} = \sigma_{nf}\left(\vec{q} \times \vec{B}_o\right)$, where σ_{nf} and $\vec{q}\left(\vec{U}, \vec{V}\right)$ show the electrical conductivity of nanofluid and bulk velocity field of the nanofluid, respectively. The magnetic force $\bar{j} \times B_o$ and magnetic dissipation $\frac{\vec{j}.\vec{j}}{\sigma_{nf}}$ are simplified to $-\sigma_{nf}B_o^2U$ and $\sigma_{nf}B_o^2U^2$, respectively.

The equations governing the incompressible nanofluid flow for the present problem are

$$\frac{\partial U}{\partial X} + \frac{\partial V}{\partial Y} = 0, \tag{1}$$

$$U\frac{\partial U}{\partial X} + V\frac{\partial U}{\partial Y} = \nu_{nf}\frac{\partial^2 U}{\partial Y^2} - \frac{\sigma_{nf}B_o^2U}{\rho_{nf}}, \tag{2}$$

$$\left(U\frac{\partial T}{\partial X} + V\frac{\partial T}{\partial Y}\right) = \left(\frac{1}{\rho C_p}\right)_{nf}\left(k_{nf}\frac{\partial^2 T}{\partial Y^2} + \mu_{nf}\left(\frac{\partial U}{\partial Y}\right)^2 + \sigma_{nf}B_o^2U^2\right) \tag{3}$$

The imposed boundary conditions are as follows:

$$\left.\begin{array}{c} U(X,0) = U_w(X) = U_oX,\ V(X,0) = V_w,\ T(X,0) = T_w(X) = T_b + C_oX^2 \\ U(X, Y \to \infty) \to 0,\ T(X, Y \to \infty) \to T_b \end{array}\right\} \tag{4}$$

The governing self-similar equations are obtained from Equations (2) and (3) by using the following dimensionless variables:

$$\eta = Y\sqrt{\frac{U_o}{\nu_{bf}}},\ U = U_oXf'(\eta),\ V = -\sqrt{U_o\nu_{bf}}f(\eta),\ \theta(\eta) = \frac{T - T_b}{T(X,0) - T_b} \tag{5}$$

Equations (2) and (3) under the transformation in Equation (5) become

$$\frac{G_1}{G_o}f''' + ff'' - f'^2 - \frac{G_3}{G_o}M^2f' = 0, \tag{6}$$

$$\frac{G_5}{G_4}\theta'' + \frac{G_1}{G_4}\mathrm{EcPr}f''^2 + \mathrm{Pr}f\theta' + \frac{G_3}{G_4}\mathrm{Ec}M^2\mathrm{Pr}f'^2 - 2\mathrm{Pr}\theta f' = 0 \tag{7}$$

The imposed boundary conditions are transformed to

$$f(0) = -\frac{V_w}{\sqrt{U_o\nu_{bf}}} = S,\ f'(0) = 1,\ f'(\eta \to \infty) = 0 \tag{8}$$

$$\theta(0) = 1,\ \theta(\eta \to \infty) = 0 \tag{9}$$

where $G_o = (1-\phi) + \phi\left(\frac{\rho_s}{\rho_{bf}}\right)$, $G_1 = (1-\phi)^{-2.5}$, $G_3 = \frac{\sigma_{nf}}{\sigma_{bf}}$, $G_4 = 1 - \phi + \phi\left(\frac{(\rho C_p)_s}{(\rho C_p)_{bf}}\right)$, $G_5 = \frac{k_{nf}}{k_{bf}}$, and $Ec = \frac{U_w^2}{(C_p)_{bf}\,(T(X,0)-T_b)}$ (Eckert number), and the subscripts bf and s are used for base fluid and nanoparticles, respectively. $\Pr = \frac{\nu_{bf}}{\alpha_{bf}}$ (Prandtl number); α_{bf} indicates base fluid thermal diffusivity; $M^2 = \frac{\sigma_{bf} B_o^2}{\rho_{bf} U_0}$; $S = -\frac{V_w}{\sqrt{U_o \nu_{bf}}}$ and shows the dimensionless mass-transfer parameter; and ν_{nf}, σ_{nf}, ρ_{nf}, k_{nf}, and $(\rho C_p)_{nf}$ are defined in Table 1. The thermophysical properties of CuO, Fe_3O_4, and working fluid (H_2O) are shown in Table 2.

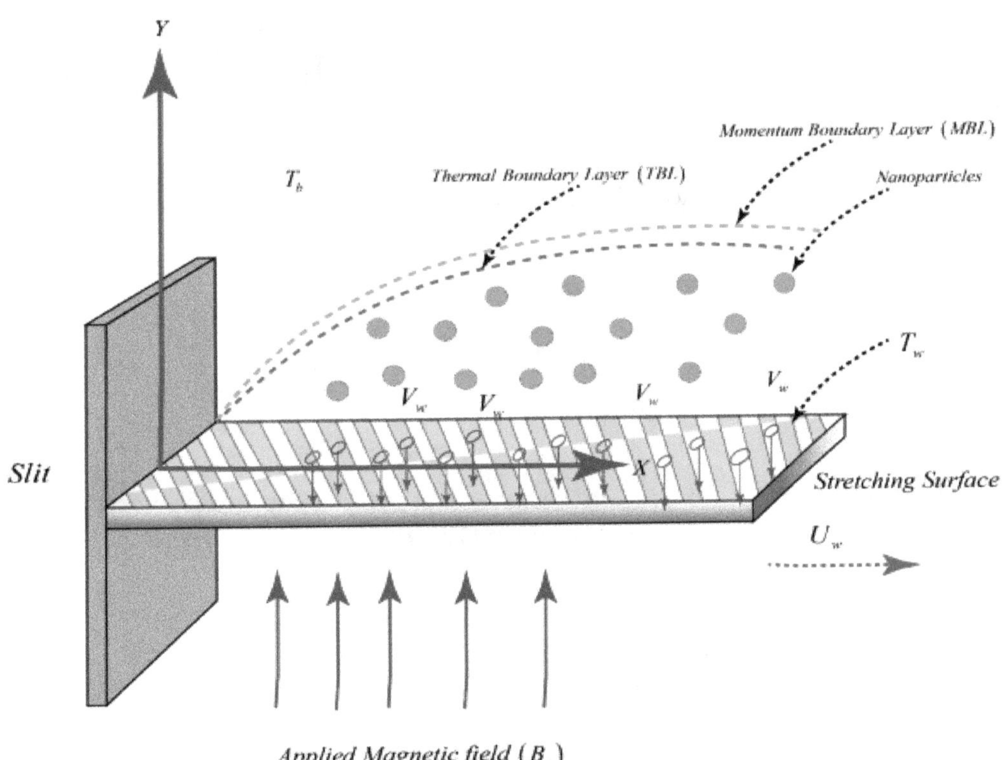

Figure 1. Physical flow model and coordinate system.

Table 1. Effective thermophysical properties of nanofluid [47–52].

Thermophysical Property of Nanofluid	Symbol	Defined
Thermal conductivity	k_{nf}	$k_{nf} = \frac{(k_s+2k_{bf})-2\phi(k_{bf}-k_s)}{(k_s+2k_{bf})+\phi(k_{bf}-k_s)} k_{bf}$ here, ϕ represents sold volume fraction of nanoparticles.
Viscosity	μ_{nf}	$\mu_{nf} = \frac{\mu_{bf}}{(1-\phi)^{2.5}}$
Electric conductivity	σ_{nf}	$\sigma_{nf} = 1 + \frac{3\left(\frac{\sigma_s}{\sigma_{bf}}-1\right)\phi}{\left(\frac{\sigma_s}{\sigma_{bf}}+2\right)-\left(\frac{\sigma_s}{\sigma_{bf}}-1\right)\phi} \sigma_{bf}$
Heat capacitance	$(\rho C_p)_{nf}$	$(\rho C_p)_{nf} = (1-\phi)(\rho C_p)_{bf} + \phi(\rho c_p)_s$
Density	ρ_{nf}	$\rho_{nf} = (1-\phi)\rho_{bf} + \phi\rho_s$

Table 2. Thermophysical properties of CuO, Fe_3O_4, and working fluid (H_2O).

Physical Properties	H_2O	CuO	Fe_3O_4
C_p (J/kgK)	4179	531.8	670
k (W/mK)	0.613	76.5	6.0
$\rho\left(\text{kg/m}^3\right)$	997.1	6320	5200
$\sigma\left(\text{S}\times\text{m}^{-1}\right)$	5180	2.7×10^{-8}	25,000
Pr (-)	6.8	-	-

3. Solution Methodology

3.1. Closed-Form Solution of Momentum Balance Equation

The closed-form exact solution of Equation (6) with associated boundary conditions of Equation (8) is supposed as follows:

$$f(\eta) = C_1 + C_2 e^{-\beta\eta},\ \beta > 0 \tag{10}$$

Using the first two boundary conditions defined in Equation (8), the computed arbitrary constants C_1 and C_2 are

$$C_1 = S + \frac{1}{\beta},\ C_2 = -\frac{1}{\beta} \tag{11}$$

Putting Equation (11) into Equation (10), we get

$$f(\eta) = S + \frac{1}{\beta}\left(1 - e^{-\beta\eta}\right) \tag{12}$$

The above closed-form solution trivially satisfies the far-field boundary condition as defined in Equation (8) for $\beta > 0$. To find β, we insert Equation (12) into Equation (6) and get

$$\frac{G_1}{G_o}\beta^2 - S\beta - 1 - \frac{G_3}{G_o}M^2 = 0 \tag{13}$$

By solving the above equation, we have

$$\beta = G_o\left(\frac{S + \sqrt{S^2 + 4\frac{G_1}{G_o}\left(1 + \frac{G_3}{G_o}M^2\right)}}{2G_1}\right) > 0. \tag{14}$$

The closed-form solution of the boundary value problem (Equations (6) and (7)) is given by

$$f(\eta) = S + \frac{2G_1}{G_o\left(S + \sqrt{S^2 + 4\frac{G_1}{G_o}\left(1 + \frac{G_3}{G_o}M^2\right)}\right)}\left(1 - e^{-G_o\left(\frac{S+\sqrt{S^2+4\frac{G_1}{G_o}\left(1+\frac{G_3}{G_o}M^2\right)}}{2G_1}\right)\eta}\right) \tag{15}$$

3.2. Solution of Energy Balance Equation via Laplace Transform

Equation (7) is decoupled from Equation (6) by substituting Equation (12) into Equation (7) as follows:

$$\frac{G_5}{G_4}\theta'' + \frac{G_1}{G_4}\text{EcPr}\beta^2 e^{-2\beta\eta} + \text{Pr}\left(S + \frac{1}{\beta}\left(1 - e^{-\beta\eta}\right)\right)\theta' + \frac{G_3}{G_4}\text{EcM}^2\text{Pr}e^{-2\beta\eta} - 2\text{Pr}\theta e^{-\beta\eta} = 0 \tag{16}$$

To get rid of exponential coefficients, we define a new variable, ξ, as follows:

$$\xi = \frac{\Pr}{\beta^2} e^{-\beta\eta} \tag{17}$$

By utilizing the above transformation, Equation (7) and the related boundary conditions take the following form:

$$\xi \frac{d^2\theta}{d\xi^2} + \frac{d\theta}{d\xi}\left(K + \frac{\xi}{G}\right) + \xi L - 2\frac{\theta}{G} = 0, \tag{18}$$

$$\theta\left(\frac{\Pr}{\beta^2}\right) = 1,\ \theta(0) = 0 \tag{19}$$

with

$$K = 1 - \frac{\Pr(1+\beta S)}{G\beta^2},\ L = \frac{Ec\beta^2}{G\Pr}\left(\frac{G_1}{G_4}\beta^2 + \frac{G_3}{G_4}M^2\right) \text{ and } G = \frac{G_5}{G_4}. \tag{20}$$

By employing Laplace transform on Equation (18) and then using Equation (19), we obtain

$$\frac{d\Theta(\zeta)}{d\zeta} + \Theta(\zeta)\left[\frac{\zeta(2-K) + \frac{3}{G}}{\zeta\left(\zeta + \frac{1}{G}\right)}\right] = \frac{L}{\zeta^3\left(\zeta + \frac{1}{G}\right)} \tag{21}$$

where $\Theta(\zeta)$ is the Laplace transform of the function $\theta(\xi)$. Equation (21) is a Leibnitz first-type linear equation with integrating factor

$$e^{\int \frac{\zeta(2-K)+\frac{3}{G}}{\zeta(\zeta+\frac{1}{G})} d\zeta} = \frac{\zeta^3}{(G\zeta + 1)^{1+K}}. \tag{22}$$

Solving Equation (21) by utilizing Equation (22), we have

$$\Theta(\zeta) = \frac{L}{\zeta^3(-K-1)} + c\frac{(G\zeta+1)^{K+1}}{\zeta^3} \tag{23}$$

By taking Laplace inverse of Equation (23), we get

$$\theta(\xi) = \frac{L\xi^2}{2(-K-1)} + \frac{c}{2G^{-K-1}\Gamma(-K-1)}\left(\xi^2 * \xi^{-2-K} e^{\left(\frac{-\xi}{G}\right)}\right) \tag{24}$$

Here, an asterisk (*) indicates convolution and Γ shows a gamma function. The convolution of two functions, $F(\xi)$ and $G(\xi)$, is defined as follows:

$$F(\xi) * H(\xi) = \int_0^{\xi} F(\xi - \varepsilon) H(\varepsilon) d\varepsilon \tag{25}$$

By taking $F(\xi) = \xi^2$ and $H(\xi) = e^{\frac{-\xi}{G}} \xi^{-K-2}$, Equation (24) takes the following form:

$$\theta(\xi) = \frac{L\xi^2}{2(-K-1)} + \frac{c}{2G^{-K-1}\Gamma(-K-1)} \int_0^{\xi} (\xi - \varepsilon)^2 e^{\frac{-\varepsilon}{G}} \varepsilon^{-K-2} d\varepsilon. \tag{26}$$

By employing the transformation $\varepsilon = \xi u$, the above equation takes the following form:

$$\theta(\xi) = -\frac{L\xi^2}{2(K+1)} + \frac{c\xi^{1-K}}{2G^{-K-1}\Gamma(-K-1)} \int_0^1 (1-u)^2 e^{\frac{-u\xi}{G}} u^{-K-2} du. \tag{27}$$

By utilizing the integral form of Kummer's confluent hypergeometric function, i.e., $M_{1,1}\left(-K-1; -K+2; \frac{-\xi}{G}\right) = \frac{\Gamma(2-K)}{2\Gamma(-1-K)} \int_0^1 (1-u)^2 e^{\frac{-u\xi}{G}} u^{-K-2} d$, Equation (27) becomes

$$\theta(\xi) = -\frac{L\xi^2}{2(K+1)} + \frac{cG^{K+1}\xi^{1-K}}{\Gamma(2-K)} M_{1,1}\left(-K-1; -K+2; \frac{-\xi}{G}\right). \tag{28}$$

The boundary condition at the surface of the stretching surface $\theta(0) = 0$ is satisfied identically. However, the constant of integration c is obtained by using the far-field boundary condition $\theta\left(\xi = \frac{\Pr}{\beta^2}\right) = 1$ and is given by

$$c = \frac{\Gamma(2-m)\left(\frac{2(K+1)+L\left(\frac{\Pr}{\beta^2}\right)^2}{2(K+1)}\right)}{G^{K+1}\left(\frac{\Pr}{\beta^2}\right)^{1-K} M_{1,1}\left(-1-K\,; 2-K\,; -\frac{\Pr}{G\beta^2}\right)}. \tag{29}$$

Finally, by inserting Equation (29) into Equation (28) and using the transformation $\xi = \frac{\Pr}{\beta^2} e^{-\beta\eta}$, we obtain the exact solution of the energy equation:

$$\theta(\eta) = -\frac{1}{2}\frac{L}{(K+1)}\left(\frac{\Pr e^{-\beta\eta}}{\beta^2}\right)^2 + \frac{\left(\frac{\Pr e^{-\beta\eta}}{\beta^2}\right)^{1-K} M_{1,1}\left(-1-K\,; 2-K\,; -\frac{\Pr e^{-\beta\eta}}{G\beta^2}\right)\left(1 + \frac{L}{2(1+K)}\left(\frac{\Pr}{\beta^2}\right)^2\right)}{\left(\frac{\Pr}{\beta^2}\right)^{1-K} M_{1,1}\left(-K-1\,; 2-K\,; -\frac{\Pr}{G\beta^2}\right)}. \tag{30}$$

4. Analysis of Entropy Generation

The rate of entropy generation in the presence of heat dissipation phenomenon with magnetic heating is given by

$$\dot{E}_{Gen}''' = \frac{k_{nf}}{T^2}\left(\frac{\partial T}{\partial Y}\right)^2 + \frac{\mu_{nf}}{T}\left(\frac{\partial U}{\partial Y}\right)^2 + \frac{\sigma_{nf} B_0^2 U^2}{T}, \tag{31}$$

Using Equation (6), Equation (31) becomes

$$\frac{\dot{E}_{Gen}'''}{\left(\dot{E}_{Gen}'''\right)_0} = Ns = \underbrace{G_5\frac{\theta'^2}{(\theta+\Lambda)^2}}_{N_H} + \underbrace{\frac{G_1 \Pr Ec f''^2}{(\theta+\Lambda)}}_{N_F} + \underbrace{G_3\frac{\Pr M^2 Ec f'^2}{(\theta+\Lambda)}}_{N_M}. \tag{32}$$

Here, $\left(\dot{E}_{Gen}'''\right)_0 = \frac{k_{bf} U_o}{\nu_{bf}}$ indicates characteristic entropy generation; Ns indicates entropy production rate in dimensionless form; $\Lambda = \frac{T_b}{T_w - T_b}$ shows the temperature parameter; and N_H, N_F, and N_M represent the dimensionless form of entropy generation due to heat transfer, viscous dissipation, and magnetic heating, respectively.

By utilizing the obtained exact solutions, the three sources of entropy generation stated above take the following forms:

$$N_H = \frac{1}{(\Pr)_{eff}} \frac{\left[\left(-\frac{1}{2}\frac{L}{(K+1)}\left(\frac{\Pr e^{-\beta\eta}}{\beta^2}\right)^2 + \frac{\left(\frac{\Pr e^{-\beta\eta}}{\beta^2}\right)^{1-K} M_{1,1}\left(-1-K\,;2-K\,;-\frac{\Pr e^{-\beta\eta}}{G\beta^2}\right)\left(1+\frac{L}{2(1+K)}\left(\frac{\Pr}{\beta^2}\right)^2\right)}{\left(\frac{\Pr}{\beta^2}\right)^{1-K} M_{1,1}\left(-K-1\,;2-K\,;-\frac{\Pr}{G\beta^2}\right)}\right)'\right]^2}{\left(\Lambda - \frac{1}{2}\frac{L}{(K+1)}\left(\frac{\Pr e^{-\beta\eta}}{\beta^2}\right)^2 + \frac{\left(\frac{\Pr e^{-\beta\eta}}{\beta^2}\right)^{1-K} M_{1,1}\left(-1-K\,;2-K\,;-\frac{\Pr e^{-\beta\eta}}{G\beta^2}\right)\left(1+\frac{L}{2(1+K)}\left(\frac{\Pr}{\beta^2}\right)^2\right)}{\left(\frac{\Pr}{\beta^2}\right)^{1-K} M_{1,1}\left(-K-1\,;2-K\,;-\frac{\Pr}{G\beta^2}\right)}\right)^2}. \tag{33}$$

$$N_F = \frac{Ec\Pr\left(\frac{-S+\sqrt{S^2+4\frac{G_1}{G_o}\left(1+\frac{G_3}{G_o}M^2\right)}}{2G_1} e^{-\eta\left(\frac{S+\sqrt{S^2+4\frac{G_1}{G_o}\left(1+\frac{G_3}{G_o}M^2\right)G_o}}{2G_1}\right)}\right)^2 G^3{}_o}{\left(\Lambda - \frac{1}{2}\frac{L}{(K+1)}\left(\frac{\Pr e^{-\beta\eta}}{\beta^2}\right)^2 + \frac{\left(\frac{\Pr e^{-\beta\eta}}{\beta^2}\right)^{1-K} M_{1,1}\left(-1-K\,;2-K\,;-\frac{\Pr e^{-\beta\eta}}{G\beta^2}\right)\left(1+\frac{L}{2(1+K)}\left(\frac{\Pr}{\beta^2}\right)^2\right)}{\left(\frac{\Pr}{\beta^2}\right)^{1-K} M_{1,1}\left(-K-1\,;2-K\,;-\frac{\Pr}{G\beta^2}\right)}\right)}. \tag{34}$$

and

$$N_H = \frac{M^2 Ec\Pr\left(e^{-\eta\left(\frac{S+\sqrt{S^2+4\frac{G_1}{G_o}\left(1+\frac{G_3}{G_o}M^2\right)G_o}}{2G_1}\right)}\right)^2}{\left(\Lambda - \frac{1}{2}\frac{L}{(K+1)}\left(\frac{\Pr e^{-\beta\eta}}{\beta^2}\right)^2 + \frac{\left(\frac{\Pr e^{-\beta\eta}}{\beta^2}\right)^{1-K} M_{1,1}\left(-1-K\,;2-K\,;-\frac{\Pr e^{-\beta\eta}}{G\beta^2}\right)\left(1+\frac{L}{2(1+K)}\left(\frac{\Pr}{\beta^2}\right)^2\right)}{\left(\frac{\Pr}{\beta^2}\right)^{1-K} M_{1,1}\left(-K-1\,;2-K\,;-\frac{\Pr}{G\beta^2}\right)}\right)}. \tag{35}$$

4.1. Bejan Number

To compare the spatial distribution of entropy generation in a flow field due to various sources, an irreversibility ratio parameter known as Bejan number (Be) is defined as given below

$$Be = \frac{\frac{k_{nf}}{T^2}\left(\frac{\partial T}{\partial Y}\right)^2 \Rightarrow (Entropy\ generation\ due\ to\ heat\ transfer)}{\left(\frac{k_{nf}}{T^2}\left(\frac{\partial T}{\partial Y}\right)^2 + \frac{\mu_{nf}}{T}\left(\frac{\partial U}{\partial Y}\right)^2 + \frac{\sigma_{nf}B_o^2U^2}{T}\right) \Rightarrow (Total\ entropy\ generation)} \tag{36}$$

After the utilization of similarity variables, Equation (36) takes the following form:

$$Be = \frac{G_5\frac{\theta'^2}{(\theta+\Lambda)^2} \Rightarrow N_H}{\left(G_5\frac{\theta'^2}{(\theta+\Lambda)^2} + \frac{G_1\Pr Ecf''^2}{(\theta+\Lambda)} + G_3\frac{\Pr M^2Ecf'^2}{(\theta+\Lambda)}\right) \Rightarrow (N_H + N_F + N_M)} \tag{37}$$

5. Results and Discussion

The nondimensional complicated differential equations (momentum and energy equations) are solved by taking into consideration the exponential form solution and the Laplace transform. The exact expressions are obtained for entropy generation via heat transfer, magnetic heating, and frictional heating. The dimensionless entropy production (Ns), velocity $f'(\eta)$, and temperature $\theta(\eta)$ are plotted against η by taking various values of relevant parameters. The Bejan number (Be) profile is also plotted against the similarity variable η by considering different values of the relevant embedded parameters. All the figures are plotted by taking water as a base fluid. Nanoparticles of Fe_3O_4/CuO are dispersed in H_2O.

Figure 2a demonstrates the impact of mass suction (S) on the velocity of $Fe_3O_4 - H_2O$ and $CuO - H_2O$ nanoliquids. The decrement in motion is seen for both $Fe_3O_4 - H_2O$ and $CuO - H_2O$ nanoliquids with increasing (S). For a fixed value of (S), the velocity of the $CuO - H_2O$ nanoliquid is higher than the velocity of the $Fe_3O_4 - H_2O$ nanoliquid. Furthermore, the velocity of both nanoliquids satisfies the boundary condition at $\eta \to \infty$ asymptotically. Figure 2b demonstrates the influence of the magnetic parameter $\left(M^2\right)$ on $f'(\eta)$. It is seen that $f'(\eta)$ reduces as M^2 increases. It is a well-known fact that the Lorentz force acts as a decelerating force for fluid flow and varies directly as M^2 increases. Due to this fact, $f'(\eta)$ varies inversely with M^2. Furthermore, the velocity of the $Fe_3O_4 - H_2O$ nanoliquid is lower than the velocity of the $CuO - H_2O$ nanoliquid, and this is because of the low density of $Fe_3O_4 - H_2O$ compared to $CuO - H_2O$. Figure 3a shows the variation of temperature $\theta(\eta)$ with S by taking $M = 1$, $\phi = 0.1$, $Ec = 0.5$, and $\Pr = 6.8$. The temperature drop is observed with increasing values of S. The width of the thermal boundary layer (TBL) of the $Fe_3O_4 - H_2O$ nanoliquid is greater than that of the $CuO - H_2O$ nanoliquid. Furthermore, the difference in TBL thickness reduces as S increases. The effects of M^2 on temperature $\theta(\eta)$ are presented in Figure 3b. It is seen that $\theta(\eta)$ is augmented as M^2 increases. The rising behavior of temperature is because of magnetic heating. The effective thermal conductivity of nanoliquids is directly related to the solid volume fraction of nanoparticles (ϕ), and this augments the temperature of nanoliquids, as shown in Figure 3c. Furthermore, the width of TBL is smaller for base fluid H_2O and larger for $Fe_3O_4 - H_2O$. This is due to the low thermal conductivity of water and the high effective thermal conductivity of the $Fe_3O_4 - H_2O$ nanoliquid. Figure 3d reveals the influence of the Eckert number (Ec) on $\theta(\eta)$. It is found that increasing Ec leads to a rising temperature. The dissipation function implies that frictional heating varies directly with velocity gradients, and the velocity gradients are high in the vicinity of stretching surface. Due to this fact, the temperature shoots up suddenly, resulting in a higher Eckert number in the vicinity of the stretching plate, as shown in Figure 3d.

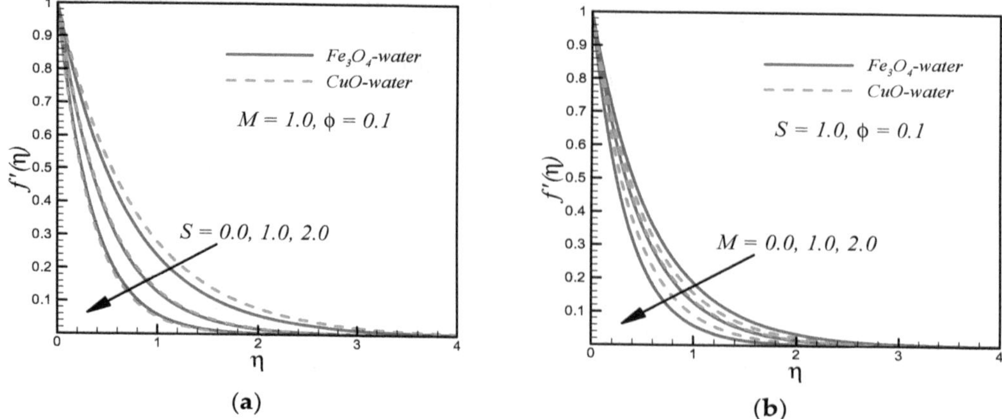

Figure 2. Variation of the velocity profile $f'(\eta)$ with (**a**) S and (**b**) M.

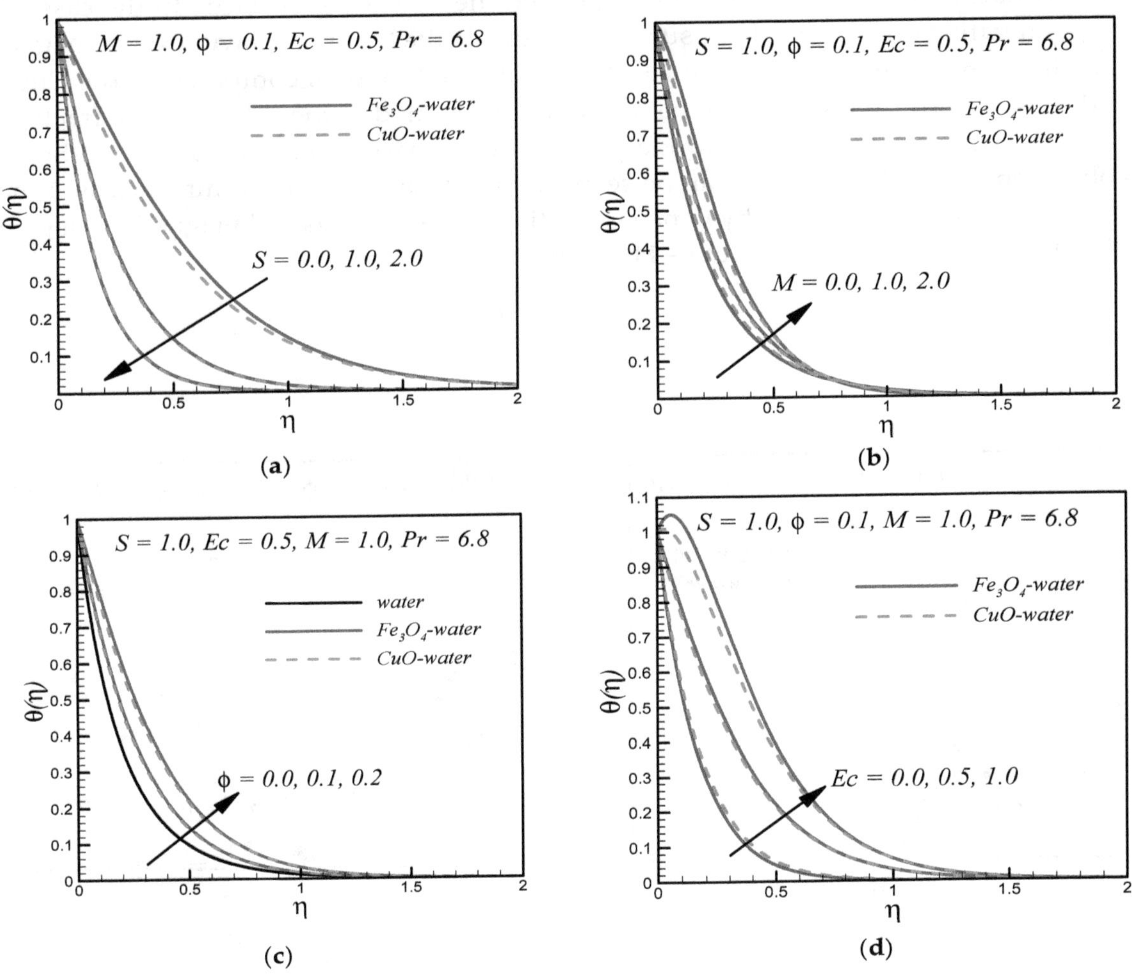

Figure 3. Variation of the temperature profile $\theta(\eta)$ with (**a**) S, (**b**) M, (**c**) ϕ,and (**d**) Ec.

Figure 4a portrays the effects of the Eckert number (Ec) on the entropy generation number (Ns). As seen from the plot, Ns is directly related to the Eckert number. This happens since frictional heating increases with the increasing Eckert number. The entropy generation in the $Fe_3O_4 - H_2O$ nanoliquid than that in the $CuO - H_2O$ nanoliquid. Furthermore, the surface of the solid boundary is the region where maximum entropy is generated. The features of mass suction (S) on Ns are revealed in Figure 4b. As S increases, entropy generation rises at the solid wall and its vicinity, but the opposite trend is observed to start at a certain distance away from the boundary. Furthermore, entropy generation is higher in the $Fe_3O_4 - H_2O$ nanoliquid at the solid boundary and its neighborhood as compared to the $CuO - H_2O$ nanoliquid, but the trend becomes the opposite at a certain distance from the boundary. The nature of entropy generation (Ns) with disparate values of the solid volume fraction of nanoparticles (ϕ) is shown in Figure 4c. From this plot, it can be seen that Ns increases as ϕ increases. This increase in Ns is due to the boost of heat transfer with increasing ϕ. It is well known that the magnetic force is nonconservative. The entropy generation is directly related to the nonconservative forces, and this fact is depicted in Figure 4d. The variations of Ns with temperature difference function (Λ) are presented in Figure 4e. The Ns decreases with increasing values of Λ. Figure 5a shows that the Bejan number (Be) has a maximum value at the surface of the stretching boundary for a nonzero suction parameter (S). In the case of an impermeable stretching boundary, the entropy generation in the $Fe_3O_4 - H_2O$ nanoliquid is due to dissipative forces (viscous and magnetic) near and on the boundary, which are high in comparison to those of the $CuO - H_2O$ nanoliquid. An opposite trend is observed to start at a certain vertical distance from the stretching surface. In the case of $S > 0$, the entropy generation on the stretching surface and inside the boundary layer due to magnetic and viscous heating is more dominant in the $Fe_3O_4 - H_2O$ nanoliquid as compared to the $CuO - H_2O$ nanoliquid. It is noticed from Figure 5b that *Be* is directly related to the solid volume fraction (ϕ) in the region away from the stretching boundary. In the vicinity of an elastic boundary, the opposite trend is observed. From Figure 5c, it can be seen that the Bejan number diminishes as Λ increases. Furthermore, the entropy generation by nonconservative forces (viscous and magnetic) is higher in the $Fe_3O_4 - H_2O$ nanoliquid than in the $CuO - H_2O$ nanoliquid.

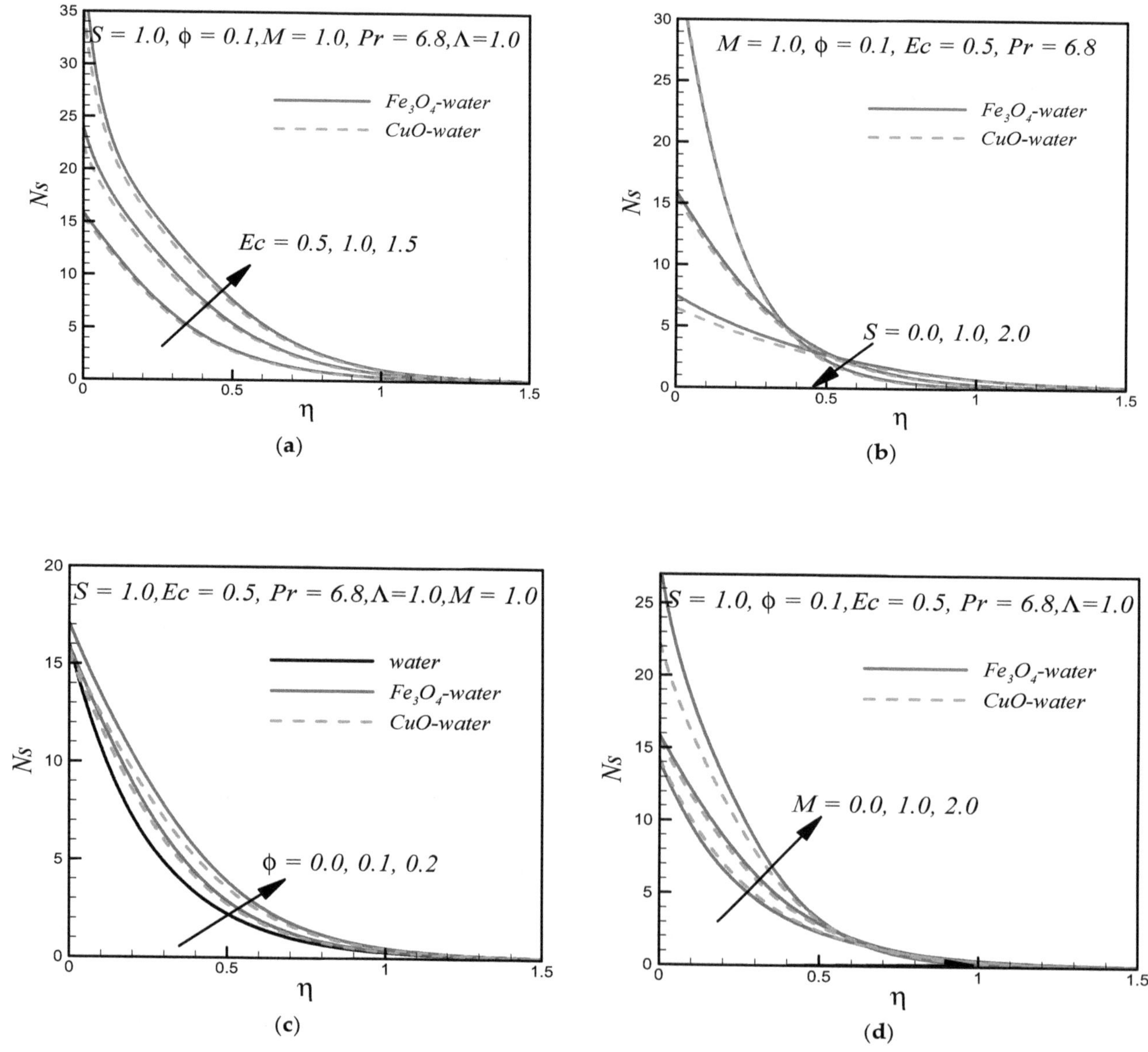

Figure 4. *Cont.*

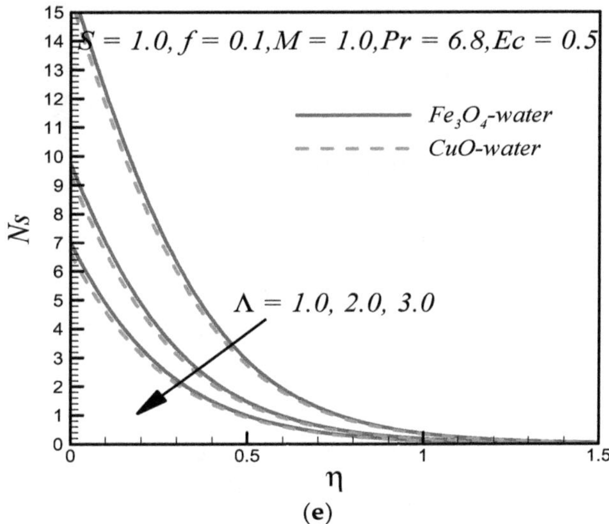

(e)

Figure 4. Variation of entropy generation profile $Ns(\eta)$ with (**a**) Ec, (**b**) S, (**c**) ϕ, (**d**) M, and (**e**) Λ.

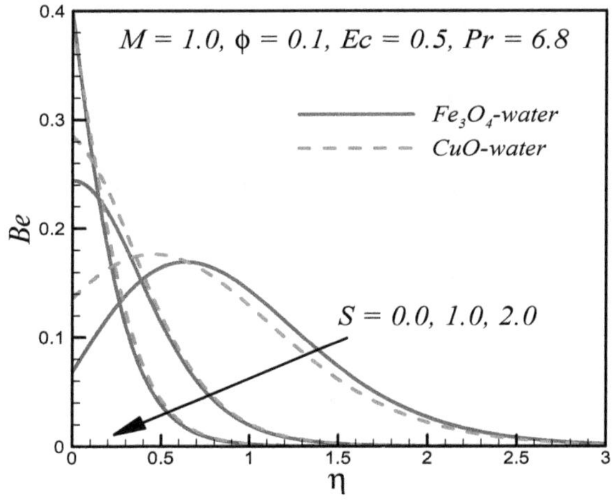

(**a**)

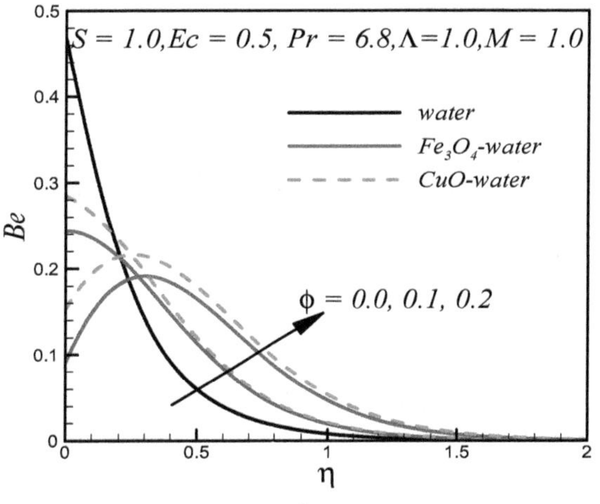

(**b**)

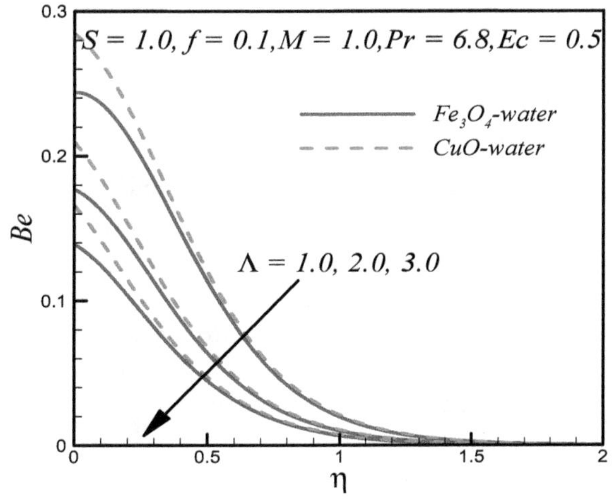

(**c**)

Figure 5. Variation of Bejan number $Be(\eta)$ with (**a**) S, (**b**) M and (**c**) ϕ.

6. Concluding Remarks

In this study, we investigated flow, heat transfer, and entropy production in a dissipative nanofluid flow under the influence of a magnetic field. The following findings can be drawn from the exact results:

- The decrement in motion is seen for both $Fe_3O_4 - H_2O$ and $CuO - H_2O$ nanofluids with increasing S and M^2.
- The velocity of the $CuO - H_2O$ nanofluid is higher than that of the $Fe_3O_4 - H_2O$ nanofluid.
- The temperature $\theta(\eta)$ is observed to decrease with increasing values of S.
- The temperature $\theta(\eta)$ increases as M^2, ϕ, and Ec increase.
- The thermal boundary layer (TBL) width of the $Fe_3O_4 - H_2O$ nanoliquid is greater than that of the $CuO - H_2O$ nanoliquid.
- The entropy generation number (Ns) is directly related to the Eckert number (Ec) and solid volume fraction (ϕ).
- Entropy generation (Ns) by nonconservative forces is higher in the $Fe_3O_4 - H_2O$ nanoliquid than in the $CuO - H_2O$ nanoliquid.

Author Contributions: M.I.A. and M.Q. formulated the problem. D.L. and U.A. solved the problem. U.F. and M.I.A. computed the results. All the authors equally contributed in writing and proofreading the paper. All authors have read and agreed to the published version of the manuscript.

Acknowledgments: The authors would like to acknowledge the Department of Mathematics, Faculty of Science, Jiangsu University, Zhenjiang China for technical support.

Nomenclature

C_o	(KL^{-2})	Dimensional constant
Be	(Dimensionless)	Bejan number
B_0	$(MT^{-2}I^{-1})$	The applied magnetic field. ("I" shows electric current)
$(C_p)_{bf}$	$(L^2T^{-2}K^{-1})$	Specific heat at a constant pressure of a base fluid
$(C_p)_{nf}$	$(L^2T^{-2}K^{-1})$	Specific heat at a constant pressure of nanofluid
Ec	(Dimensionless)	Eckert number
$f(\eta)$	(Dimensionless)	Velocity normal to the solid surface
$f'(\eta)$	(Dimensionless)	Velocity along the solid surface
$\vec{j}$	$L^{-2}I$	Current density
k_{nf}	$(MLT^{-3}K^{-1})$	Thermal conductivity of nanofluid
k_{bf}	$(MLT^{-3}K^{-1})$	Thermal conductivity of the base fluid
k_s	$(MLT^{-3}K^{-1})$	Thermal conductivity of nanoparticle
M^2	(Dimensionless)	Magnetic parameter
N_H	(Dimensionless)	Entropy generation due to heat transfer
N_F	(Dimensionless)	Entropy generation due to viscous dissipation
N_M	(Dimensionless)	Entropy generation due to the magnetic field
N_s	(Dimensionless)	Entropy generation number
Pr	(Dimensionless)	Prandtl number
S	(Dimensionless)	Mass transfer parameter
$\dot{E}'''_{Gen}$	$(ML^{-1}K^{-1}T^{-3})$	Rate of volumetric entropy generation
$(\dot{E}'''_{Gen})_o$	$(ML^{-1}K^{-1}T^{-3})$	Characteristic entropy generation
T	(K)	The temperature inside the boundary layer

$T_w(x)$	(K)	The temperature at the solid boundary
T_b	(K)	The temperature of fluid outside the thermal boundary layer
$U_w(x)$	$\left(LT^{-1}\right)$	The velocity of a stretching sheet
U	$\left(LT^{-1}\right)$	Velocity component along the surface of the solid body
U_o	(T^{-1})	Constant
V	$\left(LT^{-1}\right)$	Velocity component normal to the surface of the solid body
V_w	$\left(LT^{-1}\right)$	Normal velocity component at the boundary
X, Y	(L)	Cartesian coordinates

Greek Symbols

η	(Dimensionless)	Similarity variable
μ_{bf}	$\left(ML^{-1}T^{-1}\right)$	Dynamic viscosity of a base fluid
μ_{nf}	$\left(ML^{-1}T^{-1}\right)$	Dynamic viscosity of nanofluid
ν_{nf}	(L^2T^{-1})	Kinematic viscosity of nanofluid
ρ_{nf}	$\left(ML^{-3}\right)$	Nanofluid density
ρ_{bf}	$\left(ML^{-3}\right)$	The density of a base fluid
ρ_s	$\left(ML^{-3}\right)$	Density of nanoparticles
σ_{nf}	$\left(M^{-1}L^{-3}T^3I^2\right)$	Electric conductivity
σ_{bf}	$\left(M^{-1}L^{-3}T^3I^2\right)$	The electric conductivity of a base fluid
σ_s	$\left(M^{-1}L^{-3}T^3I^2\right)$	The electric conductivity of nanoparticle
$\theta(\eta)$	(Dimensionless)	Temperature
ϕ	(Dimensionless)	The solid volume fraction of nanoparticles
Λ	(Dimensionless)	Temperature difference parameter

References

1. Wang, C.Y. Exact solutions of the steady-state Navier-Stockes equations. *Annu. Rev. Fluid Mech.* **1991**, *23*, 159–177. [CrossRef]
2. Hui, W.H. Exact solutions of the unsteady two-dimensional Navier-Stokes equations. *J. Appl. Math. Phys.* **1987**, *38*, 689–702. [CrossRef]
3. Polyanin, A.D. Exact solutions to the Navier-Stokes equations with generalized separation of variables. *Dokl. Phys.* **2001**, *46*, 726–731. [CrossRef]
4. Al-Mdallal, Q.M. A new family of exact solutions to the unsteady Navier–Stokes equations using canonical transformation with complex coefficients. *Appl. Math. Comput.* **2008**, *196*, 303–308. [CrossRef]
5. Daly, E.; Basser, H.; Rudman, M. Exact solutions of the Navier-Stokes equations generalized for flow in porous media. *Eur. Phys. J. Plus* **2018**, *133*, 173. [CrossRef]
6. Crane, L.J. Flow past a stretching plate. *Zeitschrift für Angewandte Mathematik und Physik* **1970**, *4*, 645–647. [CrossRef]
7. Fang, T.; Zhang, J.; Yao, S. Slip MHD viscous flow over a stretching sheet—An exact solution. *Commun. Nonlinear Sci. Numer. Simul.* **2009**, *14*, 3731–3737. [CrossRef]
8. Fang, T.; Zhang, J. Closed-form exact solutions of MHD viscous flow over a shrinking sheet. *Commun. Nonlinear Sci. Numer. Simul.* **2009**, *14*, 2853–2857. [CrossRef]
9. Liu, I.-C. Exact Solutions for a Fluid-Saturated Porous Medium with Heat and Mass Transfer. *J. Mech.* **2011**, *21*, 57–62. [CrossRef]
10. Turkyilmazoglu, M. Mixed convection flow of magnetohydrodynamic micropolar fluid due to a porous heated/cooled deformable plate: Exact solutions. *Int. J. Heat Mass Transf.* **2017**, *106*, 127–134. [CrossRef]
11. Khan, Z.H.; Qasim, M.; Ishfaq, N.; Khan, W.A. Dual Solutions of MHD Boundary Layer Flow of a Micropolar Fluid with Weak Concentration over a Stretching/Shrinking Sheet. *Commun. Theor. Phys.* **2017**, *67*, 449. [CrossRef]
12. Bejan, A. A Study of Entropy Generation in Fundamental Convective Heat Transfer. *J. Heat Transf.* **1979**, *101*, 718–725. [CrossRef]

13. Bejan, A. The thermodynamic design of heat and mass transfer processes and devices. *Int. J. Heat Fluid Flow* **1987**, *8*, 258–276. [CrossRef]
14. Butt, A.S.; Ali, A.; Mehmood, A. Entropy analysis in MHD nanofluid flow near a convectively heated stretching surface. *Int. J. Exergy* **2016**, *20*, 318–342. [CrossRef]
15. Das, S.; Chakraborty, S.; Jana, R.N.; Makinde, O.D. Entropy analysis of unsteady magneto-nanofluid flow past accelerating stretching sheet with convective boundary condition. *Appl. Math. Mech.* **2015**, *36*, 1593–1610. [CrossRef]
16. Hakeem, A.K.A.; Govindaraju, M.; Ganga, B.; Kayalvizhi, M. Second law analysis for radiative MHD slip flow of a nanofluid over a stretching sheet with non-uniform heat source effect. *Sci. Iran.* **2016**, *23*, 1524–1538. [CrossRef]
17. Rashidi, M.M.; Freidoonimehr, N. Analysis of Entropy Generation in MHD Stagnation-Point Flow in Porous Media with Heat Transfer. *Int. J. Comput. Methods Eng. Sci. Mech.* **2014**, *15*, 345–355. [CrossRef]
18. Makinde, O.D. Entropy analysis for MHD boundary layer flow and heat transfer over a flat plate with a convective surface boundary condition. *Int. J. Exergy* **2012**, *10*, 142. [CrossRef]
19. Butt, A.S.; Ali, A.; Mehmood, A. Numerical investigation of magnetic field effects on entropy generation in viscous flow over a stretching cylinder embedded in a porous medium. *Energy* **2016**, *99*, 237–249. [CrossRef]
20. Ding, H.; Li, Y.; Lakzian, E.; Wen, C.; Wang, C. Entropy generation and exergy destruction in condensing steam flow through turbine blade with surface roughness. *Energy Convers. Manag.* **2019**, *196*, 1089–1104. [CrossRef]
21. Al-Odat, M.Q.; Damseh, R.A.; Al-Nimr, M.A. Effect of Magnetic Field on Entropy Generation Due to Laminar Forced Convection Past a Horizontal Flat Plate. *Entropy* **2004**, *4*, 293–303. [CrossRef]
22. Rashidi, M.M.; Mohammadi, F.; Abbasbandy, S.; Alhuthali, M.S. Entropy Generation Analysis for Stagnation Point Flow in a Porous Medium over a Permeable Stretching Surface. *J. Appl. Fluid Mech.* **2015**, *8*, 753–765. [CrossRef]
23. Das, S.; Jana, R.N.; Makinde, O.D. Entropy generation in hydromagnetic and thermal boundary layer flow due to radial stretching sheet with Newtonian heating. *J. Heat Mass Transf. Res.* **2015**, *2*, 51–61.
24. Ajibade, A.O.; Jha, B.K.; Omame, A. Entropy generation under the effect of suction/injection. *Appl. Math. Model.* **2011**, *35*, 4630–4646. [CrossRef]
25. Govindaraju, M.; Ganga, B.; Hakeem, A.K.A. Second law analysis on radiative slip flow of nanofluid over a stretching sheet in the presence of lorentz force and heat generation/absorption. *Front. Heat Mass Transf.* **2017**, *8*, 1–8. [CrossRef]
26. Milanese, M.; Iacobazzi, F.; Colangelo, G.; Risi, A. An investigation of layering phenomenon at the liquid–solid interface in Cu and CuO based nanofluids. *Int. J. Heat Mass Transf.* **2016**, *103*, 564–571. [CrossRef]
27. Iacobazzi, F.; Milanese, M.; Colangelo, G.; Lomascolo, M.; Risi, A. An explanation of the Al_2O_3 nanofluid thermal conductivity based on the phonon theory of liquid. *Energy* **2016**, *116*, 786–794. [CrossRef]
28. Colangelo, G.; Milanese, M.; De Risi, A. Numerical simulation of thermal efficiency of an innovative Al_2O_3 nanofluid solar thermal collector: Influence of nanoparticles concentration. *Therm. Sci.* **2017**, *21*, 2769–2779. [CrossRef]
29. Colangelo, G.; Favale, E.; Milanese, M.; de Risi, A.; Laforgia, D. Cooling of electronic devices: Nanofluids contribution. *Appl. Therm. Eng.* **2017**, *127*, 421–435. [CrossRef]
30. Colangelo, G.; Favale, E.; Miglietta, P.; Milanese, M.; de Risi, M. Thermal conductivity, viscosity and stability of Al_2O_3-diathermic oil nanofluids for solar energy systems. *Energy* **2016**, *95*, 124–136. [CrossRef]
31. Colangelo, G.; Favale, E.; Milanese, M.; Starace, G.; De Risi, A. Experimental Measurements of Al_2O_3 and CuO Nanofluids Interaction with Microwaves. *J. Energy Eng.* **2017**, *143*, 04016045. [CrossRef]
32. Iacobazzi, F.; Milanese, M.; Colangelo, G.; de Risi, A. A critical analysis of clustering phenomenon in Al_2O_3 nanofluids. *J. Therm. Anal. Calorim.* **2018**, *135*, 371–377. [CrossRef]
33. Hsiao, K.-L. Stagnation electrical MHD nanofluid mixed convection with slip boundary on a stretching sheet. *Appl. Therm. Eng.* **2016**, *98*, 850–861. [CrossRef]
34. Ma, Y.; Mohebbi, R.; Rashidi, M.; Yang, Z.; Sheremet, M. Nanoliquid thermal convection in I-shaped multiple-pipe heat exchanger under magnetic field influence. *Phys. A Stat. Mech. Its Appl.* **2020**, *550*, 124028. [CrossRef]

35. Wakif, A.; Chamkha, A.; Thumma, T.; Animasaun, I.L.; Sehaqui, R. Thermal radiation and surface roughness effects on the thermo-magneto-hydrodynamic stability of alumina–copper oxide hybrid nanofluids utilizing the generalized Buongiorno's nanofluid model. *J. Therm. Anal. Calorim.* **2020**, 1–20. [CrossRef]
36. Hsiao, K.-L. Nanofluid flow with multimedia physical features for conjugate mixed convection and radiation. *Comput. Fluids* **2014**, *104*, 1–8. [CrossRef]
37. Ma, Y.; Mohebbi, R.; Rashidi, M.M.; Yang, Z. MHD convective heat transfer of Ag-MgO/water hybrid nanofluid in a channel with active heaters and coolers. *Int. J. Heat Mass Transf.* **2019**, *137*, 714–726. [CrossRef]
38. Prasad, K.V.; Vaidya, H.; Makinde, O.D.; Vajravelu, K.; Wakif, A.; Basha, H. Comprehensive examination of the three-dimensional rotating flow of a UCM nanoliquid over an exponentially stretchable convective surface utilizing the optimal homotopy analysis method. *Front. Heat Mass Transf.* **2020**, *14*. [CrossRef]
39. Gebhart, B. Effects of viscous dissipation in natural convection. *J. Fluid Mech.* **1962**, *14*, 225–232. [CrossRef]
40. Desale, S.; Pradhan, V.H. Numerical Solution of Boundary Layer Flow Equation with Viscous Dissipation Effect Along a Flat Plate with Variable Temperature. *Procedia Eng.* **2015**, *127*, 846–853. [CrossRef]
41. Vajravelu, K.; Hadjinicolaou, A. Heat transfer in a viscous fluid over a stretching sheet with viscous dissipation and internal heat generation. *Int. Commun. Heat Mass Transf.* **1993**, *20*, 417–430. [CrossRef]
42. Alsabery, A.; Saleh, H.; Hashim, I. Effects of Viscous Dissipation and Radiation on MHD Natural Convection in Oblique Porous Cavity with Constant Heat Flux. *Adv. Appl. Math. Mech.* **2017**, *9*, 463–484. [CrossRef]
43. Mohamed, M.K.A.; Sarif, N.M.; Noar, N.A.Z.M.; Salleh, M.Z.; Ishak, A. Viscous dissipation effect on the mixed convection boundary layer flow towards solid sphere. *Trans. Sci. Technol.* **2016**, *3*, 59–67.
44. Jamaludin, A.; Nazar, R.; Khan, I. *AIP Conference Proceedings*; Boundary Layer Flow and Heat Transfer in a Viscous Fluid Over a Stretching Sheet with Viscous Dissipation, Internal Heat Generation and Prescribed Heat Flux; AIP Publishing LLC: New York, NY, USA, 2017; Volume 1870, p. 040029. [CrossRef]
45. Makinde, O.D. Effects of viscous dissipation and Newtonian heating on boundary-layer flow of nanofluids over a flat plate. *Int. J. Numer. Methods Heat Fluid Flow* **2013**, *23*, 1291–1303. [CrossRef]
46. Motsumi, T.G.; Makinde, O.D. Effects of thermal radiation and viscous dissipation on boundary layer flow of nanofluids over a permeable moving flat plate. *Phys. Scr.* **2012**, *86*, 1–8. [CrossRef]
47. Minkowycz, W.J.; Sparrow, E.M.; Abraham, J.P. *Nanoparticle Heat Transfer and Fluid Flow*; CRC Press: Boca Raton, FL, USA, 2012.
48. Bianco, V.; Manca, O.; Nardini, S.; Vafai, K. *Heat Transfer Enhancement of Nanofluids*; CRC Press: Boca Raton, FL, USA, 2015.
49. Kleinstreuer, C.; Feng, Y. Experimental and Theoretical Studies of Nanofluid Thermal Conductivity Enhancement: A Review. *Nanoscale Res. Lett.* **2011**, *6*, 229. [CrossRef]
50. Bashirnezhad, K.; Bazri, S.; Safaei, M.R.; Goodarzi, M.; Dahari, M.; Mahian, O.; Dalkılıça, A.S.; Wongwises, S. Viscosity of nanofluids: A review of recent experimental studies. *Int. Commun. Heat Mass Transf.* **2016**, *73*, 114–123. [CrossRef]
51. Sheikholeslami, M. *Nanofluid Heat and Mass Transfer in Engineering Problems*; Intech Open: Rijeka, Croatia, 2017.
52. Sajid, M.U.; Ali, H.M. Recent advances in application of nanofluids in heat transfer devices: A critical review. *Renew. Sustain. Energy Rev.* **2019**, *103*, 556–592. [CrossRef]

Permissions

All chapters in this book were first published in MDPI; hereby published with permission under the Creative Commons Attribution License or equivalent. Every chapter published in this book has been scrutinized by our experts. Their significance has been extensively debated. The topics covered herein carry significant findings which will fuel the growth of the discipline. They may even be implemented as practical applications or may be referred to as a beginning point for another development.

The contributors of this book come from diverse backgrounds, making this book a truly international effort. This book will bring forth new frontiers with its revolutionizing research information and detailed analysis of the nascent developments around the world.

We would like to thank all the contributing authors for lending their expertise to make the book truly unique. They have played a crucial role in the development of this book. Without their invaluable contributions this book wouldn't have been possible. They have made vital efforts to compile up to date information on the varied aspects of this subject to make this book a valuable addition to the collection of many professionals and students.

This book was conceptualized with the vision of imparting up-to-date information and advanced data in this field. To ensure the same, a matchless editorial board was set up. Every individual on the board went through rigorous rounds of assessment to prove their worth. After which they invested a large part of their time researching and compiling the most relevant data for our readers.

The editorial board has been involved in producing this book since its inception. They have spent rigorous hours researching and exploring the diverse topics which have resulted in the successful publishing of this book. They have passed on their knowledge of decades through this book. To expedite this challenging task, the publisher supported the team at every step. A small team of assistant editors was also appointed to further simplify the editing procedure and attain best results for the readers.

Apart from the editorial board, the designing team has also invested a significant amount of their time in understanding the subject and creating the most relevant covers. They scrutinized every image to scout for the most suitable representation of the subject and create an appropriate cover for the book.

The publishing team has been an ardent support to the editorial, designing and production team. Their endless efforts to recruit the best for this project, has resulted in the accomplishment of this book. They are a veteran in the field of academics and their pool of knowledge is as vast as their experience in printing. Their expertise and guidance has proved useful at every step. Their uncompromising quality standards have made this book an exceptional effort. Their encouragement from time to time has been an inspiration for everyone.

The publisher and the editorial board hope that this book will prove to be a valuable piece of knowledge for researchers, students, practitioners and scholars across the globe.

List of Contributors

Marcin Stasiak
Institute of Mathematics, Poznań University of Technology, 60-965 Poznań, Poland

Grzegorz Musielak and Dominik Mierzwa
Institute of Technology and Chemical Engineering, Poznań University of Technology, 60-965 Poznań, Poland

Riccardo Tesser
NICL—Naples Industrial Chemistry Laboratory, Department of Chemical Science, University of Naples Federico II, 80126 Naples, Italy

Elio Santacesaria
CEO of Eurochem Engineering Ltd., 20139 Milan, Italy

Krzysztof Górnicki, Radosław Winiczenko and Agnieszka Kaleta
Department of Fundamental Engineering, Warsaw University of Life Sciences, Nowoursynowska 164 St., 02-787 Warsaw, Poland

Rui Quan
Hubei Key Laboratory for High-efficiency Utilization of Solar Energy and Operation Control of Energy Storage System, Hubei University of Technology, Wuhan 430068, China
Hubei Collaborative Innovation Center for High-efficiency Utilization of Solar Energy, Hubei University of Technology, Wuhan 430068, China

Yufang Chang
Hubei Collaborative Innovation Center for High-efficiency Utilization of Solar Energy, Hubei University of Technology, Wuhan 430068, China

Tao Li and Yousheng Yue
Hubei Key Laboratory for High-efficiency Utilization of Solar Energy and Operation Control of Energy Storage System, Hubei University of Technology, Wuhan 430068, China

Baohua Tan
School of Science, Hubei University of Technology, Wuhan 430068, China

Seok Yoon, WanHyoung Cho, Changsoo Lee and Geon-Young Kim
Division of Radioactive Waste Disposal Research, Korea Atomic Energy Research Institute (KAERI), 989-111, Daedeok-daero, Yuseong-gu, Daejeon 34057, Republic of Korea

Hyun Sung Kang
Eco-Friendly Vehicle R & D Division, Korea Automotive Technology Institute, 303 Pungse-Ro, Pungse-Myeon, Cheonan-Si 330-912, Korea
Department of Mechanical Engineering, Korea University, 409 Innovation Hall Building, Anam-Dong, Sungbuk-Gu, Seoul 02841, Korea

Myong-Hwan Kim
Hydrogen Fuel Cell Mobility R&D Center, Korea Automotive Technology Institute, 303 Pungse-Ro, Pungse-Myeon, Cheonan-Si 330-912, Korea

Yoon Hyuk Shin
Eco-Friendly Vehicle R & D Division, Korea Automotive Technology Institute, 303 Pungse-Ro, Pungse-Myeon, Cheonan-Si 330-912, Korea

Miha Kovačič and Robert Vertnik
Štore Steel Ltd., Železarska cesta 3, SI-3220 Štore, Slovenia
Faculty of Mechanical Engineering, University in Ljubljana, Aškerčeva 6, SI-1000 Ljubljana, Slovenia

Klemen Stopar
Štore Steel Ltd., Železarska cesta 3, SI-3220 Štore, Slovenia

Božidar Šarler
Faculty of Mechanical Engineering, University in Ljubljana, Aškerčeva 6, SI-1000 Ljubljana, Slovenia
Institute of Metals and Technology, Lepi pot 11, SI-1000 Ljubljana, Slovenia

Xiaolong Ma, Zhongchao Zhao, Pengpeng Jiang, Shan Yang, Shilin Li and Xudong Chen
School of Energy and Power, Jiangsu University of Science and Technology, Zhenjiang, Jiangsu 212000, China

Dongxu Han and Bo Yu
School of Mechanical Engineering, Beijing Key Laboratory of Pipeline Critical Technology and Equipment for Deepwater Oil & Gas Development, Beijing Institute of Petrochemical Technology, Beijing 102617, China

Qing Yuan
National Engineering Laboratory for Pipeline Safety, Beijing Key Laboratory of Urban Oil and Gas Distribution Technology, China University of Petroleum, Beijing 102249, China

Danfu Cao and Gaoping Zhang
Storage and Transportation Company, Sinopec Group, Xuzhou 221000, China

Liangyu Wu, Yingying Chen, Weibo Yang and Fangping Tang
School of Hydraulic, Energy and Power Engineering, Yangzhou University, Yangzhou 225127, China

Suchen Wu and Mengchen Zhang
Key Laboratory of Energy Thermal Conversion and Control of Ministry of Education, School of Energy and Environment, Southeast University, Nanjing 210096, China

Yeon Je Shin and Woo Jun You
Department of Architecture & Fire Safety, Dongyang University, Yeongju 36040, Korea

Yong Li
School of Electrical Engineering, Beijing Jiaotong University, Beijing 100044, China
China North Vehicle Research Institute, Beijing 100072, China

Weili Li and Ying Su
School of Electrical Engineering, Beijing Jiaotong University, Beijing 100044, China

Xikang Cheng, Wei Liu, Ziliang Tan, Zhilong Zhou, Binchao Yu, Wenqi Wang, Yang Zhang and Sitong Liu
Key Laboratory for Precision and Non-Traditional Machining Technology of the Ministry of Education, Dalian University of Technology, Dalian 116024, China

Dianchen Lu, Umer Farooq and Muhammad Qasim
Department of Mathematics, Faculty of Science, Jiangsu University, Zhenjiang 212013, China

Muhammad Idrees Afridi
Department of Computing, Abasyn University, Islamabad 45710, Pakistan

Usman Allauddin
Department of Mechanical Engineering, NED University of Engineering & Technology, Karachi 75270, Pakistan

Index

Printed in the USA
CPSIA information can be obtained
at www.ICGtesting.com
JSHW051410091023
49903JS00006B/365